Disaster Risk Reduction

Methods, Approaches and Practices

For further volumes:
http://www.springer.com/series/11575

About the Series

Scope of the Series

Disaster risk reduction is a process, which leads to the safety of community and nations. After the 2005 World Conference on Disaster Reduction, held in Kobe, Japan, the Hyogo Framework for Action (HFA) was adopted as a framework of risk reduction. The academic research and higher education in disaster risk reduction has made/is making gradual shift from pure basic research to applied, implementation oriented research. More emphasis is given on the multi-stakeholder collaboration and multi-disciplinary research. Emerging university networks in Asia, Europe, Africa and Americas have urged for the process-oriented research in disaster risk reduction field. Keeping this in mind, this new series will promote the outputs of action research on disaster risk reduction, which will be useful for a wider range of stakeholders including academicians, professionals, practitioners, and students and researchers in the related field. The series will focus on some of emerging needs in the risk reduction field, starting from climate change adaptation, urban ecosystem, coastal risk reduction, education for sustainable development, community based practices, risk communication, human security etc. Through academic review, this series will encourage young researchers and practitioners to analyze field practices, and link it to theory and policies with logic, data and evidences. Thus, the series emphasizes evidence based risk reduction methods, approaches and practices.

Series Editor

Editorial Advisory Group

Rajib Shaw · Fuad Mallick · Aminul Islam
Editors

Disaster Risk Reduction Approaches in Bangladesh

Springer

Editors
Rajib Shaw
Kyoto University, Kyoto, Japan

Aminul Islam
Sustainable Development
United Nations Development Programme [UNDP]
Dhaka, Bangladesh

Fuad Mallick
Department of Architecture
BRAC University
Dhaka, Bangladesh

ISSN 2196-4106 ISSN 2196-4114 (electronic)
ISBN 978-4-431-54251-3 ISBN 978-4-431-54252-0 (eBook)
DOI 10.1007/978-4-431-54252-0
Springer Tokyo Heidelberg New York Dordrecht London

Library of Congress Control Number: 2013939620

Printed on acid-free paper

Springer is part of Springer Science+Business Media (www.springer.com)

Preface

Disasters are increasing, and their impacts on people have become more conspicuous in recent years. In this respect, Bangladesh is known for its innovations in disaster risk reduction at the national, local, and community levels. Globally, the concept of disaster management has changed from response to pre-disaster risk reduction. It is often said that US$1 investment in risk reduction saves US$7 in recovery. The Hyogo Framework for Action (HFA: 2005–2015) has become an accepted mechanism of risk reduction globally, nationally, and locally, and the mid-term report of the HFA emphasizes the need for local approaches and initiatives.

Bangladesh is prone to floods and cyclones, and the risk of other disasters such as drought, earthquakes, and tornados is increasing. The impacts are becoming more visible at the local level, with greater impacts on poor and vulnerable communities. The Comprehensive Disaster Management Program has generated a wealth of resources over the last 10 years. There exist other innovations in community-based risk reduction, conducted by nongovernmental organizations and civil society bodies. Several universities in Bangladesh have started master's programs in disaster risk reduction; however, there are few resource books available. There is an increasing demand for resources that utilize field-based knowledge and for linking those to the curriculum of the master's and diploma programs. With those facts in mind, this book is a modest attempt to compile and analyze some of the existing practices on disaster risk reduction.

The contents of this book were developed through a consultation workshop with different universities and related organizations over the past year. The book is one of two volumes prepared for disaster risk reduction and climate change adaptation. Thus, readers are encouraged to look at the contents of the other volume, titled *Climate Change Adaptation Actions of Bangladesh*. The editors acknowledge the support of the United Nations Development Programme (UNDP) Dhaka office in this regard. Several eminent authors were able to fulfill their commitments to make contributions in spite of their busy schedules, and we are thankful to all of them.

All the statements and opinions in this book are those of the authors and editors, and do not officially represent the organizations to which they belong.

This book is written for students and young researchers aspiring to a career in environmental studies and/or disaster risk reduction. We hope that they will find the book useful and relevant to their work.

Kyoto, Japan
Dhaka, Bangladesh

Rajib Shaw
Fuad Mallick
Aminul Islam

Message: The Challenging Road to Disaster Resilience in Bangladesh

Bangladesh is a disaster risk hotspot, ranked fifth in the top 15 countries with highest risks.[1] This unfortunate ranking reflects the very profound, multi-layered challenges that face the country. Population density and poverty run hand in hand to increase vulnerability. Poor people in a crowded country like Bangladesh have limited options: they live in areas prone to storms, flooding or landslides, their homes no match against the forces of nature. Climate change adds another layer of risk and its threat to Bangladesh is easy to grasp: in a densely populated least developed country (LDC), we have the melting Himalayas to the north, sending ever more unpredictable river-flow through the country, squeezed against the real potential of a rising Bay of Bengal to the south. Bangladesh has the highest number of people exposed to flooding in the world[2] and nearly 20 % of the land mass is covered by floodwaters in a normal flood year—up to 68 % during periodic disastrous floods, such as those of 1998! Simply stated, climate change in Bangladesh means natural disasters are more frequent, more intense and more destructive. The cost of not being prepared for future disasters and mitigating against climate change impacts is unacceptably high for the people of Bangladesh.

Within this incredibly challenging context, let me share the good news. Bangladesh is one of the best examples in the world of real achievements in disaster management, a result of long-term focus on the issue. The Government, United Nations and humanitarian partners have worked over 40 years to save lives through improved disaster management systems. Bangladesh fundamentally transformed its approach, long before the global consensus on the risk reduction agenda was agreed under the Hyogo Framework of Action (HFA) in 2005. In my view, Bangladesh has

[1] World Disaster Report 2012 (page-9: World Risk Index).

[2] Global Assessment Report 2011, UNISDR.

moved from "reactive humanitarian relief" after a disaster to "proactive risk management" before the disaster and an emphasis on the combination of response and an early return to the development trajectory, immediately after a crisis:

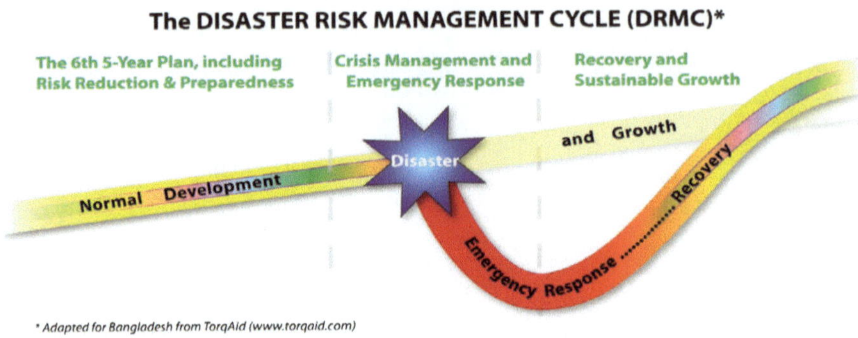

In the long term, these collective efforts have clearly paid off in the reduction of mortality associated with Cyclones: from almost half a million deaths in the 1970 Cyclone Bhola, to several hundred in Cyclone Aila in 2009. Cyclones Sidr (2007) and Aila (2009) offer specific examples of effective response. Improved institutional capacities and infrastructure enabled the activation of cyclone early warning systems and the national preparedness system. Coordination and leadership—both local and national, facilitated the evacuation of 3.5 million people within 48 h. Stockpiled relief and rescue items reached most affected areas within 24 h. More recently, Government has passed a significantly improved Disaster Management Act which aims to secure legal empowerment for citizens vulnerable to disaster. Building on lessons learned from humanitarian coordination in the past, the Local Consultative Group[3] on Disaster & Emergency Response (LCG DER) has been revitalized. LCG DER has enabled effective use of assessment tools and created a tailor made framework around existing development coordination structures to rapidly and systematically respond to emergencies. With UN support, training has been given to 27,000 government officials on disaster planning and management! Disaster management subjects have been introduced at primary and secondary schools and dedicated degree programmes are now available across 22 universities and tertiary institutions.

Innovation is a crucial element of how Bangladesh responds to disasters. Early warnings for cyclones and floods are disseminated through mobile phone alerts from the Disaster Management Information Centre (DMIC)—imagine the importance of this approach for remote communities! New technology has been applied by UNDP in the construction of more than 16,000 disaster-resilient homes in highly vulnerable communities. Critically important, the communities themselves have

[3]Multilateral and bilateral Bangladesh based donors and UN agencies are organized under the umbrella of the Local Consultative Group (LCG), to engage in dialogue with the Government of Bangladesh on 18 thematic areas aligned with Sixth Five Year Plan.

been engaged in the process of fine-tuning building designs, to better protect their livelihoods in the event of a disaster. The UN system, particularly UNDP, has supported the development and adoption of saline tolerant and drought-resistant crop varieties in communities that are increasingly vulnerable to drought and salinity intrusion due to climate change. Similarly, mangrove afforestation, first piloted in 2006 to trap sediment along the coastal belt, is now guarding vulnerable communities from the increasing severity and intensity of climatic events, such as cyclones and tidal surges, as well as creating greater areas of useable land in the long run that can provide livelihood options for poor and vulnerable coastal people.

But even with all this good news, it is very clear that Bangladesh faces a real challenge in managing and reducing risk, in sustainably addressing the threats posed by climate change, and in ensuring highest quality preparedness and response, after any disaster. Even when response to a disaster is well coordinated and effective, real leadership is required to transition the country from the humanitarian approach, back to a sustainable development path.

And if the country's high vulnerability isn't enough, new and/or emerging problems add further complexity. Slow-onset disasters, such as water logging, which persist for months or even years, have a pervasive impact on livelihoods, access to food, shelter, water, sanitation and healthcare. Rapid, unplanned urbanization exponentially increases the number of people and property at risk. Low and mid-level tremors in recent months remind us of the seldom addressed earthquake risk in urban centers such as Dhaka city. There is much work to be done to reduce the scale and scope of urban risks in Bangladesh. It is heartening to note that a UN supported Urban Community Volunteer Program, 64,000 volunteers from the Fire Service and Civil Defense (FSCD) were trained, nationwide, to better serve as the "frontline" in the face of disasters. Another program supported by UNDP and UKAid, the national Urban Partnership for Poverty Reduction (UPPR) supports three million urban poor, particularly women and girls, to improve their living conditions by building secure surroundings and reducing risk factors.

As important as these efforts are, they are little compared to what is urgently needed to reduce disaster risks to an acceptable level in Bangladesh. Towards that lofty objective, Government must provide leadership, not only in terms of a long-term focus on managing and reducing the risks associated with natural disasters and climate change, but in creating an enabling environment for innovation in the private sector, for research and development coming out of Bangladesh's highly regarded think-tanks and universities, as well as for NGO's that have traditionally been a pro-active force for sustainable development in Bangladesh. With a renewed sense of urgency and determination, I encourage all walks of people in this great country to get involved, to work together in making Bangladesh a strong, resilient and disaster-ready nation.

United Nations Resident Coordinator
Dhaka, Bangladesh

Neal Walker

been engaged in the process of fine-tuning building designs to better protect their livelihoods in the event of a disaster. The UN system, particularly UNDP, has supported the development and adoption of saline tolerant and drought-resistant crop varieties in communities that are increasingly vulnerable to drought and salinity [illegible] due to climate change. Similarly, mangrove afforestation has [illegible] [illegible] vulnerable communities [illegible] and tidal surges, as well as creating greater areas of livable land in the long run that can provide livelihood options for poor and vulnerable coastal people.

[illegible] Bangladesh [illegible] to manage and reduce risks [illegible] the threats posed by climate change, and in ensuring higher quality preparedness and responses [illegible] effective [illegible] transition [illegible] from the humanitarian approach [illegible].

[illegible]

Contents

Part I Outline and Basics of DRR

1 **Disaster, Risk and Evolution of the Concept** 3
Rajib Shaw, Aminul Islam, and Fuad Mallick

2 **Understanding Vulnerability and Risks** 23
A.K.M. Mamunur Rashid

3 **National Perspectives of Disaster Risk Reduction in Bangladesh** 45
Rajib Shaw, Aminul Islam, and Fuad Mallick

Part II Hazard and Risk Profiles of Bangladesh

4 **Flood Risks and Reduction Approaches in Bangladesh** 65
Rezaur Rahman and Mashfiqus Salehin

5 **Cyclone and Tornado Risk and Reduction Approaches in Bangladesh** 91
Fuad Mallick and Aminur Rahman

6 **Earthquake Risk and Reduction Approaches in Bangladesh** 103
A.S.M. Maksud Kamal

7 **Drought Risk and Reduction Approaches in Bangladesh** 131
Umma Habiba, Rajib Shaw, and Abu Wali Raghib Hassan

8 **Landslide and Flashflood in Bangladesh** 165
Abdul Awal Sarker and A.K.M. Mamunur Rashid

9 Arsenic Contamination in Bangladesh: Contemporary Alarm and Future Strategy 191
Md. Anwarul Abedin and Rajib Shaw

10 Sea-Level Rise Along the Coast of Bangladesh 217
Md. Golam Mahabub Sarwar

Part III Approaches and Issues

11 Urban Risk Reduction Approaches in Bangladesh 235
Gulsan Ara Parvin, S.M. Reazul Ahsan, and Rajib Shaw

12 Community-Based Disaster Risk Reduction Approaches in Bangladesh 259
Umma Habiba, Rajib Shaw, and Md. Anwarul Abedin

13 Crop Insurance as Risk Management Strategy in Bangladesh 281
Umma Habiba and Rajib Shaw

14 Disaster Education in Bangladesh: Opportunities and Challenges 307
Umma Habiba, Md. Anwarul Abedin, and Rajib Shaw

15 Disaster Risk Reduction Investment and Reduction of Response Cost in Bangladesh 331
Ahmadul Hassan, Aminul Islam, Nityananda Chakravorty, and Bhuiya Md. Tamim Al Hossain

16 Gender and Social Exclusion Analysis in Disaster Risk Management 343
A.K.M. Mamunur Rashid and Hasan Shafie

About the Editors 365

Part I
Outline and Basics of DRR

Chapter 1
Disaster, Risk and Evolution of the Concept

Rajib Shaw, Aminul Islam, and Fuad Mallick

Abstract Disaster is a function of hazard, vulnerability and capacity. The subject of disaster has evolved over time, and the concept of risk reduction is an accepted trend globally in this regard. A number of recent literatures and reports supported the accelerated paradigm shift from response to disaster risk reduction in different countries. United Nations International Strategy for Disaster Reduction (UNISDR) has promoted the integrated concept of disaster risk reduction through five priority areas, known as Hyogo Framework for Action (HFA: 2005–2015). HFA is considered as the global framework of risk reduction, agreed by UN member states, with specific targets. Since the adoption of the HFA in 2005, a certain progress has been made in HFA implementation by national governments with support from international and regional agencies. At the same time, the need for comprehensive DRR approach, thus HFA implementation at local level, has been strongly recognized. Effective DRR relies on the efforts of many different stakeholders, including UN agencies, regional and international organizations, CSOs, private sectors, media and academics. The collaboration and cooperation among all stakeholders is crucial in order to improve the resilience of communities. Thus, local level implementation and multi-stakeholder collaboration are considered as the key aspects of disaster risk reduction.

Keywords Disaster risk • HFA • ISDR • Resilience • Vulnerability

---

R. Shaw (✉)
Graduate School of Global Environmental Studies, Kyoto University, Kyoto, Japan
e-mail: shaw.rajib.5u@kyoto-u.ac.jp

A. Islam
Sustainable Development, United Nations Development Programme [UNDP], Dhaka, Bangladesh

F. Mallick
BRAC University, Dhaka, Bangladesh

R. Shaw et al. (eds.), *Disaster Risk Reduction Approaches in Bangladesh*,
Disaster Risk Reduction, DOI 10.1007/978-4-431-54252-0_1, © Springer Japan 2013

1.1 Introduction

The world continues to experience dramatic suffering and loss of life due to natural hazards. Disasters caused by natural hazards seriously undermine the result of development gains and investment, and remains a major impediment to sustainable development.

> Over the last 20 years, it is conservatively estimated that disasters have killed 1.3 million people, affected 4.4 billion and resulted in economic losses of USD 2 trillion. These are staggering numbers when you consider what it means in terms of missed opportunities, shattered lives, lost housing, schools and health facilities destroyed, cultural losses and roads washed away (UNISDR 2012a).

The Asia Pacific region is the most disaster prone and the most affected by disasters in the world. Although the Global Assessment Report 2011 suggests that deaths due disasters is declining globally, the concentration of human losses has been enormous in the region. Seventy-five percent of all deaths due to disasters from 1970 to 2011 are concentrated in the Asia pacific region (UNISDR and UNESCAP 2012).

The region also bares the largest proportion of disaster losses globally. For example, in 2011 the economic losses in the region were close to 80 % of the global losses. The most frequent hazard in the region is hydro-meteorological in nature. This means the region is more susceptible to the effects of climate extremes and climate variations. For example, 1.2 billion people have been exposed to hydro-meteorological risks through 1,215 events since 2000, compared to about 355 million people exposed to 394 climatological, biological and geo-physical disaster events during the same period (UNISDR and UNESCAP 2012; UNISDR 2012b).

Also, human exposure to hydro-meteorological hazards still continues to rise. Population almost doubled from 2.2 to 4.2 billion people between 1970 and 2010, and the average number of people exposed to yearly flooding more than doubled from 29.5 to 63.8 million. In addition, the population's resident in cyclone-prone areas grew from 71.8 to 120.7 million (UNISDR and UNESCAP 2012).

1.2 Evolution of the Disaster Paradigm

1.2.1 Recognition Towards Disaster Risk Reduction

In the 1970s and early 1980s, the vulnerability approach to disasters began with a rejection of the assumption that disasters are "caused" in any simple way by external natural events, and a revision of the assumption that disasters are "normal" (Wisner et al. 2004). In addition, since determinant of disaster risk was more focused than the existence of hazards themselves, the key problem of vulnerability was less understood. There is a danger in treating disasters as something peculiar, as events that deserve their own special focus. For example, it is to risk separating "natural" disasters from the social frameworks that influence how hazards affect people, thereby putting too much emphasis on the natural hazards themselves, and not nearly enough on the surrounding social environment (Wisner et al. 2004). Hence global trends shows increasing losses from disasters and human activities are related to the increasing losses. Thus, people and societies are becoming more vulnerable. Although the frequency of dramatic natural events may be constant, human activities contribute to their increased intensity. Impact depends on development practices, environmental protection, regulated growth of cities, distribution of people and wealth and government structures (UNISDR 2002). If a disaster occurred in a megacity, the aftermath is spread all over the world because of the large information system and network, complexity and globalization. Therefore, international and comprehensive disaster risk reduction is needed. Development gains were being jeopardized by the increasing losses from severe disasters despite advances in science and technology. Because of these changes, a global movement concerned with a world safer form disasters was starting to develop from late 1980s (Kyoto University and UNISDR 2010).

For the last several years, Disaster Risk Reduction (DRR) has gained its strong recognition due to the increased loss and damages of human lives and economic assets caused by the impact of natural hazards and through the evolution of the international discussion on DRR, which will be explained in the next section Yodmani (2001). Within the UNISDR Terminology 2009, DRR is defined as "the concept and practice of reducing disaster risk through reduced exposure to hazards, lessened vulnerability of people and property, wise management of land and the environment and improved preparedness for adverse events". DRR has a broad context including governance, technical, education and awareness, infrastructure, mitigation and preparedness issues. As Twigg (2007) considers, there are different definitions of the term of DRR in the technical literature but it is generally understood to mean the broad development and application of policies, strategies and practices to minimize vulnerabilities and disaster risks through society (Matsuoka 2013).

A number of recent literatures and reports supported the accelerated paradigm shift from response to disaster risk reduction in different countries. The below are the several examples in their national policies recognizing the need and the ongoing paradigm shift (Shaw and Okazaki 2004).

- The Pakistan's National Disaster Management Ordinance was promulgated in December 2006. The National Disaster Management Authority (NDMA) was established and assigned to manage complete spectrum of all types of disasters through a paradigm shift by moving away from response and relief oriented approach and by adopting a disaster risk reduction perspective from local government level upwards (Government of Pakistan 2006).
- The Bangladesh National Plan for Disaster Management 2005–2006 specified "Our future direction is to ensure we achieve a paradigm shift in disaster management from conventional response and recovery to a more comprehensive risk reduction culture" (Government of Bangladesh 2005).
- The Philippine's National Disaster Risk Reduction and Management Plan 2011–2028 (NDRRMP) mentioned that "the enactment of Republic Act 10121 otherwise known as the Philippines Disaster Risk Reduction and Management

Act 2010 has laid the bases for a paradigm shift from disaster preparedness and response to disaster risk reduction and management" (Government of the Philippines 2011).

- Lao PDR's the latest draft National Disaster Management Plan 2012–2015 specified that "the plan has been developed on the basis of a national vision and mission to reduce the vulnerability of all the people of the Lao PDR to the effects of natural, environmental and human induced hazards to a manageable and acceptable humanitarian level by a bringing a paradigm shift in disaster management from conventional response and relief practice to a more comprehensive risk reduction culture" (Government of Lao PDR 2012).

As Yodmani (2001) discussed, it is important to note that a paradigm shift in the development sector from income poverty to human poverty has been paralleled in the disaster management sector by a shift from setting disasters as extreme evens created by natural forces, to viewing them as manifestations of unresolved development problems.

1.2.2 Resilience

The DRR community used the terms prevention, preparedness, resistance, mitigation, response and so on to describe various risk-reduction efforts. However, recently, building resilience against disasters has become one of the important concepts within DRR. As Surjan et al. (2011) discussed, resilience was first talked about in the 1970s in terms of defining ecosystem, and has gained stronger reorganization for the last decade or so in the socioeconomic regimes through the advanced holistic understanding of ecological, sociological, and economic systems. In addition, social scientist uses the term resilience to explain how human capabilities to return to normalcy after absorbing stress or surviving negative changes. For the last decade or so, the DRR community started paying attention to the notions of resilience and examining ways to build, nature and increase resilience (Surjan et al. 2011).

Resilience is defined as " the ability of a system, community or society exposed to hazards to resist, absorb, accommodate to and recover from the effects of a hazard in a timely and efficient manger, including through the preservation and restoration of its essential basic structures and function" (UNISDR 2009).

As Joerin (2012) pointed out, there were a number of scholars (Adger 2000; Allen 2006; Bruneau et al. 2003; Paton and Johnston 2001; Twigg 2007) regard the extent of people's abilities to respond to a disturbance (e.g. disaster) to be shaped by the political, economic, physical and natural context of their environment where they are embedded in. Twigg (2007) suggested the three capacities as a system or community resilience: (1) capacity to absorb stress or destructive forces through resistance or adaptation; (2) capacity to manage, or maintain certain basic functions and structures, during disastrous events; (3) capacity to recover or bounce back after an event. Surjan et al. (2011) considers that resilience has four main elements:

Redundancy, Flexibility, Capacity to reorganize, and Capacity to learn. Yodmani (2001) presented a disaster risk formula as below:

$$\text{Disaster Risk} = \frac{\text{Hazard} \times \text{Vulnerability}}{\text{Capacity}}$$

Therefore, considering these discussions, resilience has a broader concept than capacity because it goes beyond the coping or managing capacity. As Twig discussed, a focus on resilience means putting greater emphasis on what communities can do for themselves and how to strengthen their capacities, rather than concentrating on their vulnerability to disaster or their needs in a emergency.

World Disaster Report 2010 pointed out that the ultimate objective of DRR and climate change adaptation is to produce resilient cities. In the Asia Pacific Disaster Report 2012, Heads of the UNESCAP and UNISDR acknowledges "we are still working to identify the ways in which different components of risk consisting of hazards, vulnerability and exposure, which interact to increase to total risk and trigger damage" (UNISDR and UNESCAP 2012). Therefore, efforts to decompose risks in a community are important to understand what kind of hazards, vulnerability and exposure they are faced with to build their resilience.

1.2.3 *Vulnerability and Exposure*

As Twigg (2007) discussed, the terms "resilience" and "vulnerability" are opposite sides of the same coin, but both are relative terms since one has to ask what individuals, communities and systems are vulnerable or resilience to, and to what extent. Vulnerability is defined by UNISDR (2009) that "the characteristics and circumstances of a community, system or asset that make it susceptible to the damaging effects of a hazard". Exposure is defined by UNISDR (2009) that "People, property, systems, or other elements present in hazard zones that are thereby subject to potential losses" (UNISDR 2009). Exposure to hazards has multiplicd as urban centers grow and economic activities expand into increasingly exposed and hazard-prone land (UNISDR and UNESCAP 2012).

$$\text{Natural Hazards} \times \frac{\text{Vulnerability and Exposure}}{\text{Resilience}} = \text{Disaster Risk}$$

Considering the recent discussions on resilience, exposure and vulnerability, the formula suggested by Yodmani (2001) can be modified as the other one as above, which reflects better the purpose of Disaster Risk Reduction to reduce vulnerability and exposure to hazards and to build resilience from impacts of disasters. As Asia Pacific Disaster Report 2012 pointed out, exposure to disaster risk is growing faster than our ability to build resilience and the shared challenge is to control both thc growing exposure and rising vulnerability (UNISDR and UNESCAP 2012).

1.3 Evolution of the International Agenda on DRR

As various scholars (Twigg 2007; Joerin 2012; Surjan et al. 2011) discussed, disaster risk reduction is a relatively new concept and evolved only over the past two to three decades. The international agenda on Disaster Risk Reduction (DRR) advanced significantly during the last two decades. In the late 1980s, increasing losses in development gains from disasters prompted a global movement towards disaster risk reduction.

1.3.1 International Decade for Natural Disaster Reduction

The United Nations declared 1990s as International Decade for Natural Disaster Reduction (IDNDR) to contribute to technical and scientific buy-in and to make DRR agenda imperative. "Yokohama Strategy and Plan of Action" which was adopted at the first United Nations World Conference on Disaster Reduction (WCDR) in 1994 through the mid-review of IDNDR provided the first international blueprint for disaster reduction policy guidance focusing on social and community orientation while largely encouraging technical solutions to lessen the probability of disasters.

1.3.2 International Strategy for Disaster Reduction

At the end of the IDNDR in 1999, the United Nations General Assembly established the secretariat of the United Nations International Strategy for Disaster Reduction (UNISDR) to facilitate the implementation of the International Strategy for Disaster Reduction, as the successor mechanism of IDNDR within the United Nations to promote increased commitment to DRR and strong linkages to sustainable development. UNISDR was mandated in the United National General Assembly Resolution (56/195) "to serve as the focal point in the United Nations system for the coordination of disaster reduction and to ensure synergies among the disaster reduction activities of the United Nations system and regional organizations and activities in socio-economic and humanitarian fields" (UNISDR 2011a).

The second UN WCDR was held 3 weeks after the catastrophic event of the Indian Ocean tsunami in January 2005 in Kobe City, Hyogo Prefecture, Japan. With stronger political commitment on DRR, the "Hyogo Framework for Action (HFA) 2005–2015: Building the Resilience of Nations and Communities to Disasters" (UNISDR 2011a, b) was adopted by 168 Member States and endorsed unanimously by all UN Member States in the General Assembly. The discussions at the second UN WCDR and the HFA as the outcome document were based on the consultation process through the Inter Agency Task Force on Disaster Reduction facilitated by

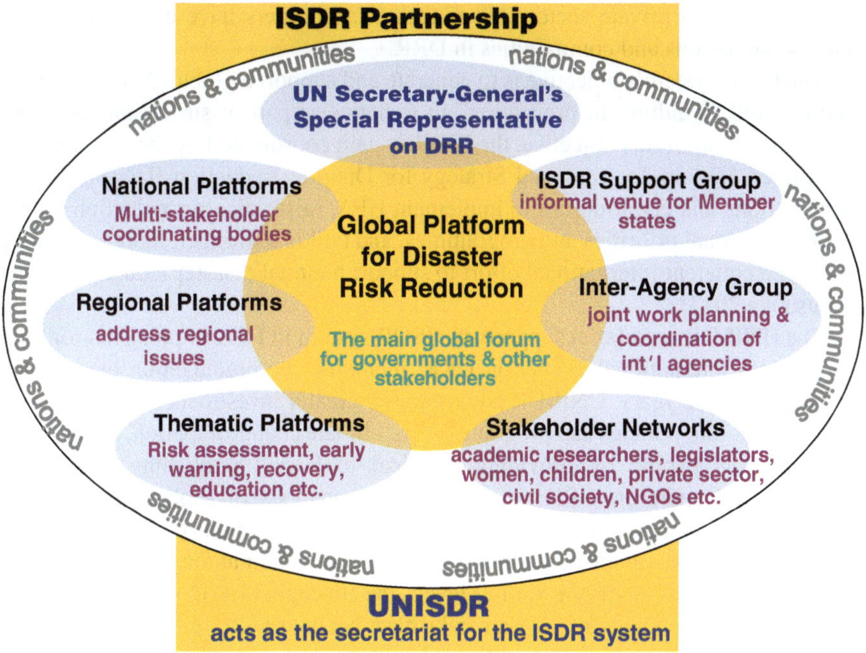

Fig. 1.1 Elements consisting ISDR system (*Source*: UNISDR 2005)

UNISDR. The consultation process was building on the review of the Yokohama Strategy, which identified key challenges. International DRR agenda shifted from technical and scientific work among experts to political commitment by decision makers backed by such experts. This shift and increased political commitment and recognition on DRR led to the development of HFA as the comprehensive DRR policy guidance to all stakeholders. With the adoption of HFA, the United Nations General Assembly tasked UNISDR with supporting its implementation and monitoring the progress of its implementation. UNISDR is the UN office dedicated entirely to disaster risk reduction, as an entity of the UN Secretariat led by the Special Representative of the Secretary-General for Disaster Risk Reduction (UNISDR 2011a).

Adoption of the HFA at the second UN WCDR and its follow up has created and fostered the movement on DRR. This period corresponds to the period of realization of the people's vulnerability and emerging comprehensive approach on disaster management discussed by Surjan et al. (2011), leading to a more comprehensive approach by considering the inter-related components: hazard assessment; vulnerability analysis; and enhancement of management capacity.

UNISDR mobilizes and coordinates a vibrant network (Fig. 1.1) and partnership called ISDR system, which consists of numerous organizations, governments, inter-governmental and non-governmental organizations, international financial institutions, scientific and technical bodies and specialized networks, UN agencies, civil society and private sector. All of these stakeholders have essential roles in supporting nations and communities in DRR.

The ISDR system's objective is to generate and support a global DRR movement and to build "a culture of prevention" in societies as part of sustainable development. In pursuit of this objective, the ISDR system coordinated by the secretariat of the United Nations International Strategy for Disaster Reduction (UNISDR) supports nations and communities to implement HFA, raises disaster reduction profile in organizational priorities and programmes, and builds a stronger, more systematic and more coherent international effort to support national disaster reduction efforts (Matsuoka 2013).

The UNISDR secretariat supports the ISDR system in HFA implementation, and coordinates international efforts on DRR, including the organization of a Global Platform every 2 years. UNISDR advocates for greater investment and the integration of DRR into policies and programmes for climate change adpation, and informs and connects people by providing practical tools and publishing the Global Assessment Report on DRR, an authoritative analysis of global disaster risk. UNISDR also supports the monitoring of HFA implementation (UNISDR 2012a).

Throughout the enhanced partnership and collaboration among a wide range of partners to implement HFA at various levels, the concept of resilience, vulnerability, exposure as components of disaster risks have become more informed concept in order to take concrete actions directly addressing to these components.

1.4 Progress on DRR at Different Levels

As Twigg (2007) discussed, no single group or organization can address every aspect of DRR which sees disasters as complex problems demanding a collective response from different disciplinary and institutional group—in other words, partnerships. The level of follow up actions of the ISDR system to promote HFA implementation includes global, regional, national, and thematic levels by emphasizing the multi-stakeholder partnership.

1.4.1 Global Level

The Global Platform on Disaster Risk Reduction is the main global forum being organized by UNISDR every 2 years in Geneva, Switzerland where the Headquarters of the UNISDR is located. The Global Platform brings together all parties and stakeholders involved in DRR, and campaigns to build global awareness. The first session of the Global Platform on DRR was organized in 2007, the second session in 2009, the third session in 2011, and the fourth session will be held in 2013. The Global Platform allows key actors to assess HFA implementation progress, enhance awareness of DRR, share experiences and learn from good practice, and identify remaining gaps to accelerate national and local implementation. The outcomes of the session of the Global Platform are captured in a Chair's summary, which become key guidance for multi-stakeholders working on DRR in their own priority setting. Analysis of Chair's summaries of the Global Platform will be discussed later in this chapter to explore the trends within the recent discussions. The Global Platform is complemented by national, regional, and thematic platforms (UNISDR 2011b). Since the adoption of the HFA in 2005, the Global Platform has been playing a crucial role as the global mechanism to bring multi-stakeholders on DRR to discuss the progress and challenges in the implementation of HFA.

1.4.2 Regional Level

Regional Platforms, being coordinated by regional presences of UNISDR with a wide range of regional partners, are regional-focus mechanisms which exist in all the regions (Africa, Asia, Pacific, Americas, and Europe). Through the regional platforms, including regional ministerial conferences, regional DRR actors get together and discuss regional progress and challenges on DRR, including participations from representatives from states, national platforms, NGOs, scientific and technical organizations, and regional intergovernmental organizations, UN offices, economic commissions, development banks, inter-governmental organizations, committees, associations and networks. In Asia and Pacific Region, the regional ministerial conference on DRR is organized every 2 years to promote the regional implementation of HFA and aims at addressing common regional challenges and priorities. ASEAN Agreement on Disaster Management and Emergency Response (AADMER) was entered into force in 2009 as a regional agreement that legally binds ASEAN Member States to promote regional cooperation and collaboration in reducing disaster losses and in testifying joint emergency response to disasters in the region. In Africa, African Regional Strategy for DRR was endorsed by the African Union. Pacific region has the Pacific Disaster Risk Reduction and Disaster Management Framework for Action: 2005–2010 (the Madang Framework). The Medium Term Plan 2007–2011 was adopted by the Ministerial Session of the European and Mediterranean Major Hazards Agreement (EUR-OPA). The Arab Strategy for DRR 2010–2015 was adopted by the League of Arab States Council of Ministries Responsible for the Environment.

The HFA has brought about a significant momentum for change at the regional level, as these regional adopted strategies are modeled on HFA (UNISDR 2011b). As appreciated within the external evaluation conducted by AusAID, the regional platforms and ministerial conference have improved the ability of UNISDR to coordination among various actors (AusAID 2012). Also considering the increased importance and priority to DRR in each region witnessed for the last several years in particular since the adoption of HFA, the regional actors on DRR have been increasing which requires a regional coordinating mechanism.

1.4.3 National Level

The need to systematically reduce the increased impact of disaster has gained strong recognition especially after the December 2004 Indian Ocean tsunami disaster (UNISDR 2007a). As discussed earlier in this chapter, DRR is the cross cutting issues and requires political and legal commitment, public understanding, scientific knowledge, careful development planning, responsible enforcement of policies and legislation, people-centered early warning systems, and effective disaster preparedness and response mechanism. A multi-stakeholder National Platform for DRR can help provide and mobilize the required knowledge, skills and resources (UNISDR 2007a).

A National Platform for Disaster Risk Reduction is a nationally owned and led forum or committee for advocacy, coordination, analysis and advice on DRR. This mechanism of national DRR coordination has been promoted by UNISDR for the last two decades including IDNDR period. HFA reinforced the call for national platforms for DRR as well (UNISDR 2011b). The number of countries which established its national platform on DRR has been increased to 81 countries as of the end of 2011 (UNISDR 2012a). Ideally, National Platforms are comprised of various stakeholders to combine different expertise. Stakeholders include government, non-governmental organizations, academic and scientific institutions, professional associations, Red Cross/Red Crescent Societies, private sector, media, etc.

Harvey (2010) stressed that the role of the national government in DRR is the responsibility to set the laws and regulations. Asia Pacific Disaster Report 2012 pointed out the challenge in terms of development of DRR legislation. There is no linearity in the development of DRR legislation, policy on its subsequent integration into development planning. This means that countries develop and adopt instruments that fit their needs without necessarily going through a sequential and comprehensive process. Out of 47 countries analyzed in the Asia and Pacific region, only 10 countries have available laws and policies on DRR and development plans that cover DRR and CCA. Of these ten countries, only one country, Viet Nam have DRR legislation, a DRR plan that is long term, and both DRR and CCA fully integrated into its national development plan (UNISDR and UNESCAP 2012). The Philippines is also one of the countries which advanced the DRR legislation and policies, including the enactment of "Philippines Disaster Risk Reduction and Management Act of 2010" as well as the Act No. 9729 "Climate Change Act of 2009" (Office of Civil Defense 2011). In order to monitor these progresses made at national levels, the monitoring mechanism through National Progress Reporting for HFA implementation being facilitated by UNISDR, which will be discussed in detail in the next section in this chapter.

1.4.4 Thematic Level

DRR is a cross-cutting issue and need to be integrated into various thematic areas and be taken acted upon. Thematic Platforms are independent groups in the disaster risk reduction community focused on supporting the implementation of the HFA on a specific area focus. They aim to integrate specific global technical expertise with the concerns of policy makers and practitioners in their thematic areas. Existing thematic platforms within ISDR system includes the themes such as recovery, early warning, earthquake, flood, wind related hazards, land slide, education, capacity development etc. A number of self-organized thematic platforms mainly composed of technical and scientific bodies have been established. They integrate global technical expertise, regional concerns, and national capacities within the thematic areas. By working with these thematic platforms, UNISDR developed guidelines for integrating thematic issues such as gender issues for disaster risk reduction, which was appreciated in the external evaluation on UNISDR conducted by AusAID (2012).

Considering the broad expertise requiring for taking comprehensive DRR policies and actions, expertise inputs through thematic platforms are useful for the international community. Together with the scientific and technical committee, these expert thematic platforms contribute to the thematic technical analysis for the Global Assessment of DRR.

As UNISDR is a relatively young organization being established in 2000, early years of UNISDR focused heavily on advocacy to ensure DRR to be more widely recognized. In recently years, the coordination function has significantly advanced through its convening role including the organization of the Global Platform, Regional Platform, facilitation of these thematic platforms, monitoring of the HFA implementation, and the issuance of the Global Assessment Report. United Nations General Assembly Resolution in 2011 (United Nations 2011) requested UNISDR to facilitate the development of a post-2015 framework for disaster risk reduction.

1.5 Hyogo Framework for Action and Its Five Priorities

"Hyogo Framework for Action 2005–2015: Building the Resilience of Nations and Communities to Disasters" was formulated as a comprehensive, action-oriented response to international concern about the growing impacts of disasters on individuals, communities and national development. Based on careful study of trends in disaster risks and practical experience in DRR, and subjected to intensive negotiations during 2004 and early 2005, the HFA was finally brought to fruition and adopted at the second United Nations World Conference on Disaster Reduction (January 2005 in Kobe, Hyogo, Japan), and was endorsed unanimously by all the United Nations Member States at the United Nations General Assembly in the same year (Matsuoka et al. 2011, 2012; Matsuoka 2013). The expected outcome of the HFA is substantive reduction of disasters losses in lives and in the social, economic and environment assets of communities and countries. This is further elaborated into the below three Strategic Goals and five Priorities for Action (UNISDR 2005).

HFA three strategic goals:

1. The more effective integration of disaster risk considerations into sustainable development policies, planning and programming at all levels, with a special emphasis on disaster prevention, mitigation, preparedness and vulnerability reduction;

2. The development and strengthening of institutions, mechanisms and capacities at all levels, in particular at the community level, that can systematically contribute to building resilience to hazards; and,
3. The systematic incorporation of risk reduction approaches into the design and implementation of emergency preparedness, response and recovery programmes in the reconstruction of affected communities.

HFA five priorities for action:

HFA priority 1: making disaster risk reduction a priority,
HFA priority 2: improving risk information and early warning,
HFA priority 3: building a culture of safety and resilience,
HFA priority 4: reducing the risks in key sectors, and
HFA priority 5: strengthening preparedness for response.
Through the priorities 1–5, HFA recommends a set of actions to help implement comprehensive DRR.

1.6 Mid-Term Progress Review of the HFA

1.6.1 Mid-Term Progress Review of the HFA

During the year 2010, the Mid-term review of the HFA implementation was conducted by UNISDR. Since the adoption of the HFA in 2005, a certain progress has been made in HFA implementation by national governments with support from international and regional agencies. At the same time, the need for comprehensive DRR approach, thus HFA implementation at local level, has been strongly recognized. The Report for the HFA Mid-term review (UNISDR 2011b) admitted that there was still insufficient level of implementation of the HFA at the local level. In addition, GAR 2011 mentions that the strong recognition to the central role of local governance in DRR and DRM acknowledge by most countries, and also added that a failure to strengthen local governments and make progress in community participation means that the gap between rhetoric and reality is widening (UNISDR 2011d). Such gap is being targeted to address through international initiatives such as the ISDR World Campaign for DRR "Making Cities Resilient 2010–2015" (UNISDR 2010), which promotes local governments from around the world to take action in implementing DRR activities.

1.6.2 Comparison Between Global and Asia-Pacific Regional Progress

This section provides an overview analysis of general trends of the HFA implementation in the Asia-Pacific region, as compared to broader global trends towards improved DRR. First, the section briefly outlines the background of the recent 2011 publication, the Global Assessment Report (GAR) on Disaster Risk Reduction. Then, it presents a comparative bar chart with a written commentary on the position of the Asia-Pacific region, in relation to global progress over the past 4 years of 2007–2011, consisting of two reporting cycles.

In the reporting period of 2007–2009, 102 countries participated in the reporting process. The Global Assessment Report 2009 considered 78 countries for its analysis. In the reporting period of 2009–2011, 133 countries participated in the reporting process. The Global Assessment Report 2011 considered 82 countries for its analysis (delayed submissions were not included in the analysis by the GAR). The Asia-Pacific regional synthesis report reflects the review and analysis of information from the national progress reports of 27 countries in the region, covering the 2009–2011 reporting period (UNISDR 2011c). Except for Priority Area 2, in general, HFA progress in the Asia-Pacific was either less than, or equal to, the global progress during the 2009–2011.

1.7 Localizing HFA Implementation

The HFA appeals to national governments, while acknowledging the enabling support of international and regional players, to take action so that disaster losses, in terms of lives, social, economic and environmental assets, are substantially reduced by 2015. To help attain that outcome, it identifies five specific Priorities for Action. The five priorities are not mutually exclusive, especially when focusing on the processes. HFA implementing guideline for national governments titled "Words Into Action: A Guide for Implementing the Hyogo Framework" (UNISDR 2007b) was produced by UNISDR together with partners to be used as a guideline on what processes governments can take in order to take actions and accomplish the five priority areas.

There has been progress in implementation of HFA at the national level; however, a strong need for a comprehensive DRR action at the local level has arisen. This is because impacts of disasters are most immediately and intensely felt at the local levels; therefore, the most effective process in which the HFA would be implemented is at the local level, adapted and owned by the citizens and officials of the local government. Through this process, the decentralized local/city governance in DRR activities is strengthened, and stakeholder roles and responsibilities are identified, clarified, and eventually carried out (Shaw 2009).

Each local entity is unique in its immediate and long-term needs for DRR. All people and entities have a stake in DRR to protect their lives and livelihoods; therefore, not only should their voices be heard, but also they should be able to participate actively. The HFA will greatly increase in its importance and impacts if implemented by local/city governments who have access to those citizens and entities. To facilitate this process, development of the HFA implementation guideline for local governments called "A Guide for Implementing the Hyogo Framework for

Table 1.1 20 Tasks drawn from five HFA priorities to be implemented by local stakeholders (*Source*: Kyoto University, UNISDR 2010)

Local/city governance (HFA priority 1 related)	
Task 1.	Engage in multi-stakeholder dialogue to establish foundations for disaster risk reduction
Task 2.	Create or strengthen mechanisms for systematic coordination for DRR
Task 3.	Assess and develop the institutional basis for disaster risk reduction
Task 4.	Prioritize disaster risk reduction and allocate appropriate resources
Risk assessment and early warning (HFA priory 2 related)	
Task 5.	Establish an initiative for community risk assessment to combine with country assessments
Task 6.	Review the availability of risk-related information and the capacities for data collection and use
Task 7.	Assess capacities and strengthen early warning systems
Task 8.	Develop communication and dissemination mechanisms for disaster risk information and early warning
Knowledge management (HFA priority 3 related)	
Task 9.	Raise awareness of disaster risk reduction and develop education programme on DRR in schools and local communities
Task 10.	Develop or utilize DRR training for key sectors based on identified priorities
Task 11.	Enhance the compilation, dissemination and use of disaster risk reduction information
Vulnerability reduction (HFA priority 4 related)	
Task 12.	Environment: incorporate DRR in environmental management
Task 13.	Social needs: establish mechanisms for increasing resilience of the poor and the most vulnerable
Task 14.	Physical planning: establish measures to incorporate disaster risk reduction in urban and land-use planning
Task 15.	Structure: strengthen mechanisms for improved building safety and protection of critical facilities
Task 16.	Economic development: stimulate DRR activities in production and service sectors
Task 17.	Financial/economic instruments: create opportunities for private sector involvement in DRR
Task 18.	Emergency and public safety; disaster recovery: develop a recovery planning process that incorporates DRR
Disaster preparedness (HFA priority 5 related)	
Task 19.	Review disaster preparedness capacities and mechanisms, and develop a common understanding
Task 20.	Strengthen planning and programming for disaster preparedness

Action by Local Stakeholders" (Kyoto University, UNISDR 2010) emerged under the initiative called ISDR Asia Regional Task Force on Urban Risk Reduction (RTF-URR), which is one of regional thematic platforms of the ISDR system (Table 1.1).

"A Guide for Implementing the Hyogo Framework for Action by Local Stakeholders" (referred hereon as the Guide) interprets "Words into Action" to use for local level implementation by customizing the guidelines made for national level. The Guide is not for contingency planning alone, but it is a tool for development as well as local/city governance. By using this guideline, stakeholders may identify the gaps in its DRR plans and activities, which will allow them to then seek

Table 1.2 Tools listed in the guide for implementing HFA by local stakeholder (Source: Kyoto University, UNISDR 2010)

HFA priority 1 related tools
Focal point for disaster risk reduction
Multi-stakeholder dialogue
Disaster risk reduction framework and action plan
Stakeholder engagement/coordination mechanisms
HFA priority 2 related tools
Risk communication and dissemination mechanisms for disaster risk information
Early warning systems
Community risk assessment
Gap analysis (including risk-related information)
HFA priority 3 related tools
Disaster information system
Public disaster awareness raising programme/strategy
Training programmes and networks in support of DRR
HFA priority 4 related tools
Disaster recovery plan
Environmental impact assessment
Financial/economic instruments
Poverty reduction programme/strategy
Promoting building safety and protection of critical facilities
Risk-sensitive urban and land-use planning
Sectoral sub-work groups to stimulate DRR activities in production and service sectors
HFA priority 5 related tools
Disaster preparedness planning and programming
Capacity assessment of disaster preparedness and mechanisms

appropriate partnerships and networks to work together for safer communities. Thus, putting this Guide to use requires an arena or forum at local level by which people of different backgrounds and affiliations can share experiences, uncertainties, knowledge, and success stories of others. This forum is referred to as a "local platform." The local platform of multi-stakeholders will thus serve as an advocacy tool of DRR in the local context. It will facilitate coordination and participatory process engaged in problem-solving based on evidence. Resources from various areas will be combined. Also, it will streamline the planning process so that DRR can be accepted as a public value and be mainstreamed into local/city plans as well as day-to-day operations of constituted authorities and businesses.

Breaking down a problem into its components often helps to simplify and understand the situation. Each component may have its unique solution, and different tools can be used to reach such solution. Tools are best utilized by help from trained professionals to different sectors. Table 1.2 lists a sample of tools mentioned in the Guide that would be helpful in accomplishing the tasks given in each HFA priority area. The Guide also gives detailed descriptions of these tools, including its purpose, relevance, and use.

In the guideline "Words into Action", 22 tasks are identified to implement HFA Priority for Action. According to the "Words into Action," each task is a primary area of effort for implementing DRR and can be used to monitor achievement by using them as indicators of progress. The 22 tasks of "Words into Action" were adapted to be used at local/city levels, and a slightly modified version of the list of 20 tasks was presented in the Guide for local/city government's use (Table 1.2).

1.8 DRR Stakeholders

No single agency or actor can deal with DRR issues alone (Izumi 2012). Effective DRR relies on the efforts of many different stakeholders, including UN agencies, regional and international organizations, CSOs, private sectors, media and academics. The collaboration and cooperation among all stakeholders is crucial in order to improve the resilience of communities (UNISDR 2007a; Izumi and Shaw 2011, 2012a, b). In the rest of this section, the roles of stakeholders in DRR at international, regional, national and local levels are discussed. This chapter focuses on the traditional stakeholders in DRR that have been working for DRR as programme implementer for many years.

"International organizations" mainly consist of two major categories—international governmental organizations (IGOs) such as UN agencies, and international non-governmental organizations (INGOs) (Archer 2011). Also, there are some organizations that belong to neither such as the International Committee of the Red Cross (ICRC) and the International Federation of Red Cross and Red Crescent Societies (IFRC) that hold memberships that have their own set of legal rights defined by international conventions (Roeder 2011). ICRC and IFRC together with the Red Cross and Red Crescent National Societies comprise the International Red Cross and Red Crescent Movement. IFRC has more a strong priority on DRR.

Alliances and networks also play an important role in DRR. The alliance and network do not consist of only international entities, but rather, of national and local entities invited from all over the world. DRR is deeply linked to climate change, rapid population growth, environmental degradation, and overall increased conditions of vulnerability, and these complex nature of the subject requires discussion, learning, and information exchange beyond cities, countries, and regions that all the entities belong to.

The discussion and conference to form the Local Government Alliance for DRR has taken place since 2008 at the international level with the leadership of UNISDR. This initiative is linked to the 2010–2011 World Disaster Reduction Campaign on Making Cities Resilient by UNISDR. The Alliance aims for knowledge and information sharing, encouraging the active role of local and regional authorities in DRR, improving local governments' understanding of DRR strategies and implementation by the central government, and ensuring the coordination of DRR actions among relevant stakeholders (UNISDR 2009). Through the Alliance, it is expected to promote capacity development for local governments, localize HFA, and share lessons learned from climate change adaptation programs (UNISDR 2009).

In addition, GNDR was formed in 2008 to bring together CSOs committed to influencing and implementing DRR policy and practice at the local, national, and international levels. The activity and achievement of the Network is highlighted by a global survey called "Views from the Frontline (VFL)" that targeted local authorities, CSOs, and communities in 48 countries about the implementation of DRR at the local level. For the second VFL survey conducted in 2009, 69 countries participated (GNDR 2009, 2011). The report results clearly showed there is a gap in the progress of the HFA implementation between the national and the local levels, and the progress in establishing national policies and legislation had not generated widespread challenges in local practices.

References

Adger WN (2000) Social and ecological resilience: are they related? Prog Hum Geogr 24(3):347–364

Allen KM (2006) Community-based disaster preparedness and climate adaptation: local capacity-building in the Philippines. Disasters 30(1):81–101

Archer C (2011) International organizations. Routledge, London

AusAID (2012) Australian Multilateral Assessment (UNISDR) March 2012. Australian AID. Available at: www.ausaid.gov.au/publications. Accessed on Dec 24, 2012

Bruneau M, Chang SE, Eguchi RT, Lee GC, O'Rourke TD, Reinhorn AM, Shinozuka M, Tierney K, Wallace WA, Von Winterfeldt D (2003) A framework to quantitatively and enhance the seismic resilience of communities. Earthquake Spectra 19(4):733–752

GNDR (2009) Views from the frontline: a local perspective of progress towards implementation of the Hyogo framework for action. GNDR, London

GNDR (2011) Views from the frontline: local reports of progress on implementing the Hyogo framework for action, with strategic recommendations for more effective implementation. GNDR, London

Government of Bangladesh (2005) Disaster Management Bureau Strategic Plan 2005–2006. Disaster Management Bureau, Bangladesh Government

Government of Lao PDR (2012) National Disaster Management Plan 2012–2015, Government of Lao PDR

Government of Pakistan (2006) National Disaster Management Ordinance. Available at: http://ndma.gov.pk

Government of the Philippines (2011) National disaster risk reduction and management plan 2011–2028, the Office of the Civil Defense. Available at: http://www.preventionweb.net/english/professional/news/v.php?id=28964. Accessed on Sep 15, 2012

Harvey P (2010) The role of national governments in international humanitarian response to disasters. In: 26th ALNAP meeting in Kuala Lumpur, Meeting Background Paper, ALNAP, London, 16–17 November 2010

Izumi T (2012) Effectiveness of partnership between Civil Society Organizations and local governments for disaster risk reduction in Asia: perspectives from Indonesia and Malaysia. PhD Theris, International Environment and Disaster Management Laboratory, Graduate School of Global Environment Studies, Kyoto University, Kyoto

Izumi T, Shaw R (2011) Roles of Civil Society in climate and disaster resilience of cities and local governments. In: Shaw R, Sharma A (eds) Climate and disaster resilience in cities. Emerald Publisher, Bingley, pp 261–280

Izumi T, Shaw R (2012a) Role of NGOs in community-based disaster risk reduction. In: Shaw R (ed) Community-based disaster risk reduction. Emerald Publisher, Bingley, pp 35–54

Izumi T, Shaw R (2012b) Roles of stakeholders in disaster risk reduction under the Hyogo framework for action: an Asian perspective. Asian J Environ Disaster Management 4(2):165–182

Joerin J (2012) Enhancing climate-related disaster resilience of urban communities in Chennai, India. PhD Thesis, International Environment and Disaster Management Laboratory, Graduate School of Global Environment Studies, Kyoto University, Kyoto

Kyoto University, UNISDR (2010) A guide for implementing the Hyogo framework for action by local stakeholders. Available at: http://www.preventionweb.net/english/professional/publications/v.php?id=13101. Accesed on Apr 4, 2011

Matsuoka Y (2013) Enhancing local government's capacities to cope with urban disasters in Makati, Philippines. PhD Thesis, International Environment and Disaster Management Laboratory, Graduate School of Global Environment Studies, Kyoto University, Kyoto

Matsuoka Y, Sharma A, Shaw R (2011) Linking resilience planning to Hyogo framework for action at local level. In: Climate disaster resilience in local governments, community, environment and disaster risk reduction, vol 6. Emerald Group Publishing Limited

Matsuoka Y, Takeuchi Y, Shaw R (2012) Implementation of Hyogo framework for action in Makati. Int J Disaster Resilience Built Environ (Emerald Group Publishing Limited)

Office of Civil Defense (2011) Philippines: national progress report on the implementation of the Hyogo framework for action (2009–2011) – interim. National Disaster Risk Reduction and Management Council, Manila

Paton D, Johnston D (2001) Disasters and communities: vulnerability, resilience and preparedness. Disaster Prev Manag 4:270–277

Roeder LW (2011) Diplomacy, funding and animal welfare. Springer, Berlin

Shaw R (2009) Role of local actors in community based disaster risk reduction. Middle-east Technical University Disaster Management Implementation and Research Center, Ankara

Shaw R, Okazaki K (2004) Sustainable Community Based Disaster Management (CBDM) practices in Asia: a user's guide. Disaster Management Planning Hyogo Office, United Nations Center for Regional Development (UNCRD), Kobe

Surjan A, Sharma A, Shaw R (2011) Understanding urban resilience. In: Climate and disaster resilience in cites, community, environment and disaster risk reduction, vol 6. Emerald Group Publishing Limited, pp 17–45

Twigg J (2007) Characteristics of disaster-resilient community, a guidance note. Report prepared for the DFID Disaster Risk Reduction Interagency Coordination Group. Available at: http://www.benfieldhrc.org/disaster_studies/projects/communitydrrindicators/community_drr_indicators_index.htm. Accessed on Nov 15, 2012

UNISDR (2002) Living with risk, pp 1–5

UNISDR (2005) Hyogo framework for action 2005–2015: building the resilience of nations and communities to disasters. UNISDR, Geneva. Available at: http://www.unisdr.org/eng/hfa/docs/Hyogo-framework-for-action-english.pdf. Accessed on Mar 12, 2010

UNISDR (2007a) Guidelines national platforms for disaster risk reduction. Available at: http://www.unisdr.org/we/inform/publications/601. Accessed on Apr 17, 2010

UNISDR (2007b) Words into action: a guide for implementing the Hyogo framework for action. Available at: http://www.preventionweb.net/english/professional/publications/v.php?id=594. Accessed on Feb 1, 2011

UNISDR (2009) 2009 UNISDR terminology on disaster risk reduction. UNISDR, Geneva. Available at: http://www.unisdr.org/we/inform/publicaitons/7817. Accessed on Dec 15, 2012

UNISDR (2010) Information kit for ISDR 2010–2011 World Disaster Reduction Campaign. Available at: http://www.unisdr.org/english/campaigns/campaign2010-2011/documents/campaign-kit.pdf. Accessed on Mar 5, 2011

UNISDR (2011a) Strategic framework 2025, Work Programme 2012–2015. Available at: http://www.unisdr.org/files/23291_1101657inteng.pdf. Accessed on Mar 17, 2012

UNISDR (2011b) Hyogo framework for action 2005–2015 building the resilience of Nations and Communities to Disasters, Mid-term review 2010–2011. UNISDR, Geneva. Available at: http://www.unisdr.org/files/18197_midterm.pdf. Accessed on Mar 17, 2012

UNISDR (2011c) HFA progress in Asia-Pacific: Regional Synthesis Report 2009–2011. Available at: http://www.unisdr.org/files/21158_hfaprogressinasiapacific20092011.pdf. Accessed on Mar 17, 2012

UNISDR (2011d) Global Assessment Report on Disaster Risk Reduction 2011. UNISDR, Geneva. Available at: http://www.preventionweb.net/gar. Accessed on Mar 17, 2012

UNISDR & UNESCAP (2012) The Asia-Pacific Disaster Report 2012 "Reducing Vulnerability and Exposure to Disasters", Bangkok. Available at: http://www.unisdr.org/we/inform/publications/29288. Accessed on Oct 14, 2012

UNISDR (2012a) UNISDR counts the cost of 20 years of inaction on climate change and risk reduction. Press release on 13 June 2012, UNISDR, Geneva. Available at: http://www.unisdr.org/archive/27162. Accessed on Apr 1, 2012

UNISDR (2012b) Annual Report (2011): UNISDR secretariat Work Programme 2010–2011. Available at: http://www.unisdr.org/files/23291_1101657inteng.pdf. Accessed on Apr 1, 2012

United Nations (2011) United Nations General Assembly Resolution A/RES/66/199. Available at: http://www.unisdr.org/files/resolutions/N1147076.pdf. Accessed on Apr 4, 2012

Wisner B, Blaikie P, Cannon T, Davis I (2004) At risk: natural hazards, people's vulnerability and disasters. Routledge, London

Yodmani S (2001) Disaster risk management and vulnerability reduction: protecting the poor. In: Paper presented at the Asia and Pacific Forum on Poverty Organized by the Asian Development Bank

Chapter 2
Understanding Vulnerability and Risks

A.K.M. Mamunur Rashid

Abstract Disaster Risk Reduction has become a major development agenda in Bangladesh since 2002. The global concept of "Reducing Risk" has also become very popular in Bangladesh, not only because Bangladesh is a disaster-prone country but also Bangladesh has contributed in shaping the knowledge about disaster risk. The concepts of 'vulnerability' and 'risk' has also emerged since 2002 and widely used in disaster risk reduction. The concept of resilience is relatively new, and conveniently used as reverse of the concept 'vulnerability'. Therefore, the understanding of these three concepts are very important and matured the whole discourse of disaster risk reduction. National Alliance for Risk Reduction and Response (NARRI) is a consortium in DRR in Bangladesh has measured the resilience of their community based preparedness work since 2002 in Bangladesh. Resilience analysis consisting a comprehensive resilience matrix has applied a five band scale against each characteristics of resilience clustering component of resilience under five broad theme. Highest level of resilience has been achieved in the risk assessment where community members' ability and skill of assessing risk and vulnerability increased and people are using their skills for risk and vulnerability assessment, preparation of management plan and followed by implementation for their risk reduction. Conversely due to lack of access to the structure and process traditionally led by the community and cultural and knowledge gap impede the level of resiliency achieved in the governance issues. The Bangladesh case of resilience study clearly shows that governance, capacities for risk assessment, risk education, redundant risk reduction options, resourcefulness of people, society and ecosystem, and a capacity to have rapidity in recovery has potential in building resilience. The risk reduction in many cases could be seen as "element" focused, where "resilience" offers a more robust scope to whole of society into the risk reduction framework.

A.K.M.M. Rashid (✉)
Poverty Environment and Climate Mainstreaming Project, UNDP, Dhaka, Bangladesh

PhD fellow of Department of Anthropology in Jahangirnagar University, Dhaka, Bangladesh
e-mail: nemoibsru@gmail.com

R. Shaw et al. (eds.), *Disaster Risk Reduction Approaches in Bangladesh*,
Disaster Risk Reduction, DOI 10.1007/978-4-431-54252-0_2, © Springer Japan 2013

Keywords National alliance • Preparedness and management plan • Resilience • Risk reduction • Vulnerability

2.1 Introduction

There is a very popular slogan in disaster management community, which was disseminated by ISDR extracted different cost–benefit analysis findings across world, "if you invest one dollar in DRR you will be able to save 7 dollar in emergency". There are ample evidences that natural disasters are affecting more of the world's population, and are projected to increase in severity and intensity under climate change. The cost of disasters, both in terms of lost GDP as well as spending on relief and rehabilitation, is significant. Investment in DRR is, however, limited. There is uncertainty around where and when disasters will strike, and what harm they will cause (particularly in the case of events such as cyclones, earthquakes, floods). Meanwhile, governments and donors focus spending on immediate needs for poverty reduction, such as health, water and food security, and hesitate to invest in DRR where the immediate development outcomes are not always clear.

The global disaster record demonstrates a clear increasing trend of incidence and increasing severity of the natural disasters. This is also very clear and pertinent now that the cost of disasters, both in terms of GDP loss (including hidden GDP loss) and expenditure diverted to relief and rehabilitation has increased. The World Bank has estimated that, between 1990 and 2000, natural disasters caused damage valued at between 2 % and 15 % of an exposed country's annual GDP (Humanitarian Exchange Magazine, Issue 38, June 2007).

The recent high-profile Stern Review (Stern Review 2006) also support the World Bank's argument. The Stern Report included natural disaster as one of the core component of climate change and estimated the costs and benefits of reducing the risks associated with climate change with a central conclusion that "the benefits of strong and early action far outweigh the economic costs of not acting". The Review estimates that, if steps are not taken to combat climate change (including disaster), the overall costs and risks will be equivalent to losing at least 5 % of global GDP each year, now and for the foreseeable future. If a wider range of risks and impacts is taken into account, the damage could rise to 20 % of GDP or more. In contrast, the costs of the action required to reduce emissions to a level which would avoid the worst impacts of climate change can be held at around 1 % of global GDP each year (Humanitarian Exchange Magazine, Issue 38, June 2007).

Evidence on the costs and benefits of DRR (Willenbockel 2011; Moench and The Risk to Resilience Study Team 2008; Tearfund 2004; World Bank 2007; SOPAC 2008; IFRC 2009) consistently shows that investment brings greater benefits than costs, and therefore should be a priority for development planning. However, this evidence is limited and very location- and hazard-specific. Further work is needed to demonstrate to planning and finance decision makers and donors that mainstreaming DRR is financially and economically justified. In all cases, only cost benefit analysis is done and the financial benefits are calculated where resilience is not correlated with

Table 2.1 Damage to economic sectors

Damage to economic sectors due to major disasters					
	Floods			Cyclone	
Damage	1998	2004	2007	Sidr 2007	Aila 2009
Total damage (Crore TK)	11,419	13,450	7,253	11,557	1,885
% of GDP	5.7 %	4.04 %	1.53 %	2.45 %	0.31 %
Infrastructure loss (% of total loss)	50.6 %	74.4 %	61.6 %	63.3 %	84.3 %
Non-structural/livelihood/agriculture loss (% of total loss)	49.4 %	25.6 %	38.4 %	36.7 %	23.7 %

Source: Sixth Five Year Plan of Government of Bangladesh; Part 3: Statistical Annex and Technical Framework; pp. 26–27

other economic analysis including cost benefit analysis. We have to accept the fact that cost–benefit analysis is one of the economic tools that can be used to compare the costs and benefits of DRR interventions. Moreover, only a handful of sporadic community-level studies have been conducted, and these have used different methodologies and approaches. In most cases, large scale development projects funded by Governments are not assessed and synergic contribution of Government and CSO organizations are not very clear. Initial research suggests that DRR measures often bring greater benefits than the costs they incur, but this may vary significantly depending on factors such as the type of disaster, the country concerned and the DRR measures employed. However, the questions demands critical analysis of the benefits and require further evidence of the diversified financial and economic benefits of DRR.

Disaster Risk Reduction has become a major development agenda since 2002, in Bangladesh. The global knowledge on "Reducing Risk" has also become very popular in Bangladesh, not only because Bangladesh is a disaster-prone country but also Bangladesh has contributed in shaping the knowledge about disaster risk. There is a history of strong "disaster preparedness" backbone in Bangladesh, which earned justified attention of the world in shaping the knowledge and paradigm in disaster management discipline. In Bangladesh, Bangladesh Institute of Development Studies has conducted an expenditure tracking study (Islam 2011) on disaster risk reduction, which shows a clear consequence of floods on GDP. Considering the cyclone Sidr and Cyclone Aila and the size of the GDP in 2007 and 2009, we can see that that only major floods and cyclones have caused average 3.5 % negative influence on national GDP (Table 2.1).

Despite this historical capacity and epistemology in disaster management, there is no such validated inference drawn in Bangladesh that to what extent the disaster-risk-reduction interventions contributed in building resilience. Moreover, the economic and human benefits are inadequately captured from where lessons can be replicated. The scales of interventions are different, which has not adequately linked through micro–macro analysis.

In this given context, the understanding of vulnerability, risk and resilience is key for Bangladesh and economic, social and political analysis of disaster risk reduction will be very helpful to calibrate a set of DRR interventions into solid clustered economic, social and human benefits and to understand the level of resilience at community and national level in Bangladesh.

2.2 Concepts of Risk and Vulnerability

2.2.1 Vulnerability: A Conceptual Discussion

The phenomenon "vulnerability" in the lexicon of disaster management is different from the arbitrary uses of the term for convenience in daily life. Over the last 18 years "vulnerability" has become a key concept used in the analysis of risk and disaster. Although the notion is not hard to grasp intuitively when used in the context of empirical realities, a clear specification of the nature and types of vulnerability, as well as the processes behind its construction are not always well established. The term "vulnerability" is used by different agencies and disciplines from its' own "term of convenience", and this is also true that little attempt is made to accurately define the notion, to detail its different components or to consider the processes that lead to the construction of vulnerability.

The term "natural disaster" is often used to refer to natural events such as earthquakes, hurricanes or floods. However, the phrase "natural disaster" suggests an uncritical acceptance of a deeply engrained ideological and cultural myth. Blaikie et al. (1994) in their book titled *At Risk* questions this myth and argues that extreme natural events are not disasters until a vulnerable group of people is exposed. The vulnerability is then become a thought or a study phenomenon in disaster management. After 10 years of the book published first, in its second edition Wisner et al. (2004) they accounted more expensive and deadly disasters and discusses disaster not as an aberration, but as a signal failure of mainstream "development". Two analytical models are provided as tools for understanding vulnerability. One links remote and distant "root causes" to "unsafe conditions" in a "progression of vulnerability". The other uses the concepts of "access" and "livelihood" to understand why some households are more vulnerable than others.

There is much confusion around the term; does "vulnerability" refer a state of coping capacity (Kumpulainen 2006) or a potential state for loss (Cutter et al. 2003) or a hypothetical concept (Cardona 2004; Cardona et al. 2005, 2012). The multifaceted meaning of the term demonstrates an importance of a clear definition, specification of different types and indications as to the processes through which vulnerability is constructed, is critical for the programming and planning of risk reduction strategies and instruments. The processes leading to vulnerability are inevitably linked to the development models prevalent in different country contexts and can only be permanently reduced or modified in the context of changes in the parameters of these models. A clear specification of vulnerability types and causation is required if we are to bring the risk and development problems together in a consequent manner.

The usual and common definitions are outlined in UNDP's Emergency Response Division in Geneva in 2000. The document identified several definitions of vulnerabilities.

- The predisposition of a society, or a component of society, to suffer damage or loss when exposed to natural or socially induced physical events, and to face difficulties in recovering from this loss.

- The degree of susceptibility and resilience of the community and environment to hazards (Emergency Management Australia).
- The characteristics of a person or group in terms of their capacity to anticipate, cope with, resist, and recover from the impact of a hazardous event (adapted from Blaikie et al. 1994).
- The predisposition to suffer damage due to external events. Vulnerability is about susceptibility (i.e. exposure and proximity to external events) and resilience (i.e. access to resources, capacities and capabilities). Vulnerability "factors" include underlying causes, dynamic pressures and unsafe conditions (see Blaikie et al. 1994). Vulnerability includes the following dimensions: physical, emergency management, demographic, health, economic, communications, psychological, societal/cultural and organizational (WHO/EHA document).

Seen the definitions we can define "vulnerability" as a state of resilience of people, its belongings, livelihoods, and elements to the susceptible threats to be caused by hazards. Here, this is very clear that vulnerability is defined from the perspectives of human being. The term is defined not from any other perspective but only from human point view and therefore this is very anthropocentric. Therefore, vulnerability could be determined by:

- Location
- Age, sex, physical and mental ability
- Education, Experience, Skill
- Resource Backup
  - Physical
  - Technological
  - Financial
  - Social
  - Human
- Ethnicity
- Constructions
- Flexibility/adaptability/coping ability
- Intellectual ability/intuition power
- Access to information
- Exposure and experience to past hazards

Considering the above determinants, we can assume that vulnerability is not a state of single factor determination but it belongs to the cognitive level interpreting the multiple determinants with relative value scores in a certain context.

Many different schemes exist for classifying the components or factors of vulnerability. These range from rather detailed schemes that identify many specific levels or components of vulnerability (economic, social, organisational, institutional, ideological, educational, cultural, physical, locational etc.; see Wilches-Chaux 1989) through to more condensed schemes identifying vulnerability types (economic/social, organisational/institutional, motivational etc.; see, Anderson and Woodrow 1998). Both types of scheme are relevant and recognise that the "global vulnerability" of a particular subset or group of society is the result of the combination or interaction of different generic types of vulnerability. Vulnerability involves a combination of factors that determine the degree to which someone's life and livelihood is put at risk by a discrete and identifiable event in nature or society.

2.2.2 *Risk: A Conceptual Discussion*

The world risk is also used very conveniently in daily life. The term is generally used to refer the probability of negative consequences. In many cases, the term is used refer occurrences of hazards: flood risk, earthquake risk, cyclone risk, fire risks. All these hazard risks refer the probability of hazard onset based on certain condition. However, this meaning does not refer the probability of loss or damage clearly though these are hidden under the connotation of the hazard risk. The above use of the term "risk" is different from the connotation used in referring life risk, risk of diseases, risk of houses, risk of crop, risk of livelihoods, risk of food and many other risks. The later set of risk refers clearly the probability of consequence. In risk management, these two dimensions are taken into consideration by two terms in a matrix: likelihood of occurrence, and level of consequences. This structure is popularly used in the "insurance" business, when they calculate the risk of insured element: life, property or asset. The term "element at risk" also drives from the view point of the economic perspectives of risks, which is popular perspective in insurance business to determine the premium of the insurer. This framework of risk analysis has predominantly influenced the whole risk management framework in disaster management.

As the disaster management yet to emerge as discipline as there is no structured framework of knowledge (theory, methodology, category and analytical framework) to study the subject matter and no such object of study, therefore, the confusions and ambiguity of the term "risk" is obvious in the disaster studies. "Risk" comprises the essential complex concept used in disaster studies but the existence of risk is recognized as an indispensable requisite in order for disasters to occur. Therefore, definition of risk and the identification of fundamental notions as regards the conformation of risk in society are indispensable for deliberate programming and planning purposes. The available literatures (Adam and van Loon 2000; Beck 2000; Brooks 2003; Cardona 2011; Covello and Mumpower 1985; DKKV 2011; Douglas and Wildavsky 1982; ISO 2009a, b; Blaikie, et al. 1994; Wisner, et al. 2004) in different agencies and organizations are rarely explicit in terms of the definition of risk, assuming a common understanding of its significance. This leads to confusions in many instances. Hazard and risk continue to be confused, as do risk and vulnerability.

The usual and common definitions of "risk" are outlined in UNDP's Emergency Response Division in Geneva in 2000. The document identified several definitions of risk.

- The likelihood of harmful consequences that derive from the interaction of hazards, societal vulnerability and the environment. Risk is potential, a measure of future possible harm under determined conditions, expected loss (adapted from Emergency Management Australia).
- The social loss expected due to the interaction of hazards and vulnerability in a particular time and space.
- A statistical concept relating to the probability that a negative event or condition will affect an individual in a given time and space (WHO/EHA document).

Analysing the above definitions and the initial discussions on risk we can understand that the term "risk" is used to describe the likelihood of harmful consequences, arising from the interaction of hazards, vulnerable elements and the environment. However, different people think about risk in different ways. The generic equations refers that risk is a function of hazard and vulnerability. The risk in specific terms refers the specific element risk in the context of multiple hazards and the cumulative vulnerability of the element in relation to different hazards. The INGOs and Bangladeshi NGOs use the equation of "risk" as follows.

$$\text{Risk} = \text{Hazard} \times \frac{\text{Vulnerability}}{\text{Capacity}}$$

This is also confusing in nature, as this contradicts with the definition of vulnerability. The least capacity refers more vulnerability, then how capacity has got separate variable than vulnerability is not clear at all. Perhaps, they try to refer "potential damage" under exposure to hazards. Therefore, in most calculation vulnerability is assumption based and equal to all people exposed to the hazards. The capacity then become key variables, and such equations ended with "capacity building" discourse.

2.2.3 *Risk and Vulnerability Assessment: Micro and Macro Perspectives*

Risk Assessment is considered to be the most critical element in determining the community risk and also in defining the risk environment. Generally two different approaches are followed for defining the disaster risk in Bangladesh. The engineering and scientific institutions, both from the government and autonomous organizations, define the risks from the viewpoint of more physical, hydro-meteorological and spatial aspects with very little human dimensions. The organization that follows this approach includes Water Related Planning and Research Organizations like Bangladesh Water Development Board (BWDB), Water Resources Planning Organization (WARPO), Institute of Water Modeling (IWM), Center for Environmental and Geographic Information Services (CEGIS), Local Government Engineering Department (LGED) etc. Clearly, these organizations put considerable emphasis on the opinions and decisions of thereby demonstrate a strong bias to outsider's viewpoint termed as etic culture. Although they make occasional interactions with the vulnerable communities, their assessment is generally governed by their scientific analysis bases on geo-physical considerations. In most cases, it is observed that the assessment of the risks in this approach remains within the scientific communities and to some extent to the very senior officials of the Government and a few research based organizations. Recently, few organizations like CEGIS have slowly started to incorporate human dimensions into their risk assessment process. The value of the etic viewpoint is extremely important because it provides a broader and future dimension from more of technology-based observations. For example people hardly can perceive the climate change predications while they map out the hazard scenario and the physical vulnerability.

On the other hand, a large number of organizations, mostly NGO's are very keen to define the risks from the viewpoint of "people at risk". A number of methods are generally practiced by these organizations. Participatory Vulnerability Assessment (PVA) and Participatory Capacity Assessment (PCA) are the key methods of risk assessment by the NGOs. It is clearly evident from the very name that "peoples participation" is the key to this approach and thereby retains a dominance of local viewpoint. In most cases, different Participatory Rural Appraisal (PRA) tools are used in these methodologies; for example, mapping, seasonality, diagramming, group discussion, and a range of ranking tools are usually applied in these methodologies. These NGOs hardly use any modelling data to conduct risk assessment at the community level. Their approach is totally based on the risks perceived by the people and interpreted by the people from their own point of view, which is called emic approach. This approach highly focuses on human dimensions; diversity, differences, cultures, economy, political process, and inequalities with very little input from scientific or technical considerations. This approach seeks more in-depth answer to the question "Why do some people remain more vulnerable than others?" which is not limited to physical and spatial dimensions but bring many other root causes like human rights, good governance, gender and other inequality issues. ActionAid Bangladesh, CARE Bangladesh, Oxfam, Concern, World Vision, Bangladesh Disaster Preparedness Center (BDPC), Bangladesh Unnayan Parishad (BUP), and many other local NGOs are the practitioners of the approach.

In 2006, CDMP introduced a nationally driven consultative process that takes account of the strength of both approaches to standardized Community Risk Assessment (CRA) process. In doing so, CDMP bring the practitioners of both approaches together and organizes a series of consultative workshops at various stages from the community to the national level. The consultative workshops yield a new approach that combines the existing methods in a synergistic way. The beauty of this integration was the blending of technology-based assessment into participatory process. NGOs who work very closely with community people will collect the etic perspective before entering into the risk assessment process. The secondary and technical information will give more precise point of discussion and participation to bring the peoples perspective and rank the risks as people prioritise by their own criteria. The scientific maps from organizations like IWM, CEGIS, LGED and BWDB having predictions and trends of hazards, provide useful reference for the community in the preparation of participatory hazards maps. This would contribute like "which hazard produces risks to which sector, which people of which area and at which level" are adequately addressed in the new approach. The integrated Community Risk Assessment process has six key steps: (a) Scoping the Communities (pre-field visit and field visit), (b) Identification of Hazards, (c) Identification of Vulnerable Sectors and specific Risks in each sector, (d) Analysis and Evaluation of Risks, (e) Determine Specific Risk Reduction Options/Measures and (f) Monitoring and Review. In each step, etic and emic approaches are better integrated to bring the optimum results to define the risk environment of a community. In every step, gender and social exclusion analysis were done using emic approach, which helped to identify the differences of risks by gender, age, physical ability, social and economic class and ethnic identity.

2.2.4 Assessing Resilience: An Emerging Phenomenon in DRR

Resilience is relatively a new phenomenon in the disaster management discourse though the use of the term is observed since 1991 in Bangladesh to refer the bounce back capacity of people after any disaster. The whole phenomenon is developed under the niche of emergency response and disaster preparedness discourse. The "resilience" as term refers an internal strength of the people or element which in many cases similar to "capacity" and opposite to "vulnerability" in the lexicon of the disaster management. However, the ambiguity is evident by three distinct initiatives: Center for Community Enterprise (CCE) in 1999, Characteristics of a Disaster Resilient Community by John Twigg in 2007 and revised in 2009 initiated by Interagency Group and the very recent further initiative of BOND Disaster Risk Reduction Group in 2012.

If we carefully look at these three initiatives and try to understand the insight they want to give us characterising the "resilience" is not similar rather sometime confusing. The CCE defined a resilient community as "a resilient community is one that takes intentional action to enhance the personal and collective capacity of its citizens and institutions to respond to, and influence the course of social and economic change (CCE 2000)". Twigg (2009) has defined the "resilience" by a set of broad characteristics, "System or community resilience can be understood as the capacity to anticipate, minimize and absorb potential stresses or destructive forces through adaptation or resistance; manage or maintain certain basic functions and structures during disastrous events and recover or 'bounce back' after an event". Twigg (2009) also illustrated the resilience in relation to capacity in his book as "Resilience is generally seen as a broader concept than 'capacity' because it goes beyond the specific behaviour, strategies and measures for risk reduction and management that are normally understood as capacities. However, it is difficult to separate the concepts clearly. In everyday usage, 'capacity' and 'coping capacity' often mean the same as resilience" (page 8, para 8). The BOND's new additional effort in 2012 also describes the resilience in a system approach and tried to explain that "A resilient system has the capacity to respond positively to change, maintaining or improving function; this includes monitoring, anticipating and managing known risks and vulnerabilities to existing shocks and stresses whilst being able to address uncertainties in the future. Change and responses may be incremental or transformational (Characteristics of Resilience Building: A Discussion Paper, 2012, page 7)". In this latest initiative, again the resilience is referred as a set of capacities and not only based on past and current shocks and stress but can also for future uncertainties. All these initiatives also suggested a set of characteristics, which also contradicts each other. CCE (2000) has suggested 4 dimensions of resilience: people, organization, community process and resources with 23 characteristics of resilience. Twigg (2009) suggested 167 characteristics of resilience in line with Hyogo Framework for Action (HFA). BOND (2012) has suggested five characteristics; one as process characteristics and four enabling characteristics for building resilience.

When the issue of measuring the characteristics, there is also lack of clear instruction from these initiatives. However, there is no such initiatives are observe in Bangladesh, where community resilience is measured. Many other characteristics and its measurement are hypothetical in nature. In 2012, a NGO alliance has initiated a process to measure such resilience at community level with a methodology and robust set of scales under all 167 characteristics of resilience that Twigg (2009) has proposed. The particular case could be illustrated in next section.

2.3 Resilience Analysis in Bangladesh: A Case Study

National Alliance for Risk Reduction and Response (NARRI) is a consortium in DRR has taken a recent process to measure the resilience of their community based on their work since 2002 in Bangladesh. They have adopted the framework of Twigg (2009) and made conscious attempt to do an analysis of resilience of community through different DRR interventions by DIPECHO assistance in Bangladesh. Adopting John Twigg's Characteristics of Resilience (Twigg 2009), a country specific community resilience analysis framework was developed. Various DRR interventions were categorized and therefore measured and assessed applying the framework. The resilience analysis also considered the economic analysis and well as vulnerability analysis from community perspectives. The generic types of interventions for DRR in Bangladesh were assessed for resilience analysis was follows.

1. Development of Risk Protection measures through Plantation/forestation
2. Development of emergency evacuation route from risk areas to emergency shelter
3. Raising risk awareness among larger community and most vulnerable people
4. Building community volunteerism for early warning
5. Development of Alternative Livelihood Options (ALO) and Hazard Resilient Livelihoods
6. Established Alternative Options to ensure access to safe drinking water to reduce water born disease
7. Introducing and piloting of early, short duration rice variety and flood tolerant agriculture product

8. Development of alternative options to protect crops/agricultural products from natural hazards
9. Raise the plinth of houses above flood level/free/resilient from/to Hazards
10. Renovate and construction of Disaster Endurable Houses
11. Raise the plinth of common places/school fields/other community resource places above/free/resilient from/to Hazards
12. Raise the plinth of Graveyard above highest Flood Level to safe burial
13. Protection of cropland and living places from hazards by small structural intervention and strengthening the existing structures
14. Re-excavate water way for drain out surplus water
15. Emergency support to Marooned people
16. Capacity Building of TBAs to reduce new born death through ensure safe delivery using modern technique
17. Capacity Building of Persons with Disability to be Able to Enhance Their Mobility
18. Protection of Educational Institute from River Erosion hazards by constructing movable school.

Resilience analysis used a comprehensive resilience matrix using a five band scale against each characteristics of resilience under certain component of resilience clustered under five broad theme of resilience. One example of the scales is described in Table 2.2. The five bands will be interpreted as following significance.

1. Very Low Resilience (initiate a resilience process at community)
2. Low Resilience (benefits are not obvious or under observation by the initiative)
3. Medium Resilience (benefits are obvious and community are willing to continue such initiative, there is a sense of ownership)
4. High Resilience (community has fully owned the initiative and practiced themselves without external support)
5. Very High Resilience (community has scaled up the practices and stimulated many others to practice the same or scaled up)

The study (Rashid et al. 2012) has tested the resilience outcome of the past DIPECHO projects applying the Twigg (2007) characteristics for measuring community resilience in five key components identified by Hyogo Framework on Action (HFA) including governance, risk assessment, knowledge and education, risk management and vulnerability reduction, and disaster preparedness and response. The result shows that community under the DIPECHO programme has been achieved medium level of resilience means benefits are obvious and community are willing to continue such initiatives and the sense of ownership is developed among them (Fig. 2.1).

This resilience analysis also helped to align specific contribution in the country progress to HFA. The resilience framework was therefore aligned with HFA in following manner (Table 2.3).

Highest level of resilience has been achieved in the risk assessment where community people's ability and skill of risk and vulnerability assessment increased and

Table 2.2 Framework for "Resilience Analysis": an example of measuring scales against characteristics

Thematic area	Component of resilience	Characteristics of a disaster resilient community	Measuring scales	Tangible benefits	Intangible benefits
Risk management and vulnerability reduction	Sustainable livelihoods	Adoption of hazard-resistant agricultural practices (e.g. soil and water conservation methods, cropping patterns geared to low or variable rainfall, hazard-tolerant crops) for food security	1. Technology is transferred at farmers level 2. Farmers find benefit of such practices 3. Inputs and product markets are accessible to the farmers 4. Farmers are practicing new agriculture without external support 5. Practice is scaled up and has become integral part of agriculture system	Increased income Increased agricultural productivity	Mental stress of farmers decreased Farmers empowerment increased

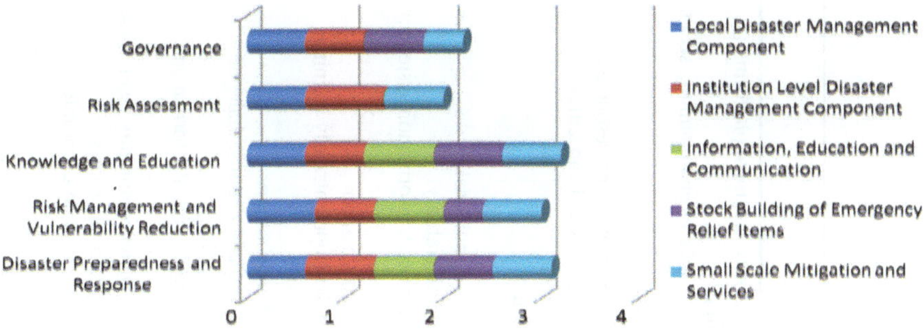

Fig. 2.1 Observed DIPECHO intervention and level of resilience. *Source*: Rashid et al. (2012, p. 53)

people are using their skills for risk and vulnerability assessment, preparation of management plan and followed by implementation for their risk reduction. Conversely due to lack of access to the structure and process traditionally led by the community and cultural and knowledge gap impede the level of resiliency achieved in the governance issues. The study also identified that the uses of Twigg's characteristics though help to report to HFA progress but limit to understand the resilience of a community. The study offered four broad characteristics and four dimensions to analyse the community resilience. The four dimensions are: People, Organization, Process and Resources; and the four characteristics are Robustness, Redundancy, Resourcefulness and Rapidity (Fig. 2.2).

Applying this alternative set of characteristics the study (Rashid et al. 2012) has identified that DIPECHO achieved wide range of changes in the Disaster Risk Reduction (DRR) approaches in Bangladesh. Among them accountability, community participation, education and training, social protection and voluntarism are important areas. A brief discussion of the key findings will be interesting to worth in illustrating the case.

2.3.1 Accountability and Community Participation

Accountability and community participation hold an important role in disaster resilience where the accountability as a social and political process ensures role and responsibilities of government and service providing agencies and empowered community, and community participation as an additional element ensuring transparency, and responsiveness feed reversing the exponential increase in disaster occurrence, build culture of safety, and ensure sustainable development. Accountability and participation empowered people and improve their capacities for identification of their own risks and vulnerabilities against hazard along with the possessed capacities and resources, and the needs for managing those. Integration of accountability and participation in disaster risk reduction intervention help

Table 2.3 Synergies between Resilience Framework and Hyogo Framework

Resilience thematic area	Components of resilience	HFA link	HFA indicators
Governance	• Policy, planning, priorities and political commitment • Legal and regulatory systems • Integration with dev. policies and planning • Integration with emergency response and recovery • Institutional mechanisms, capacities and structures; allocation of responsibilities • Partnerships • Accountability and community participation	HFA 1. Ensure that disaster risk reduction (DRR) is a national and a local priority with a strong institutional basis for implementation	• DRR institutional mechanisms (national platforms); designated responsibilities • DRR part of development policies and planning, sector wise and multi-sector • Legislation to support DRR • Decentralization of responsibilities and resources • Assessment of human resources and capacities • Foster political commitment • Community participation
Risk assessment	• Hazards/risk data and assessment • Vulnerability/capacity and impact data and assessment • Scientific and technical capacities and innovation	HFA 2. Identify, assess and monitor disaster risks and enhance early warning	• Risk assessments and maps, multi-risk • Indicators on DRR and vulnerability • Early warning: people centered; information systems; public policy • Data and statistical loss information • Scientific and technical development; early warning • Regional and emerging risks
Knowledge and education	• Public awareness, knowledge and skills • Information management and sharing • Education and training • Cultures, attitudes, motivation • Learning and research	HFA 3. Use knowledge, innovation and education to build a culture of safety and resilience at all levels	• Information sharing and cooperation • Networks across disciplines and regions; dialogue • Use of standard DRR terminology • Inclusion of DRR into school curricula, formal and informal education • Training and learning on DRR: community level, local authorities, targeted sectors; equal access • Research capacity: multi-risk; socio-economic; application • Public awareness and media

Risk management and vulnerability reduction	• Environmental and natural resource management • Health and well being • Sustainable livelihoods • Social protection • Financial instruments • Physical protection; structural and technical measures • Planning régimes	HFA 4. Reduce the underlying risk factors	• Sustainable ecosystems and environmental management • DRR strategies integrated with climate change adaptation • Food security for resilience • DRR integrated into health sector and safe hospitals • Protection of critical public facilities • Recovery schemes and social safety-nets • Vulnerability reduction with diversified income options • Financial risk-sharing mechanisms • Public–private partnerships • Land use planning and building codes • Rural development plans and DRR
Disaster preparedness and response	• Organizational capacities and coordination • Early warning systems • Preparedness and contingency planning • Emergency resources and infrastructure • Emergency response and recovery • Participation, voluntarism, accountability	HFA 5. Strengthen disaster preparedness for effective response at all levels	• Disaster Mgt. Capacities: policy–technical–institutional • Dialogue, coordination and information exchange between disaster managers and development sectors • Regional approaches to disaster response, with risk reduction focus • Review and exercise preparedness and contingency plans • Emergency funds • Voluntarism and participation

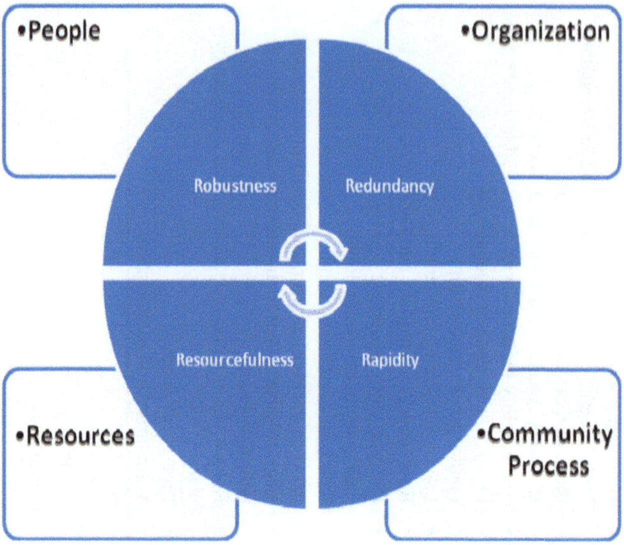

Fig. 2.2 Dimensions of resilience with characteristics

achieving disaster mitigation, preparedness, and response process through building confidence among the community people and enhance their capabilities. DIPECHO achieved high level of resiliency (4.05 out of 5.0 scales) in accountability and community participation.

DIEPCHO help forming local and community organizations dedicated to disaster risk reduction and response actions and ensured their roles and responsibilities to be performed effectively. Attaching those institutions with the project components DIEPCHO create sustainable partnership with community people and promote their accessibility in the decision making process.

2.3.2 *Partnership*

Disaster is a national challenge and become resilient against disaster is the collective responsibility of all sectors of the society, including all levels of government, private, the non-government and individuals of the community. Partnership approach in disaster risk reduction is one of the key to ensure long term sustainability through bring coordination, collaboration, sense of ownership and networking among different stakeholders. Partnership ensures effective participation and creates platform for the citizen for raising their voice to convey their views and needs. DIEPCHO achieved high level of resiliency (3.9 out of 5.0 scales) in creating sustainable partnership against disaster.

DIPECHO through disaster preparedness interventions ensure effective participation among implementing organizations, local government and national stakeholders, community based organizations and individuals. Partnership has been the core concepts behind the implementation of DRR interventions, which are reflected through following practices: ensure the community led process of implementation of risk reduction and emergency response actions through analyzing their risks and preparing risk reduction plan, enabling people and relevant institutions to understand potential hazards, associated risks and vulnerability, ensure long term sustainability of the project results through capacity building and developing sense of ownership among the community stakeholders, engaging local level government agencies in the implementation process through building their capacities and creating network with community people resulting empowered community and development of capacity of the local groups and organizations in the recruitment, train, support and motivate community volunteers for DRR and work together for achieving resilience.

2.3.3 Education and Training

Community education and training is an ongoing process in which community members acquires knowledge about their roles, responsibilities, expectations, and rights for individual preparedness as well as the way of responding collectively and recover from disaster situation. Education and training in disaster risk reduction help the community members to know where to turn for help and how to receive those, and enabling entire community to be resilient against disasters. Training community members, business, and other stakeholders in risk and vulnerability assessment and disaster preparedness and risk reduction ensure effective communication among different stakeholders and create a strong community infrastructure for resilience. DIPECHO achieved high level of resiliency (3.69 out of 5.0 scales) in providing DRR and emergency response education and training for capacity development of community.

DIPECHO improved awareness and capacity of the community people in disaster management through incorporating education system, practicing through simulation and exercise, and sharing different coping strategies. Community knowledge, self confidence, and motivation towards identification and reduction of risks and vulnerability improved through transformation of education into different practices, including: community knowledge on past disaster experiences and coping strategies increased and reflected through their DRR planning and actions, member's skills and capacity improved in disaster risk reduction and preparedness, and practices in hazard, vulnerability and risk assessment, community DRM planning, search and rescue, shelter management and relief distribution, basic and extra-curricular activities by different schools incorporated DRR issues and practiced regularly through simulation exercise, community knowledge on DRR and DP used need assessment and prioritized actions for disaster resiliency, and family preparedness plans, school and hospital contingency plan prepared and updated by community themselves and ensured regular simulation exercise.

2.3.4 Social Protection

Social protection along with risk reduction aimed to reduce vulnerability of community people against disasters. Integrated disaster risk reduction and social protection interventions helps strengthening community's resilience especially the poor and vulnerable peoples of the community. Building a resilient society needs socially inclusive, sustainable social protection and labor system that effectively coordinate different sector. This helps people managing risks, protect community from reduction of income during crisis, and provide them with tools for building skills and create employment opportunities. Integrating social protection into disaster risk reduction can potentially ensure pro-poor DRR interventions and empowers the poor to build resilient communities and livelihoods. DIPECHO through ensuring social protection aspects into different interventions achieved high level of resiliency (4.5 out of 5.0 scales).

DIPECHO ensures preservation of social capital of the community and ensure mutual relationship with different institutions. Social network has been created through creation of community organization and perform coordinated actions. Mutual assistance system, social network, and support mechanism to feed disaster risk reduction ensured through direct and indirect practices, including: alternative livelihood options has been created through research and capacity building of community people, established infrastructures and services facilitated community's economic activities and income of the household, effective cooperation established and sustained with formal and informal structures and maintained through effective communication, protection of personal property and assets has been promoted throughout the society which ensure social security and social assistance helped to shield the poor and most vulnerable people of the community from greater poverty.

2.3.5 Participation, Voluntarism, and Accountability

The concept of resilience *i*nvolves the element of risk being mitigated by high levels *of* voluntarism and participation in community activities where everyone knows and performs their roles and responsibilities. Volunteerism benefits both the community and individuals strengthening trust, solidarity, and reciprocity among the stakeholders and creates opportunity for participation in DRR intervention, where accountability ensures effective implementation. DIPECHO thorough high level of voluntarism and participation ensure high level of resiliency (4.2 out of 5.0 scales) in ensuring accountability.

DIPECHO involve wide range of volunteers and community organizations implementing different interventions to promote community resilience. Community volunteers participate in different DRR and emergency response activities and work for building awareness among people. Motivated and well organized volunteer groups created through mutual assistance from relevant government agency are actively involved through search and rescue operations. DIPECHO creates a platform for community to create self help and prepared volunteer group who support achieving resiliency. Related achievements include: justifiable community confidence in preparedness and emergency response developed along with ability to take effective action against disaster, high level of community voluntarism in all aspects of preparedness, response and recovery ensures through representations of all sector of community, self help group has been created who are prepared and alert for any hazard and active in disaster risk reduction and response actions, prepared and resilient community for dedicated actions developed and functional within the community.

2.4 Conclusion

The concept of vulnerability, risk and resilience is yet to mature in the disciplinary growth of the disaster management. The disaster management is also yet to emerge as a discipline, where a lot more core phenomena, key theories and methodology needs to be developed. In the context of a potential discipline, the concept of vulnerability, risk and resilience can be explored more through robust research and investigation. Research in disaster management is very premature in Bangladesh, though the country is paradise for the development of discipline and disaster studies. The essence is currently handled by UN agencies, INGOs and national NGOs in partnership with Ministry of Disaster Management and Relief. The academic maturity will eventually grow and we are very hopeful that different account of knowledge will be restored in upcoming institution Disaster Management Research and Training Institute.

The key concepts and its illustration will be very helpful in the formulation of a discipline. The core phenomena in disaster management is differently used and perceived by different disciplines and area of practices. These key concepts has one common point of genesis: disaster is not given, we have to have something to do to reduce the disasters either by protecting or reducing hazards and by reducing vulnerabilities or increasing resilience to shocks, which all contributes in disaster risk reduction.

The Bangladesh case of resilience study clearly shows that governance, capacities for risk assessment, risk education, redundant risk reduction options, resourcefulness of people, society and ecosystem, and a capacity to have rapidity in recovery has potential in building resilience. The risk reduction in many cases could be seen as “element” focused, where “resilience” offers a more robust scope to whole of society into the risk reduction framework.

In general perception, risks due to hazards are seen differently from ecological, technical or social risk, in many cases even health risks. Conflicts has added a questions to the whole risk reduction models and thus social vulnerability and social risks has to be addressed in the broader risk reduction framework, the framework of resilience can accommodate the disaster, climate change, conflict, and environmental issues into one umbrella. The time is matured now to think the issues from a broader perspective and develop theories, methodologies, capture the epistemology and ontology to form a discipline on "Disaster Management". Recent socio-political, economic, technological events demonstrate that any event, which have consequences on the people directly or indirectly that go beyond local damage can be also be the subject matter of the emerging discipline. The limit as to what can reasonably be expected of security in human life, economy, society and built environment, planning and efforts in future is the clear scope for the discipline. An integral and holistic approach to disaster management also means that every risk has to be considered within the context of other risks of natural, technological, biological and socio-political origin. Many of these risks might not be the current practice areas for disaster management professional but in developing a discipline these should be the integral part of a full discipline.

References

Adam B, van Loon J (2000) Repositioning risk; the challenge for social theory. In: Adam B, Beck U, van Loon J (eds) The risk society and beyond. Sage, London, pp 1–31

Anderson MB, Woodrow PJ (1998) Rising from the ashes: development strategies in times of disaster. Lynne Rienner, London, pp 338

Beck U (2000) Risk society revisited: theory, politics and research programmes. In: Adam B, Beck U, van Loon J (eds) The risk society and beyond. Sage, London, pp 211–229

Blaikie P, Cannon T, Davis I, Wisner B (1994) At risk: Natural hazards, people's vulnerability, and disasters. Routledge, London

BOND (2012) The characteristics of resilience building: a discussion paper. http://community.eldis.org/?233@@.5ad4406d!enclosure=.5ad4406e&ad=1

Brooks N (2003) Vulnerability, risk and adaptation: a conceptual framework. Tyndall Centre for Climate Change Working Paper 38, University of East Anglia, Norwich

Cardona OD (2004) The need for rethinking the concepts of vulnerability and risk from a holistic perspective: a necessary review and criticism for effective risk management. In: Bankoff G, Frerks G, Hilhorst D (eds) Mapping vulnerability: disasters, development and people. Earthscan Publishers, London, pp 37–51

Cardona OD (2011) Disaster risk and vulnerability: notions and measurement of human and environmental insecurity. In: Brauch HG, Oswald Spring U, Mesjasz C, Grin J, Kameri-Mbote P, Chourou B, Dunay P, Birkmann J (eds) Coping with global environmental change, disasters and security – threats, challenges, vulnerabilities and risks. Springer, Berlin, pp 107–122

Cardona OD, Hurtado JE, Chardon AC, Moreno AM, Prieto SD, Velásquez LS, Duque G (2005) Indicators of Disaster Risk and Risk Management. Program for Latin America and the Caribbean: Summary Report. Inter-American Development Bank, Department of Sustainable Development, Washington. Available at http://idea.unalmzl.edu.co

Cardona OD, van Aalst MK, Birkmann J, Fordham M, McGregor G, Perez R, Pulwarty RS, Schipper ELF, Sinh BT (2012) Determinants of risk: exposure and vulnerability. In: Field CB, Barros V, Stocker TF, Qin D, Dokken DJ, Ebi KL, Mastrandrea MD, Mach KJ, Plattner G-K, Allen SK, Tignor M, Midgley PM (eds) Managing the Risks of Extreme Events and Disasters to Advance Climate Change Adaptation. A special report of working groups I and II of the Intergovernmental Panel on Climate Change (IPCC). Cambridge University Press, Cambridge, pp 65–108

Canadian Centre for Community Renewal (CCE) (2000) The Community Resilience Manual. Canadian Centre for Community Renewal. Port Alberni, B.C

Centre for Community Enterprise (1999) The Community Resilience Manual. Center for Community Enterprise, British Columbia
Covello V, Mumpower J (1985) Risk analysis and risk management: an historical perspective. Risk Anal 5(2):103–120
Cutter LS, Boruff BJ, Lynn Shirley W (2003) Social vulnerability to environmental hazards. Soc Sci Q 84(2):242–261 (the Southwestern Social Science Association)
DKKV (2011) Adaptive disaster risk reduction – enhancing methods and tools of disaster risk reduction in the light of climate change. DKKV Publication Series no. 43, German Committee for Disaster Reduction, Bonn, www.dkkv.org/de/publications/schriftenreihe.asp?h=5
Douglas M, Wildavsky A (1982) Risk and culture: an essay on the selection of technological and environmental dangers. University of California Press, Berkeley
IFRC (2009) Assessing Quality and Cost Benefit: A Philippines Case Study
Islam N (2011) Public expenditure tracking on disaster management projects. BIDS, Dhaka
ISO (2009a) Risk management – principles and guidelines. ISO 31000, International Organization for Standardization, Geneva
ISO (2009b) Risk management – vocabulary. ISO Guide 73, International Organization for Standardization, Geneva
Kumpulainen S (2006) Vulnerability concepts in hazard and risk assessment. In: Schmidt-Thomé P (ed) Natural and technological hazards and risks affecting the spatial development of European regions. Geological Survey of Finland, Special Paper 42, pp 65–74
Moench M, The Risk to Resilience Study Team (2008) Understanding the costs and benefits of disaster risk reduction under changing climatic conditions. In: Moench M, Caspari E, Pokhrel A (eds) From risk to resilience Working Paper No. 9. ISET, ISET-Nepal and ProVention, Kathmandu, 38 pp
Rashid AKMM, Islam N, Alam K, Uddin SA, Sinha NC (2012) Impacts of Last Ten Years DIPECHO's Intervention in Bangladesh. Development Frontiers, Dhaka
SOPAC (2008) Economic Analysis of Flood Risk Reduction Measures for the Lower Vaisigano Catchment Area
Tearfund UK (2004) Community based disaster risk reduction in the Indian state of Bihar: A case study. Available at http://tilz.tearfund.org/Topics/Disaster+Management
The Stern Review (2006) The economics of climate change, HM Treasury. Available at http://webarchive.nationalarchives.gov.uk/+/http://www.hm-treasury.gov.uk/independent_reviews/stern_review_economics_climate_change/stern_review_report.cfm
Twigg J (2009) Characteristics of a disaster resilient community: a guidance note. Aon Benfield UCL Hazard Research Centre, London, Web link www.abuhrc.org/research/dsm/Pages/project_view.aspx?project=13
Wilches-Chaux G (1989) Desastres, ecologismo y formación profesional. SENA, Popayán
Willenbockel D (2011) A cost-benefit analysis of practical action's livelihood-centred disaster risk reduction project in Nepal. IDS, Brighton
Wisner B, Blaikie P, Cannon T, Davis I (2004) At risk: natural hazards, people's vulnerability and disasters, 2nd edn. Routledge, London
World Bank (2007) Project Appraisal Document. "Western Kenya Community Driven Development and Flood Mitigation Project

Chapter 3
National Perspectives of Disaster Risk Reduction in Bangladesh

Rajib Shaw, Aminul Islam, and Fuad Mallick

Abstract According to Climate Risk Index, Bangladesh is one of the most disaster-prone countries in the world. Bangladesh is highly vulnerable to different types of disaster because of climatic variability, extreme events, high population density, high incidence of poverty and social inequity, poor institutional capacity, inadequate financial resources, and poor infrastructure. Bangladesh initiated its actions for disaster preparedness immediately after the cyclone of 1991. At present Bangladesh has National Disaster Management Act-2012, National Disaster Management Policy, Standing Order on Disaster and National Plan for Disaster Management 2010–2015 as key documents guiding the disaster management works in Bangladesh. To implement the documents in ground works Bangladesh has Disaster Management and Relief Division, Disaster Management Bureau (DMB) and Comprehensive Disaster Management Programme (CDMP) under the Ministry of Food and Disaster Management (NPDM, National Plan for Disaster Management 2010–2015, Disaster Management Bureau, Disaster Management & Relief Division, Ministry of Food and Disaster Management, Dhaka). The country also has disaster management mechanism at both national and sub-national levels. The Bangladesh National Plan for Disaster Management is a strategic document to be effective for a certain period of time. This is an umbrella plan that provides the overall guideline for the relevant sectors and the disaster management committees at all levels to prepare and implement their area of roles specific plans (both thematic level and different levels of administrative structure). Furthermore, the Disaster Management

R. Shaw (✉)
Graduate School of Global Environmental Studies, Kyoto University, Kyoto, Japan
e-mail: shaw.rajib.5u@kyoto-u.ac.jp

A. Islam
Sustainable Development, United Nations Development Programme [UNDP], Dhaka, Bangladesh

F. Mallick
BRAC University, Dhaka, Bangladesh

R. Shaw et al. (eds.), *Disaster Risk Reduction Approaches in Bangladesh*, Disaster Risk Reduction, DOI 10.1007/978-4-431-54252-0_3,

Act is enacted by Act No. 34 of 2012 that forms the legislative basis to desirable, integrate, and strengthen the disaster management related activities and to establish an effective institutional framework for disaster management to deal with all types of disaster in Bangladesh.

Keywords Comprehensive disaster management program (CDMP) • Disaster management plan • Disaster policy • National disaster management act • Standing order

3.1 Introduction

Bangladesh is one of the most disaster vulnerable countries in the world. With adverse effects of climate change Bangladesh is expected to lose one-third of its mainland into the Bay of Bengal due to rise in sea level. Along with this rising temperature may reduce crop production significantly. Since the 1990s disaster management has become an issue of importance to the policy makers. As a result many international and local drivers have been introduced to mitigate threats of disaster.

The country is overwhelmingly an agricultural, one of the densely populated country occupying world's eighth and Asia's fifth position with 160 million people. Agriculture is the life force of the economy as it has the potential to produce multiplier effects on the growth of other sectors of the economy. Its contribution to gross domestic product (GDP) is 23.5 % of total GDP of the country (BBS, 2006 and Handbook Agricultural Statistics, Ministry of Agriculture). Almost 48 % of the populations are still dependent on agriculture for their livelihoods. Therefore, any fluctuations in agricultural productivity influence the food security and potential stability of the country (Habiba 2012; Habiba et al. 2010, 2011).

Straddling the Tropic of Cancer, Bangladesh has a tropical monsoon climate characterized by heavy seasonal rainfall, high temperatures and high humidity. Regional climatic differences in this flat country are minor. Three seasons are generally recognized: a hot, muggy summer from March to June; a hot, humid and rainy monsoon season from June to November; and a warm-hot, dry winter from December to February. In general, maximum summer temperature ranges between 38 °C and 41 °C (100.4 °F and 105.8 °F). April is the hottest month in most parts of the country. January is the coolest month, when average temperature for most of the country is 16–20 °C (61–68 °F) during the day time and around 10 °C (50 °F) at the night time.

According to Climate Risk Index, Bangladesh is one of the most disaster-prone countries in the world (Harmeling 2009). Bangladesh is highly vulnerable to different types of disaster because of climatic variability, extreme events, high population density, high incidence of poverty and social inequity, poor institutional capacity, inadequate financial resources, and poor infrastructure (Ahmed 2004). Almost every year, the country experiences disasters of one kind or another, such as tropical cyclones, storm surges, coastal erosion, floods, tornadoes and droughts, that causes heavy loss of life and property and jeopardizing development activities (Ali 1996). During the period of 1991–2000, Bangladesh has suffered from 93 large scale natural disasters which killed 0.2 million people and caused loss of properties valued about US$59 billion in the agricultural and infrastructure sector (Climate Change Cell 2009). Therefore, the international community considers Bangladesh as the country most threatened by disasters.

3.2 Disaster Scenario: Goal, Objectives and Principles

Bangladesh regularly suffers worst disaster that is tropical cyclones, often comes with storm surges. The cyclone of 1961, 1963, 1965, 1970, 1985, 1988, 1991, 1994, 1995, 1997, 2007 and 2009 were of severe in nature (Habiba 2012). Among them, cyclone in 1970 is the deadliest cyclone that hits Bangladesh coastline and took away the lives of over 300,000 people and damaged of about US$2.5 billion equivalent crops and property. In 1991, the catastrophic cyclone killed over 150,000 people and property damages were more than US$ two billion. Most recently, in 2007 cyclone "Sidr" struck in Bangladesh that killed and injured over 3,363 and 55,282 people, respectively. It is also reported that 563,877 houses were totally destroyed and 955,065 houses were partially damaged. In terms of agricultural sector, it was accounted for fully damaged of 186,883 ha of crop areas and partly damaged of 498,645 ha area (NPDM 2010). Likewise, in 2009, cyclone "Aila" caused death of 330 lives, made one million people homeless and total damage of US$40.7 million. Estimation depicts that 20 million people of Bangladesh were at risk of post disaster diseases due to Aila.

In addition, flood is a regular phenomenon in Bangladesh and tends to occur between April–May and September–November (NAPA 2005). Annually an average of 15 % of the total geographical area is inundated by floods. Even, in extreme years, two-thirds of the country can be inundated by floods (Mirza 2002). Approximately, 37 %, 43 %, 52 % and 68 % of the country is inundated with floods of return periods of 10, 20, 50 and 100 years, accordingly (NPDM 2010). The severe floods of 1822, 1854, 1922, 1955, 1966, 1974, 1987, 1988, 1998, 2002, 2004 and 2007 are worth mentioning. It has wreaked havoc in Bangladesh throughout the history. The flood of 1988 during August–September inundated 89,000 km^2 areas of 52 districts in Bangladesh and caused loss of 1,517 human lives. The 1998 flood in Bangladesh with unprecedented duration of 65 days inundated 53 districts covering about 100,000 km^2 areas and it took lives of 918 people. The 2004 flood in Bangladesh inundated 40 districts and it took lives of 747 people. According to the government statistics, 298 people died and a total of 10,211,780 people were badly affected by recent 2007 flood (Habiba 2012).

Other disaster such as drought, earthquake, tornado etc. appears in Bangladesh. But, the details of drought disaster will be mentioned in the next section. Considering earthquake, it is an emerging risk for the country, the capital Dhaka and other major cities such as Sylhet and Chittagong are extremely vulnerable. Earthquakes of 1869, 1885, 1897, 1918, 1930, 1934, 1959, 1997 and 1999 are noteworthy. River erosion is no less dangerous than other sudden and devastating calamities. Around 10,000 ha of land is eroded as a result of river bank erosion per year in Bangladesh.

3.3 National Disaster Management Plan

Bangladesh initiated its actions for disaster preparedness immediately after the cyclone of 1991 (before introduction of many of the international drivers) At present Bangladesh has National Disaster Management Act-2008, National Disaster Management Policy, Standing Order on Disaster and National Plan for Disaster Management 2010–2015 as key documents guiding the disaster management works in Bangladesh. To implement the documents in ground works Bangladesh has Disaster Management and Relief Division, Disaster Management Bureau (DMB) and Comprehensive Disaster Management Programme (CDMP) under the Ministry of Food and Disaster Management (NPDM 2010).

National Disaster Management Plan 2010–2015 has been prepared in light of the Hyogo Framework for Action 2005–2015 and SAARC framework on Disaster Management. The plan prepared to help address the disasters in a comprehensive way. The plan is prepared with the aim to reduce vulnerability of the poor to natural, environmental and human induced disaster to a manageable and acceptable level. To achieve the aims, the concerned authorities have adopted strategies like:

(a) Bringing a paradigm shift in disaster management from conventional response and relief practice to a more comprehensive risk reduction culture.
(b) Strengthening the capacity of the Bangladesh disaster management system in improving the response and recovery management at all levels.

The plan is expected to accommodate effective intervention by DM&RD for coordinating all disaster management activities within the country. The plan is also expected to contribute towards a cohesive and well-coordinated programming framework, incorporating GoB, NGOs and the private sector. Thus the key focus of the document is to establish institutional accountability in preparing and implementing disaster management plans at different levels of administration. Development Plans incorporating Disaster Risk Reduction and Hazard Specific Multi-Sectoral Plans have made this plan an exclusive tool for reducing risk and achieving sustainable development. The plan is prepared in a participatory way, having several consultations with stakeholders and established a road map of effective partnership with the organizations working at local, national and regional levels. It is expected that this plan will contribute towards developing and strengthening regional and national networks.

The objectives of this plan are to:

- Align the strategic direction of disaster management programs with national priorities and international commitments.
- Articulate the vision and goals for disaster management.
- Outline the strategic direction and priorities to guide the design and implementation of disaster management policies and programs.
- Create a cohesive and well-coordinated programming framework incorporating government, non-government and private sector.

- Ensure that disaster management has a comprehensive and all-hazards focus comprising disaster risk reduction and emergency response.
- Illustrate to other ministries, NGOs, civil society and the private sector how their work can contribute to the achievements of the strategic goals and government vision on disaster management.

The core principles of this plan have been adopted from the PRSP.

- Country-driven, promoting national ownership of strategies through broad based participation of government, NGOs and civil society.
- Result oriented and focused on outcomes that will benefit vulnerable communities, especially women, the poor and socially disadvantaged.
- Comprehensive in recognizing the multidimensional nature of risk reduction.
- Partnership oriented, involving coordinated participation of development partners (government, domestic stakeholders, and external donors).
- Based on a long-term perspective for risk reduction.

The strategic goals of the plan are drawn from the SAARC Disaster Management Framework.

- **Goal 1**: professionalizing the disaster management system
- **Goal 2**: mainstreaming risk reduction
- **Goal 3**: strengthening institutional mechanisms
- **Goal 4**: empowering at risk communities
- **Goal 5**: expanding risk reduction programming
- **Goal 6**: strengthening emergency response systems
- **Goal 7**: developing and strengthening networks

3.4 Disaster Management Structures

Bangladesh has disaster management mechanism at both national and sub-national levels. A figure illustrating the position, activity and authority is given below. Disaster Management Structure of Bangladesh (at the national level) is as follow (Fig. 3.1):

1. **National Disaster Management Council (NDMC)** headed by the Honorable Prime Minister to formulate and review the disaster management policies and issue directives to all concerns.
2. **Inter-Ministerial Disaster Management Co-ordination Committee (IMDMCC)** headed by the Hon'ble Minister in charge of the Disaster Management and Relief Division (DM&RD) to implement disaster management policies and decisions of NDMC/Government.
3. **National Disaster Management Advisory Committee (NDMAC)** headed by an experienced person having been nominated by the Honorable Prime Minister.
4. **National Platform for Disaster Risk Reduction (NPDRR)** headed by Secretary, DM&RD and DG, DMB functions as the member secretary. This

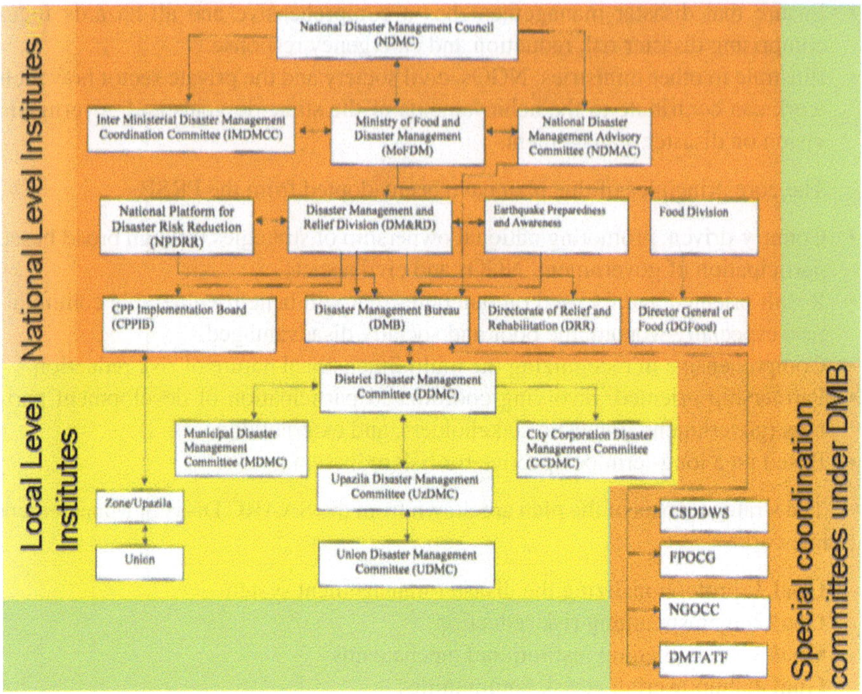

Fig. 3.1 Disaster management institutions in Bangladesh

platform shall coordinate and provide necessary facilitation to the relevant stakeholders.

5. **Earthquake Preparedness and Awareness Committee (EPAC)** headed by Honorable minister for MoFDM and DG, DMB act as member secretary.
6. **Cyclone Preparedness Program Implementation Board (CPPIB)** headed by the Secretary, Disaster Management and Relief Division to review the preparedness activities in the face of initial stage of an impending cyclone.
7. **Cyclone Preparedness Program (CPP)** Policy Committee headed by Honorable Minister, MoFDM and Secretary, DM&RD act as member secretary. Disaster Management Training and Public Awareness Building Task Force (DMTATF) headed by the Director General of Disaster Management Bureau (DMB) to coordinate the disaster related training and public awareness activities of the Government, NGOs and other organizations.
8. Focal **Point Operation Coordination Group of Disaster Management (FPOCG)** headed by the Director General of DMB to review and coordinate the activities of various departments/agencies related to disaster management and also to review the Contingency Plan prepared by concerned departments.
9. **NGO Coordination Committee on Disaster Management (NGOCC)** headed by the Director General of DMB to review and coordinate the activities of concerned NGOs in the country.

10. **Committee for Speedy Dissemination of Disaster Related Warning/Signals (CSDDWS)** headed by the Director General of DMB to examine, ensure and find out the ways and means for the speedy dissemination of warning/signals among the people.

Disaster Management Structure of Bangladesh (at sub-national levels) is as follow:

1. **District Disaster Management Committee (DDMC)** headed by the Deputy Commissioner (DC) to coordinate and review the disaster management activities at the District level.
2. **Upazila Disaster Management Committee (UZDMC)** headed by the Upazila Nirbahi Officer (UNO) to coordinate and review the disaster management activities at the Upazila level.
3. **Union Disaster Management Committee (UDMC)** headed by the Chairman of the Union Parishad to coordinate, review and implement the disaster management activities of the concerned Union.
4. **Pourashava Disaster Management Committee (PDMC)** headed by Chairman of Pourashava (municipality) to coordinate, review and implement the disaster management activities within its area of jurisdiction.
5. **City Corporation Disaster Management Committee (CCDMC)** headed by the Mayor of City Corporations to coordinate, review and implement the disaster management activities within its area of jurisdiction.

3.5 Disaster Management Regulatory Framework

Bangladesh's regulative framework for disaster management provides for the relevant legislative, policy and best practice framework under which the activity of Disaster Risk Reduction and Emergency Management in Bangladesh is managed and implemented (Fig. 3.2). The framework includes the following (NPDM 2010):

3.5.1 Disaster Management Act

A Disaster Management Act will be enacted with a view to create the legislative tool under which disaster risk and emergency management will be undertaken in Bangladesh, and the legal basis in which activities and actions will be managed. It will also create mandatory obligations and responsibilities on Ministries, committees and appointments. The objectives of the Act will be (a) to help communities to mitigate the potential adverse effects of hazard events, prepare for managing the effects of a disaster event, effectively respond to and recover from a disaster or an emergency situation, and adapt to adverse effects of climate change; (b) to provide for effective disaster management for Bangladesh; (c) to establish an institutional

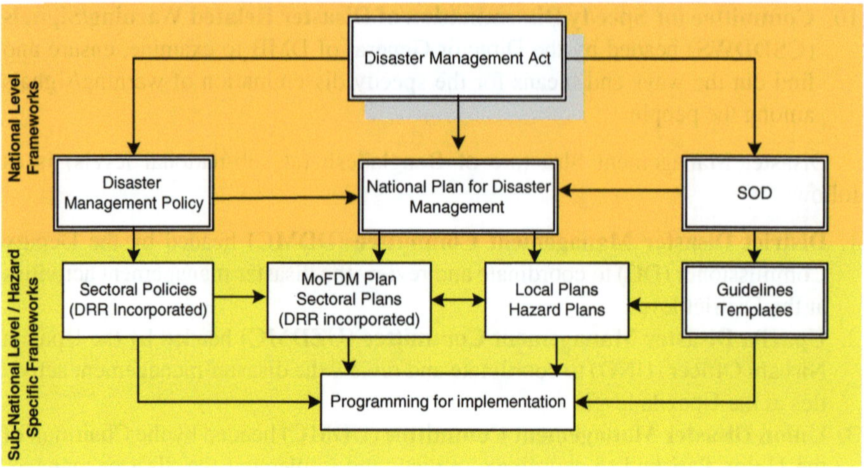

Fig. 3.2 Disaster Management Regulatory Framework

framework for disaster management; and (d) to establish risk reduction as a core element of disaster management.

3.5.2 *National Disaster Management Policy*

A National Disaster Management Policy will be formulated to define the national perspective on disaster risk reduction and emergency management, and to describe the strategic framework, and national principles of disaster management in Bangladesh. It will be of strategic in nature and will describe the broad national objectives, and strategies in disaster management.

3.5.3 *Standing Orders on Disaster*

The Standing Orders on Disaster describes the detailed roles and responsibilities of committees, Ministries and other organizations in disaster risk reduction and emergency management, and establishes the necessary actions required in implementing Bangladesh's Disaster Management Model. The Standing Orders have been prepared with the avowed objective of making the concerned persons understand their duties and responsibilities regarding disaster management at all levels, and accomplishing them. All Ministries, Divisions/Departments and Agencies shall prepare their own Action Plans in respect of their responsibilities under the Standing Orders for efficient implementation. The National Disaster Management Council (NDMC) and Inter-Ministerial Disaster Management Coordination Committee (IMDMCC) will ensure coordination of disaster related activities at the National level. Coordination at District, Thana and Union levels will be done by the respective District, Thana and Union Disaster Management Committees. The Disaster Management Bureau will render all assistance to them by facilitating the process.

3.6 Different Levels of Disaster Management Plans

The Bangladesh National Plan for Disaster Management is a strategic document to be effective for a certain period of time. This is an umbrella plan that provides the overall guideline for the relevant sectors and the disaster management committees at all levels to prepare and implement their area of roles specific plans. The Disaster Management and Relief Division (DM&RD) being the focal ministry for disaster risk reduction and emergency management will take the lead role in disaster risk reduction and emergency management planning. Additionally, there will be a few hazard specific management plans, such as Flood Management Plan, Cyclone and Storm Surge and Tsunami Management Plan, Earthquake Management Plan, Drought Management Plan, River Erosion Management Plan, etc. Moreover, there will be a detailed Disaster Management Plan for each District, Upazila, Union and Pourashava and City Corporation of the country. A District Disaster Management Plan will be the compilation of the Upazila Disaster Management Plans of the District. Similarly an Upazila Disaster Management Plan will be the compilation of the union disaster management plans of that Upazila prepared by the Union DMCs. So DMCs at Union and Pourashava levels will be mainly responsible for conducting the risk assessments and prepare the ground level plans. Once developed those will be sent to the DMCs at one level higher—Upazila DMCs, whose role will be to verify and compile the union plans and identify the resource requirements for the Upazila.

Bangladesh has various disaster management plans, aimed to work at different administrative strata (Fig. 3.3). The plans are:

1. National disaster Management Plan
2. District Disaster Management Plan (DDMP)
3. Upazila disaster Management Plan (UzDMP)
4. Union Disaster Management Plan (UDMP)
5. Pourashava/City Corporation Disaster Management Plan
6. Sectoral development plans incorporating risk reduction
7. Hazard Specific disaster Management Plan

    (a) Earthquake management plan
    (b) Cyclone Shelter Plan
    (c) Disaster Resilient Cluster Housing
    (d) Tsunami response plan etc.

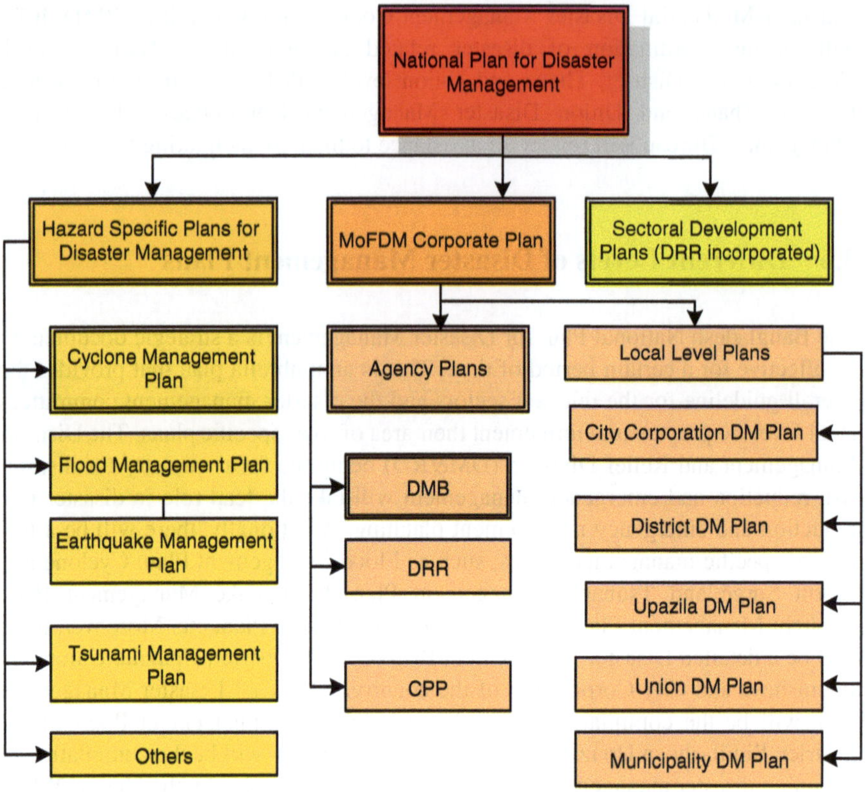

Fig. 3.3 Different levels of disaster management plans of Bangladesh

3.6.1 National Plan for Disaster Management

The National Plan for Disaster Management is prepared by the Disaster Management and Relief Division. The plan includes the following as minimum:

- Introduction
- GoB Vision for Disaster Management
- Hazards profile of Bangladesh
- Disaster development linkages: national and international drivers for change
- Aim of the plan
- Strategic goals of the plan
- Conceptualizing disaster management in Bangladesh
- Disaster management system in Bangladesh
- The roles and responsibilities of entities involved in emergency operations and risk reduction
- Disaster management regulative framework

- Action matrix for disaster risk reduction and emergency management in Bangladesh describing the priorities and the strategies
- Review and evaluation
- Implementation and follow-up
- Financing of the plan
- Other matters relating to disaster management as deemed necessary by appropriate authority for inclusion in the plan

The Plan is to be used to:

- Articulate the long-term strategic focus of disaster management in Bangladesh.
- Demonstrate a commitment to address key issues: risk reduction, capacity building, information management, climate change adaptation, livelihood security, issues of gender and the socially disadvantaged, etc.
- Show the relationship between the government vision, key result areas, goals and strategies, and to align priorities and strategies with international and national drivers for change.
- Detail a road map for the development of disaster management plans by various entities.
- Guide the DM&RD in the development and delivery of guidelines and programmes.
- Illustrate to other ministries, NGOs, civil society and the private sector how their work can contribute to the achievements of the strategic goals and government vision on disaster management.
- Provide a framework within which to report performance and success in achieving goals and strategies.

3.6.2 Sub-National Level Plans

DDMP, UzDMP, UDMP and Pourashava/City Corporation Disaster management plan has a few common key features which are given below.

1. The areas in the District vulnerable to different forms of hazards and risks.
2. Total resource requirements and the planned action for the District.
   (a) To take measures for prevention and mitigation of disasters by government agencies, NGOs, CBOs and the private sector within the District.
   (b) Capacity building and preparedness measures to be taken by government agencies, NGOs, CBOs and the private sector.
   (c) Strengthening emergency response management system plans and procedures in the event of a disaster.
3. The response plans and procedures in the event of a disaster, providing for:
   (a) Allocation of responsibilities to the departments of the government at District level and other DMC members

(b) Procedure for mobilization of resources
(c) Prompt response to disaster and relief thereof
(d) Procurement of emergency supplies
(e) Operation of disaster shelters
(f) Restoration of emergency services, such as water supply, gas supply, power, telecommunication, road links
(g) Provision of emergency medical services
(h) Burial of dead bodies
(i) Trauma counseling
(j) The dissemination of information

4. Recovery plans and procedures delineating damage assessment procedure, restoration of damaged public infrastructure, resumption of educational institutions, restoration of livelihood, rehabilitation of affected people, especially the disabled, and elderly women and children.
5. The DDMP shall be reviewed and updated annually.
6. The copies of the DDMP shall be made available to all District level stakeholders, Divisional Commissioners, etc.
7. A copy of the DDMP will be sent to the Disaster Management Bureau and all relevant ministries and divisions.
8. The DMB/NDMTI will provide technical advice and capacity building services to all DMCs.

3.6.2.1 District Disaster Management Plan (DDMP)

There is a District Disaster Management Committee (DDMC) at the District level. The DDMC consists of the Deputy Commissioner of the District as the chairperson and members comprising all District level department heads, NGO leaders and civil society members. District Relief and Rehabilitation Officer (DRRO) acts as member secretary of the committee. Members of Parliament act as advisors of the committees. The committee is required to meet bi-monthly during normal period and as and when necessary during emergency situation.

There will be a plan for each District titled "District Disaster Management Plan" comprising both disaster risk reduction and emergency response to be prepared by the District Disaster Management Committee. This is a plan to be prepared by compilation of the Upazila and Pourashava Disaster Management Plans of the District being received from the respective Upazila and Pourashava/City Corporation DMCs.

3.6.2.2 Upazila Disaster Management Plan (UzDMP)

Upazila is an important and vital administrative unit of Bangladesh. There is an Upazila Disaster Management Committee (UZDMC) at the Upazila level. The UzDMC consists of the Upazila Nirbahi Officer as the chairperson and members comprising all Upazila level department heads, NGO leaders and civil society members. The PIO acts as the member secretary of the committee. Members of Parliament act as advisors of the committees. The committee is required to meet bi-monthly during normal period and as and when necessary during emergency situation. There will be a plan for each Upazila titled "Upazila Disaster Management Plan" comprising both disaster risk reduction and emergency response to be prepared by the Upazila Disaster Management Committee by compiling all the Union Disaster Management Plans of the Upazila being received from the respective Union DMCs of the Upazila.

3.6.2.3 Union Disaster Management Plan (UDMP)

Union Parishad is the lowest administrative unit of Bangladesh. There is a Disaster Management Committee at the Union level. The UDMC is chaired by the elected Chairman of the respective Union Parishad. The Union Disaster Management Committee consists of the Union Parishad Chairman as the Chairperson and members comprising all the Government department head at Union level, members of Union Parishad, NGO leaders working in respective union and civil society members. Secretary of the respective Union Parishad acts as the member secretary of the committee. The committee is required to meet bimonthly during normal period and as and when necessary during emergency situation. There will be a plan for each Union titled "Union Disaster Management Plan" comprising both disaster risk reduction and emergency response to be prepared by the Union Disaster Management Committee following a proper community risk assessment procedure to be provided by DM&RD with the participation of vulnerable groups and the communities.

3.6.3 Sectoral Development Plans Incorporating Disaster Risk Reduction

Every Ministry/Division of the Government of Bangladesh prepares their respective Sectoral Development Plans. DM&RD with the participation of sectoral experts will prepare a general guideline to incorporate disaster risk reduction agenda for the sectors. DM&RD will also be responsible for overall monitoring and follow-up of the process to ensure that disaster risk reduction agenda are mainstreamed within the sectoral policies, plans and programs.

The development plans should address, among others, the following:

(a) Defining and redefining risk environment through hazard analysis, vulnerability assessment, risk evaluation, risk treatment options, and risk treatments.
(b) Managing the risk environment by developing programs and strategies that eliminate, or reduce the level of risk. Traditionally mitigation programs were viewed as engineering solutions to eliminate risk, but it is now accepted that all activities undertaken to eliminate or reduce risk are "mitigation" strategies (e.g.

community education and awareness, planning activities, development of warning systems). This includes activities previously described as the PPRR Model—Prevention, Preparedness, Response and Recovery.

(c) Regularly review and update the plan; and
(d) Submit a copy of the plan, and of any amendment thereto, to appropriate authority including the DM&RD.
(e) Submit a copy of its disaster management plan, and of any amendment thereto, the concerned authority.

3.6.4 Hazard Specific Multi-sectoral Disaster Management Plans

In addition to area specific disaster management plans and sector specific disaster risk reduction plans, it is envisaged that there will be a few hazard-specific management plans, such as earthquake management plan. This type of plans will be multi-sectoral and will be divided into two components: risk reduction and emergency response. This type of plans will address specific necessities to deal with a particular hazard.

3.6.4.1 Earthquake Contingency Plan

It is feared that a high intensity earthquake in these cities may result in to serious devastation and collapse the cities. Thus, a well-designed and fully coordinated plan for optimum and efficient preparedness, response and early recovery, usually known as Contingency Plan, in a systematic manner so that their capacities and resources are best utilized to fulfill the need complimenting and supplementing other agencies. Realizing the need of coordinated and comprehensive emergency response, United Nations has been promoting its humanitarian response activities in a cluster approach. This approach is proved to be effective and efficient in responding to recent disasters, for instances, the response during the earthquake on 8 October 2005 in Pakistan. Hence, it has been decided that this concept of response operations in functional clusters be applied in Bangladesh also in case of possible earthquake disaster. In this approach, under National Earthquake Contingency Plan, all response activities are grouped into nine relevant operational functional clusters based on the similarity of works, normal and disaster time mandates of different relevant organizations and possible complementarily in the resources and capacities. The clusters are as follows:

- Emergency Operations Cluster 1—Overall Command and Coordination
- Emergency Operations Cluster 2—Search, Rescue and Evacuation
- Health Cluster
- Relief Services (Food, Nutrition and other Relief) Cluster
- Shelter (Including Camp Management) Cluster

- Water Supply, Sanitation and Hygiene Cluster
- Restoration of Urban Services Cluster
- Transport (Road, Rail, Air, Sea) Cluster
- Security and Welfare Cluster

3.6.4.2 Cyclone Shelter Plan

To face the challenges particularly cyclone and tidal surges, different governmental & non-governmental organizations have constructed about 2,852 (CDMP 2009) cyclone shelters in the coastal belts of 16 districts of the country. Out of 2,852 shelters, investigation reveals that 2,590 shelters are useable while 262 are not. These shelters are insufficient in terms of necessity. So it would not be possible to provide shelter to all the affected people as well as their domestic animals. A survey team captioned as Multipurpose Cyclone Shelter Programme (MCSP) headed by Prof. Dr. Jamilur Reza Chowdhury recommended in its report of 1993 to construct 1,250 new cyclone shelter as priority no. 1 and 1,250 as priority no. 2 for providing shelters to the affected people during disasters. The report also mentioned that the total number of 2,500 cyclone shelters including primary school, madrashas and secondary schools were proposed to be constructed.

These shelters will be constructed on the government khas land/institution's land/purchased lands. There is a provision of separate latrine facilities for women. One tube-well for each shelter will be set up for supplying pure drinking water. In normal periods, these shelters will be used as educational institution.

It was decided that Bangladesh's plan of action should be inclusive to multi-hazard, all risk, and all sector approach. Therefore, following technical options are considered as critical element of the plan of action.

- Comprehensive Risk assessment (Hazard Assessment and Vulnerability Assessment), including tsunami inundation modeling and evacuation mapping;
- Warning Guidance, including seismic and sea level monitoring, data;
- Evaluation, processing and interpretation, forecasting methods and warning dissemination (a detailed plan of action is prepared);
- Mitigation and Preparedness, including education and awareness;
- Programmes, structural and non-structural mitigations, government policy and emergency management procedures;
- Development of Rescue, Relief and Rehabilitation Plan of Action based on Comprehensive Risk Assessment, and
- Existing Cyclone Preparedness Programme (CPP) should be strengthened in a way that they can prepare the community for tsunami as well as cyclone.

DMB proposes the facilitating role of local Disaster Management Committee in forming the Cyclone Center Management Committee for each center. The committee will have the following types of representation:

- A member of local Disaster Management Committee
- Locally Elected Representative (UP Member)

- Head Master of local Primary School
- Imam of Local Masjid
- NGO representative
- Women representative

DMB also propose for multipurpose use of the Cyclone Centers by local NGOs, Civil Society Groups and community people for public functions like marriage ceremonies, meetings, training sessions and other social functions under the supervision of CC Management Committee. The users will pay a minimum fee for using CC as maintenance charge. The Management Committee will be responsible for keeping financial statement of CC.

3.6.4.3 Disaster Resilient Cluster Housing

In order to provide shelter to the people of the impacted areas to the shelters, Bangladesh needs a large numbers of new shelters. Government shall develop cluster housing for a group of households that are living in marginalized hazard prone lands, initially in Khas land with necessary utilities and infrastructures that are resilient to the hazards. This set up shall have the provisions for cattle and poultry shelter, seedbeds, and schools on raised land. The architecture shall be such that it will be in a position to accommodate the adversity of the hazard impacts. Comprehensive disaster management programme (CDMP) of the DM&RD shall design, develop, pilot such disaster resilient shelters and scale up upon seeing results.

3.6.4.4 Tsunami Response Plan

Following the 2004 Tsunami, and based on several exercises, workshops, seminars and meetings, a detailed draft plan of action is prepared for Bangladesh.

3.7 Disaster Management Act

Disaster Management Act is enacted by Act No. 34 of 2012 that forms the legislative basis to desirable, integrate, and strengthen the disaster management related activities and to establish an effective institutional framework for disaster management to deal with all types of disaster in Bangladesh (DMA 2012). It is approved by the Minister of Food and Disaster Management on 12 September, 2012. The Act thereby provides a major opportunity to improve the delivery and coordination of emergency aid in Bangladesh. The objectives of this Act are:

- To help communities to:
  - Mitigate the potential adverse effects of hazard events
  - Prepare for managing the effects of a disaster event

  - Effectively respond to and recover from a disaster or an emergency situation, and
  - Adapt to adverse effects of climate change

- To make provision for an effective disaster management system in Bangladesh
- To establish an institutional framework for disaster management, and
- To establish disaster risk reduction as a core element of disaster management in Bangladesh

This Act is generally administered by the Government or by authorized bodies constituted or officers appointed by the Government. To fulfill the objectives of this Act, "Disaster Management Bureau" under the Ministry of Food and Disaster Management is predominantly responsible for coordinating and monitoring compliance of the provisions of this Act. According to the Act, the functions of Disaster Management Bureau are:

- To minimize the overall disaster impacts by adopting various disaster risk reduction activities
- To conduct emergency response, recovery and rehabilitation activity efficiently for the affected and distressed people
- To integrate, strengthen and ensure effective coordination across government, NGOs, civil society and private sectors activities and programs related to disaster risk reduction and emergency response operations
- To implement recommendations and directions of the Government related to disaster management
- To implement National Disaster Management Policy and National Plan for Disaster Management
- To establish an effective framework for disaster management that covers all types of disaster that deems to be necessary within the strategic policy framework

References

Ahmed AU (2004) Adaptation to climate change in Bangladesh: learning by doing. In: UNFCCC workshop on adaptation, Bonn, 18 June 2004

Ali A (1996) Vulnerability of Bangladesh to climate change and sea level rise through tropical cyclones and storm surges. Water Air Soil Pollut 94(1–2):171–179

CDMP (2009) Comprehensive disaster management programme, Government of Bangladesh, Dhaka, 45 pages.

Climate Change Cell (2009) Department of Environment, Ministry of Environment and Forest, Bangladesh

DMA (2012) Draft Disaster Management Act. Available at: http://www.cdmp.org.bd/cdmp_old/reports/Draft%20Disaster%20Mgt%20Act%20English%2017-02-08.pdf. Accessed 15 Jan 2013

Habiba (2012) Enhancement of drought risk management policy and actions incorporating farmer's adaptive practices in Northwestern Bangladesh. Doctorate Thesis, Kyoto University

Habiba U, Takeuchi Y, Shaw R (2010) Overview of drought risk reduction approaches in Bangladesh. In: Shaw R, Pereira J (eds) Climate change adaptation and disaster risk reduction: issues and challenges, vol 5. Emerald Group Publishing Limited, Bingley, pp 37–58
Habiba U, Shaw R, Takeuchi Y (2011) Socio-economic impact of droughts in Bangladesh. In: Shaw R, Huy N (eds) Droughts in Asian monsoon region: community, environment and disaster risk management, vol 8. Emerald Group Publishing Limited, pp 25–48
Harmeling (2009) Global Climate Risk index, Briefing Paper, Germanwatch, Bonn, 24 p
Mirza MMQ (2002) Global warming and changes in the probability of occurrence of floods in Bangladash and implications. Global Environmental Change, 12, 127–138
NAPA (2005) National Adaptation Programme of Action (NAPA) Final Report. Ministry of Environment and Forest, Government of the People's Republic of Bangladesh, Dhaka
NPDM (2010) National Plan for Disaster Management 2010–2015. Disaster Management Bureau, Disaster Management & Relief Division, Ministry of Food and Disaster Management, Dhaka

Part II
Hazard and Risk Profiles of Bangladesh

Chapter 4
Flood Risks and Reduction Approaches in Bangladesh

Rezaur Rahman and Mashfiqus Salehin

Abstract Bangladesh is one of the most flood prone areas of the world because of its unique geographical setting and physiographic features together with a massive and unique hydraulic system. The ever increasing population, ill-planned infrastructural development and massive flood control interventions in a floodplain environment have resulted in flood disasters becoming larger and more frequent in recent times. Socio-economic impact of floods is profound; the flood prone zones represent areas with the highest incidence of the extreme poor, and the number of poor living in high flood risk areas is on the rise. The damage to infrastructure constitutes the major proportion when it comes down to economic damage resulting from floods. Since mid 1960s there has been a steady growth of flood control and drainage projects in Bangladesh through the construction of embankments, drainage channels, and sluices and regulators, with the total coverage area standing at 5.37 million ha. While the projects yielded a number of positive impacts such as increase in agricultural production, increase in economic activities, and reduction of damage to infrastructure inside protected area, the projects in general could not attain the desired objectives because of lack of consideration of interdependence of land, water, ecosystems and socio-economic development. Projects were mostly formulated with single objectives, aiming to solve the immediate problems without giving adequate attention to potential, undesirable long-term consequences. Ill-planned growth of projects have impacted the hydraulic stability of the system, with projects being able to provide protection against normal floods, while largely failing to provide protection during moderate to extreme floods. The cost of environmental degradation has been rather very high. Storage functions of floodplains have been lost, impacting important hydrologic functions of floodplains such as moderation of flood peaks,

R. Rahman • M. Salehin (✉)
Institute of Water and Flood Management, Bangladesh University of Engineering and Technology, Dhaka, Bangladesh
e-mail: mashfiqussalehin@iwfm.buet.ac.bd

R. Shaw et al. (eds.), *Disaster Risk Reduction Approaches in Bangladesh*, Disaster Risk Reduction, DOI 10.1007/978-4-431-54252-0_4, © Springer Japan 2013

augmentation of dry season river flows and replenishment of groundwater storage. There has been substantial damage to capture fisheries because of disruption of hydraulic connection between river and floodplain. Disruption of links between river and floodplain by embankments has impacted country boat transports. Livelihood activities of different groups, especially the marginalized, have thus been compromised. In the light of the experiences with the flood control projects, the necessity to harness the beneficial impacts of floods especially in maintaining soil fertility and sustaining fisheries resources has become apparent. All the socio-economic and environmental concerns of earlier practices gave rise to exploration of alternative management strategies which are more resilient and environment friendly. This chapter provides a review of risks associated with different types of floods in Bangladesh. It also gives an account of historic floods and their impacts. Flood risk reduction approaches undertaken in Bangladesh and their impacts and consequences are discussed at length. The chapter also discusses three innovative flood risk reduction approaches that have evolved in Bangladesh, which tend to address the socio-economic and environmental concerns of typical flood control measures.

Keywords Flood risk • Structural options • Non-structural options • Paradigm shift • Innovative approaches

4.1 Overview of Flood Risk

Bangladesh is a deltaic country located at the lower part of the basins of the three mighty rivers—the Ganges, the Brahmaputra and the Meghna. The country is well known as one of the most flood prone areas of the world because of its unique geographical setting and physiographic features together with a massive and unique hydraulic system. About one-fifth to one-third of the country is annually flooded by overflowing rivers during monsoon (June to September) when the rainfall within the country is also very high. While normal floods are considered a blessing for Bangladesh-providing vital moisture and fertility to the soil through alluvial silt deposition in floodplains, moderate to extreme floods are of great concern, as they inundate large areas (more than 60 % of the country are inundated in large flood events), and cause widespread damage to crops and properties. Increased exposure due to growing population size and development in hazardous areas has made disasters in recent times larger and more frequent. Today's flood hazards are already difficult for Bangladesh to cope up with considering its socio-economic conditions; the climate change impacts (the country ranks high in the list of vulnerable countries in South Asia, the most vulnerable region of the world to climate change impacts), would reinforce the baseline stresses that already pose a serious impediment to the country's economic development. In the light of the experiences with flood control interventions implemented since mid 1960s, new insights have been gained and consequently new concepts of flood management have evolved. A key learning has been that flood risk reduction approaches cannot be sustainable without giving adequate attention to the interdependence of water regime, flood management interventions, other physical infrastructures, socio-economy and ecosystems in floodplain landscapes.

4.1.1 Geographical, Physiographic and Hydro-Meteorological Factors Responsible for Floods

4.1.1.1 Surrounded by Mountains

The country is surrounded by hills on its three sides, Rajmahal hills in the west, the Himalayas and the Meghalaya Plateau in the north, and Tripura–Chittagong hills in the east. The rainfall-runoff from this vast hilly area coupled with snowmelt in the Himalayas brings a huge inflow of water to Bangladesh during the wet monsoon season. A large area including Bangladesh and adjoining areas in India are under the influence of monsoons. From June to October large quantities of warm moist air travel from Indian Ocean north over Bangladesh and then to the Himalayan slopes as monsoon winds. Upper air turbulence and the long mountainous barriers stretching east–west make this moist air rise and as a result cool off and bring about enormous amounts of orographic rains which enter into Bangladesh territory in the form of runoff (Rashid 1991).

4.1.1.2 Lower Riparian Country

The country is located at the lower parts of the basins of the Ganges, the Brahmaputra and the Meghna (Fig. 4.1). The total area of the three basins stands at 1.75 million km^2 covering areas of India, Nepal, Bhutan, the Tibetan region of China and Bangladesh, of which only 7 % lies within Bangladesh. The excessive rainfall in these three river basins is the principal cause of riverine floods in Bangladesh. There are 57 rivers which originate outside the boundary of Bangladesh. About 1.18 trillion cubic meters of water flows annually to the sea, of which 1.07 trillion cubic meters or 91 % enters Bangladesh from upstream catchments and the rest are contributed by total internal rainfall (Rashid 1991). The annual volume of flow past Baruria just below the confluence of the Brahmaputra and the Ganges is 795,000 million m^3 which is equivalent to 5.52 m of depth over 14.40 million ha of land area of Bangladesh. In contrast, the annual average rainfall for the country is 2.32 m (Harza et al. 1991).

4.1.1.3 Floodplain Country

The aerial extent of flooding within Bangladesh is principally related to the unique physiographic feature of the country, which is the low-lying and extremely flat floodplain of the major rivers and their tributaries and distributaries. Floodplains cover

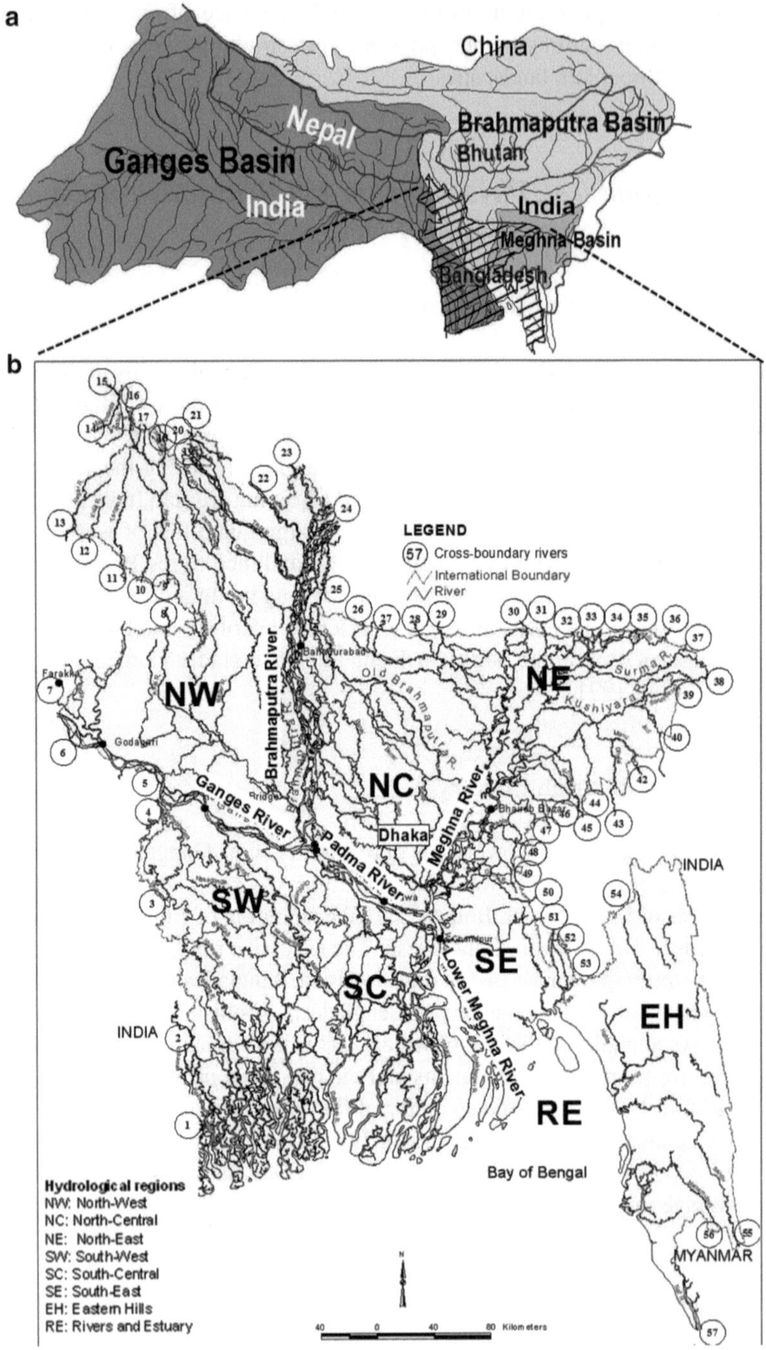

Fig. 4.1 (**a**) The Ganges–Brahmaputra–Meghna basins; (**b**) the intricate network of rivers in Bangladesh

80 % of the country, while hilly areas in the northern and eastern regions occupy about 12 % and terrace areas in the center and north-west occupy 8 %. Mean elevations range from less than 1 m on tidal floodplains, 1–3 m on the main river and estuarine floodplains, and up to 6 m in the Sylhet basin in the north-east. The capital city Dhaka which is about 225 km from the coast is within 8 m above the mean sea level (Rashid 1991). The rivers assume a minimum gradient owing to the flatness of the land surface and flood water spreads more evenly and accumulates on the plains. A relatively small increase in flood level results in inundation of a wide area in the floodplains.

4.1.1.4 Unique Hydro-Meteorological System

There is a wide spatial variation of rainfall in the basins of the Ganges, the Brahmaputra and the Meghna, as well as the whole of Bangladesh. There is also a highly skewed temporal variation of rainfall; about 80 % of the rainfall occurs during the months from May to September. As a result, unlike other deltas, the seasonal variation in river flow is highly skewed with abundant water during monsoon while very small flow during the dry season. The country, therefore, faces two major hazards: floods during the wet season and scarcity of water during the dry season. The hydrodynamic characteristics during flood flows in alluvial rivers in Bangladesh are quite different from that during low flows.

The country is crisscrossed with an intricate network of around 230 rivers. A remarkable aspect of the river system is that all the rivers, except those of the Chittagong sub-region, are hydraulically linked to each other, all rivers being either tributaries or distributaries of the three major river systems. Flood hydraulics in Bangladesh is dominated by the major rivers. High water level of the major rivers slows down the flow of their tributaries resulting in backing up of water in the tributaries. The Ganges, the Brahmaputra and the Meghna, discharge about 142,000 m^3/s into the Bay of Bengal during high-flow periods (Rahman et al. 1990). The Brahmaputra and Ganges carry about 85 % of flood flow that enter Bangladesh. The Brahmaputra has the largest flood flow followed by the Ganges and the Meghna with a flow ratio of 4.4:2.5:1. Although the Meghna has the lowest flood flow, it is by no means less important for flood processes in Bangladesh (Hofer and Messerli 1997). However, as far as area flooded is concerned, flow in the Brahmaputra has the strongest correlation with the extent of flooding in Bangladesh (Rahman et al. 2006; Salehin et al. 2007).

4.1.2 Key Determinants of Extent of Flooding

The important elements that determine the extent of flooding are the magnitude, synchronization of peaks, and duration of floods in the major rivers. The first two elements are directly related to the amount of rainfall in the upstream catchments, while the last element is related, in addition, to the downstream control provided by the coast in the form of the spring tide and the monsoon wind set-up in the Bay of Bengal.

4.1.2.1 Peak Flow

The magnitude of peak flow has a direct bearing on the extent of flooding in Bangladesh. Smaller differences in peaks of major floods can make a big difference in terms of flood affected area, since it is the spreading of floodwater evenly over a wide and flat floodplain that slows down the rate of rise in water levels. Frequency analysis of 92 hydrological stations showed that the difference in annual maximum flood levels of different return periods is not large. In general, the difference between 20- and 2-year annual flood levels remains within 2 m while the difference between 100-year and 20-year annual maximum flood levels remains within 1 m (Chowdhury et al. 1997a).

4.1.2.2 Duration of Floods

The downstream control that exacerbates the flooding condition is offered by the Lower Meghna (Fig. 4.1), which acts as the single drainage outlet of the three major river systems. Spring tide and monsoon wind setup in the Bay of Bengal cause strong back water effect in the Lower Meghna, and slow down drainage building up high water level and causing increase in the severity and duration of flood. This was the case in major flood years including 1998 and 2004, when the combined flood flow reached the Lower Meghna during spring tide, resulting in an extremely high water level and consequently severe flooding (Rahman et al. 2006).

4.1.2.3 Time of Peak Flows in Major Rivers

Synchronization of peak flows in the Brahmaputra and the Ganges is a major determinant of the extent of flooding in the country. When the peaks of the two rivers coincide, severe flooding occurs as it was the case in 1988, 1998 and also 2004 (Rahman et al. 2006). It has long been recognized that occurrence of simultaneous flood peaks in the Ganges and the Brahmaputra is not a rare event (French Engineering Consortium and BWDB 1989). Flood in 1998 is an example of simultaneous occurrence of several extreme hydrological phenomena such as coincidence of flood peaks of major rivers, high magnitude, occurrence of high flood for long duration, and arrival of peak flood at the time of spring tide.

4.1.3 Types of Floods

Five main types of natural floods occur in Bangladesh: river flood; rainfall flood; flash flood; tidal flood; and storm surge flood. Areas prone to different types of floods are shown in Fig. 4.2. In addition, some floods result from human activities. The principal sources of floods, as discussed at length in previuos sections, are the

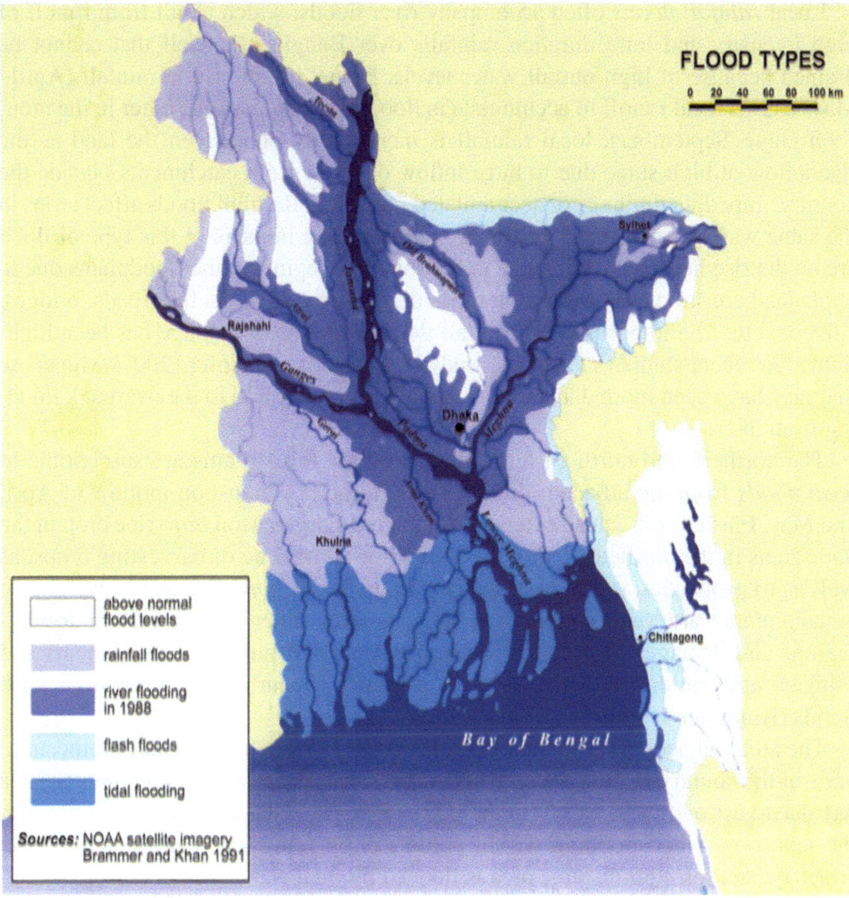

Fig. 4.2 Map of flood prone areas (Brammer and Khan 1991)

river floods from the major river systems, the Brahmaputra, the Ganges, and the Meghna, in the monsoon months. A broad strip of land, amounting to about 30 % of the country, extending beyond the active river floodplains (flooded during average normal flood) to parts of the adjoining meander floodplains (normally inundated by rainwater), is subjected to this type of flood. The timing of the flood and sometimes the duration of flooding are as important determinants of crop damage as is the absolute height reached by a particular flood. River floods in June can damage *aus* and deepwater *aman* paddy as well as jute crops, and floods in late August and September can be particularly damaging to deepwater *aman* paddy in low lands as a result of submergence at the panicle-initiation or flowering stages and to transplanted *aman* paddy on higher land because of drowning of seedlings or prevention of them from being planted (or replanted after an earlier flood loss) (Brammer 1999).

Local *rainfall floods* often accompany river floods, which result from runoff of high intensity and long duration rainfalls over Bangladesh itself that cannot be drained because of high outfall water levels. Heavy pre-monsoon rainfall (April–May) causes local runoff to accumulate in floodplain depressions. Later in the monsoon (June–September), local rainfall is increasingly ponded on the land as the rivers flow at high stage due to huge inflow of water from catchments outside the country, impeding drainage of accumulated rainwater. Rainfall floods affect crops in the same way as do the river floods in monsoon. The impacts of this type of flood are on the rise because of a change in hydrological regime in the floodplains due to unplanned construction of different types of infrastructures, such as roads, bridges, culverts, etc. The importance of rainfall floods in terms of damage has been highlighted in recent studies such as the study by Messerli and Hofer (2007) who argue that they have been more damaging in Bangladesh compared to the overbank flooding from the rivers.

The northern and north-eastern trans-boundary hill streams are susceptible to *flash floods* from the adjacent hills in India in the pre-monsoon months of April and May. Flash floods cause extensive damages to dry-season *boro* rice crop in the *haor* areas in the northeast region just before or at the time of harvesting crops, as well as to properties and infrastructures. Damages to *boro* rice and breaching of embankments are very common in some part or other of the eastern foothill regions and damages to property, especially road and railway embankments and bridges, and buildings alongside river channels, occur during very high flash floods (Brammer 1999).

The areas adjacent to estuaries and tidal rivers in the southwest and southcentral parts of the country (where they are not empoldered) experience *tidal floods* twice a day due to astronomical tide from the Bay of Bengal. Tidal water is mainly fresh in the monsoon, when flooding within polders is by rainwater. During spring tide, which occurs fortnightly, large area is flooded by tidal water, which can be damaging, especially if the water is then saline. Tide is experienced upto 225 km inland in the wet season and 325 km inland during the dry season. A considerable area in the southwest region is below the high water level of spring tide.

Approximately 12,000 km^2 of coastal land is prone to occasional cyclonic *storm-surge floods* due to tropical cyclones in the Bay of Bengal during April to June and September to November. Cyclones have the most dramatic consequences among the different hazards in Bangladesh. In comparison, the number of deaths during monsoon floods, even during extraordinary events, is small (Hofer and Messerli 1997).

Anthropogenic activities in the form of construction of infrastructure (mainly road) without sufficient drainage capacity through them, road alignments transverse to the main drainage paths, blocked drainage channels due to siltation, cross-dams or fishing activities and inadequately sized drainage sluices are increasing flood hazards (WARPO 2001a). For example, ill-planned construction of roads was one of the main reasons for rapid increase in flood depth and excessive duration of flood in September to October of 1995 in the north-west region of Bangladesh (Chowdhury et al. 1997a). Reduction of flood storage area due to filling up of low-lying floodplains and natural depressions is also exacerbating the flooding condition during major floods. Another cause of concern is the damage caused by sudden floods due to failure of flood control embankments. Flooding due to breaching of embankment does much harm to the agricultural land by depositing coarse sand, while natural flooding of agricultural land by overflowing river contributes to fertility of the land by depositing silt.

4.2 Historic Floods and Impacts

That the country has experienced floods since ancient times is quite evident from historical evidences in published literature on ancient flood management practices in the Bengal. A translation of a Persian writing during the second half of the eighteenth century by Akbaruddin (1974) showed that people used to construct earthen embankment around cultivated land to protect from floodwater, the local rulers of the Bengal built their homesteads by raising earthen mounds, and people used to grow long-stem rice which rises with the rise of floodwater. Construction of dikes along the rivers and irrigation by planned excavation of canals were the practices during the Mughal rule in the sixteenth to eighteenth century, as reported by Wilcocks (1930). However, the first systematic study of floods in Bengal was done by Mahalanobis (1927), a member of the North Bengal Flood Committee, formed by the Government of Bengal after a devastating flood in the northern part of Bengal in September 1922. It was reported that flood occurred in 25 of the years between 1870 and 1922, and were severe in 8 of these years. In 1922, the flood was caused by rainfalls of unprecedented magnitude. The rainfall in 1 week was about 10 times the normal weekly precipitation. The report observed that railway embankments hampered quick draining away of the flood water, and thus served to prolong the duration of flood.

No dependable data are available from 1923 to 1953, except a report prepared by the Irrigation Department of the Government of Bengal on the flood of 1931. The area affected by the 1931 floods was reported to be around 2,000 km^2. The flood data are, however, well recorded since 1954. A historic overview of variability of annually flooded area since 1954 is presented in Fig. 4.3. The figure also includes the growth of flood control and drainage projects in Bangladesh over the last 50 years. There is a decreasing trend of approximately 80,000 ha per year in the flooded area during the period from 1954 to 2004. This is in keeping with the increased coverage of flood protected areas, a growth of about 120,000 ha of area brought under flood control and drainage projects since 1964. However, despite implementation of all these flood control projects, the flooded area increased during major floods. There is clearly an increasing trend in year-to-year variability in the annually flooded area from mid 1970s.

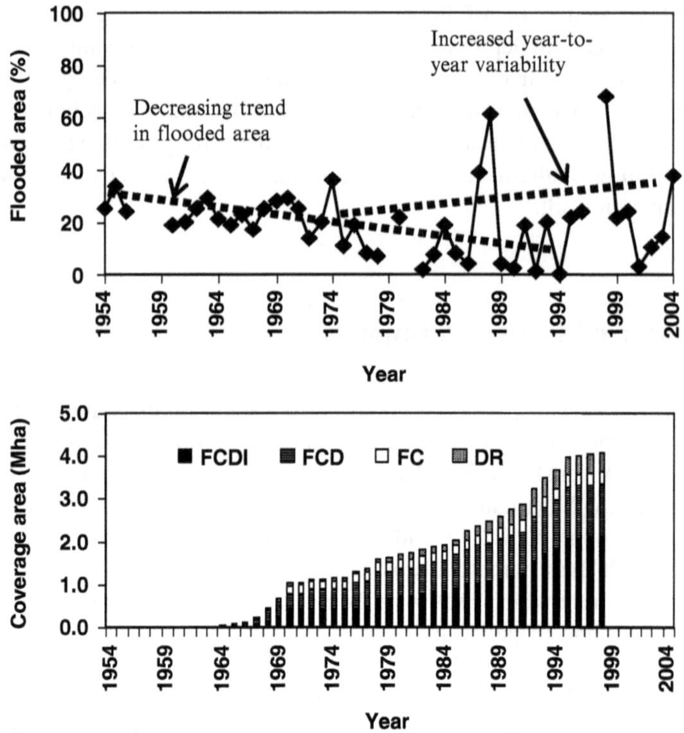

Fig. 4.3 Growth of flood control projects and variability of annually flooded area (*Source of data*: Bangladesh Water Development Board)

4.2.1 Flood Damage

The main objective of flood control projects was to protect the crop land from flood damage. However, there occurred considerable crop damage even in years of moderate flood because the embankments were also damaged. The earthen embankment was unable to give protection against severe floods in 1987, 1988, 1998 and 2004 and even against some medium floods in 1991, 1993 and 1995 (Salehin et al. 2007). However, the crop loss due to flood damage does not seem to have substantial effect upon the total rice production of the country. This is because the crops are replanted and a part or even the total amount of loss may be caught up during the later part of the season. It also happens that the crop damage by flood is largely compensated by substantially higher yields due to higher residual moisture available to the following crops (Chowdhury et al. 1997a, b). For example, *aman* production in 1998 fell short by about 21 % below the normal trend production, while the production of dry season *boro* crop was 10 % above the normal trend in the same year (Islam 2006).

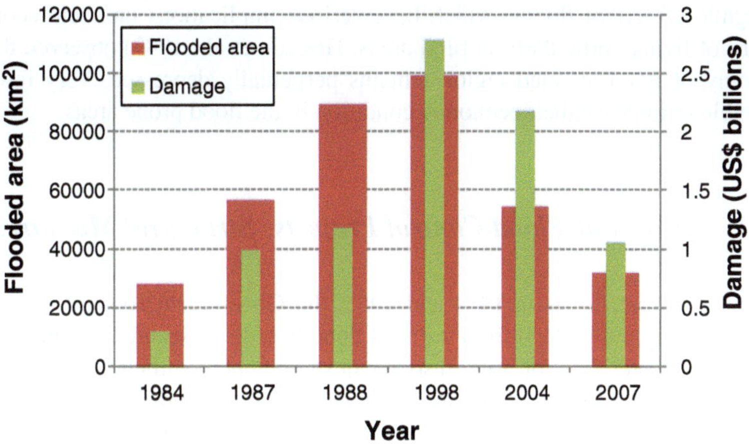

Fig. 4.4 Flood damage in moderate to major flood years (*Source of data*: Ministry of Diaster and Relief)

However, it is the damage to properties and infrastructure that is of major concern. When converted to monetary units, the damage to infrastructure outweighs the damage to crops (Chowdhury et al. 1997a, b). As can be seen in Fig. 4.4, there was substantial damage in 2004 even though the area flooded was much lower (38 %) compared to that in 1998 (67 %). The non-agricultural sector suffered loss for as high as 74 % of the total loss, with the remaining 26 % accrued to the agriculture (crop plus non-crop) (Islam 2006).

4.2.2 *People at Risk in Flood Prone Area*

With the increase of population, more and more people are settling in the flood-prone areas, making them more vulnerable to floods. An analysis conducted with 2001 population census data revealed that some 45.5 million people were exposed to severe and moderate floods (including river flood, flash flood and tidal flood), of which 22 million were male and 23.5 million were female (CCC 2009). Studies have also shown that flood-prone zones are the worst off among different disaster-prone areas in terms of food shortages, the incidence of extreme poor, insufficient income, illiteracy, and a high concentration of wage laborers (BIDS 2006). Vulnerability of the poor to floods has increased as a result of environmental deterioration and the increase in the number of poor living in high flood risk areas (Rahman 2004). Poor households suffer proportionately more from flood than richer households. Islam (1996) reports that an average poor household suffered (in terms of proportionate losses to values) four, five and three times as much compared to that suffered by a richer household, in a river flood, flash flood and tidal flood, respectively. Traditional flood damage estimates cannot capture numerous intangible effects of floods, which have serious implications on socio-economic pattern of living particularly in rural areas. Hossain et al. (1987) observed that the economy of flood affected region remains perpetually depressed. Very few have been able to improve their economic condition in the flood prone areas.

4.2.3 Growth of Flood Control Projects: Structural Measures

Following severe floods in 1954 and 1955, the Government adopted a policy of protecting agricultural land from the river flood in order to secure agricultural "production. A Master Plan for water resource development identified flood control and drainage measures as key requirements to increase agricultural production, by providing more stable conditions and reducing the depth and duration of normal and unusual flooding. Hence, since mid 1960s there has been a steady growth of flood control and drainage projects in Bangladesh through the construction of 12,850 km of embankments, 25,580 km of drainage channels and 4,190 sluices and regulators as per BWDB list of completed projects as of 1998 (WARPO 2001a). Currently the total coverage area stands at 5.37 million ha, which is about 37 % of the total area of the country and 56 % of the total cultivable lands.

Many conventional flood mitigation measures like flood control reservoirs, flood diversion or flood by-passes are infeasible inside Bangladesh because of its extreme flat topography. The main approaches that have been exercised are: (1) full protection of agricultural lands and urban areas against river flooding by constructing embankment along the rivers and providing appropriate drainage structures to minimize internal flooding; (2) partial protection against river flooding by constructing low height submersible embankments which are designed to delay pre-monsoon floods so as to ensure a safe winter crop (*boro* rice) harvest and to be overtopped during monsoon and remain submerged during the entire monsoon season; (3) evacuating unwanted rain water from behind the embankment or from within the poldered area by gravity through drainage regulators and sluices incorporated with the embankments (regulators and sluices prevent back flow from high river levels into the low-lying areas during monsoon and drain the water from the area to be protected when the river water level gets below than that inside); and (4) providing drainage by pumps in selected large projects to pump out accumulated water from the project area. The types of implemented flood control projects include flood control (FC), flood control and drainage (FCD), flood control, drainage and irrigation (FCDI), and drainage (D) projects. Other structural measures included dredging of rivers and canals at critical locations such as offtakes and confluences to remove sand bars, and hard and soft recurrent measures for bank protection and river training works.

Urban flood protection involved conventional method of constructing embankment and drainage regulator, with the exception of pumped drainage facilities in three drainage zones of the capital city Dhaka, with major part of the city dependent on detention storage-based storm drainage system. Approximately 123 FCD polders have been constructed since 1960s by Bangladesh Water Development Board under the Coastal Embankment Project (CEP) covering approximately 1.5 million ha of area in southwest, south central and Chittagong regions of the coastal zone. The objective was to create favorable condition for monsoon rice cultivation by preventing inundation of agricultural lands by saline water during high tides.

4.2.4 *Non-structural Measures*

Non-structural measures were considered as a means for mitigating flood damages. The Flood Forecasting and Warning Center (FFWC) of Bangladesh Water Development Board (BWDB) established in 1972, is responsible for making flood forecasts and flood warning during the flood seasons. The monitoring of floods and issue of flood forecasts are carried out in relation to Danger Levels (above which the flood is likely to cause damage to crops and homesteads) specified for each river gauging stations. The FFWC collects real-time data of water level (3 hourly) from 55 observation stations and rainfall from 56 observation stations. Currently a hydrodynamic mathematical model (MIKE-11) is used to forecast water levels. At present, the FFWC issues river stages forecast for 50 stations in the flood prone areas, formulated for lead time of 24 h and 48 h. The forecasts in the form of daily water level bulletins are transmitted to national radio, television, news agencies, newspapers, concerned ministries and government offices and field wireless stations.

Flood proofing of homestead is a tradition of the rural settlements in Bangladesh. Homesteads are generally raised above unusual flood level. Ponds are dug in the homestead area and banks are raised to prevent intrusion of flood water. They are the source of domestic water supply. They are also used for fish culture and for conservation of water for irrigation during droughts. Schools or health centers built on high mounds serve as flood shelters where vulnerable people can take refuge. Cyclone shelters are constructed in the coastal zone where human lives are at high risk due to cyclonic storm surge floods. Currently there are a total of 2,583 cyclone shelters in the coastal districts (CEGIS 2009). Over the years, there has been a significant change of emphasis from designing the shelters for single purpose use as flood shelter to designing them for multipurpose use. Shelters are now designed as schools, health centers and other community service centers during normal life.

4.2.5 *Experiences with Flood Control Projects*

Although there have been a number of positive impacts of the flood control projects in terms of increase in agricultural production, increase in economic activities, reduction of damage to infrastructure inside protected area, increased opportunities for culture fisheries, and generation of employment, the projects in general could not attain the desired objectives because of lack of consideration of interdependence of land, water, ecosystems and socio-economic development. Many projects were implemented with a view to solving the immediate problems without giving adequate attention to potential, undesirable long-term consequences. Project design and implementation largely ignored people's participation. There was lack of consultation across sectoral and institutional boundaries, and the projects did not fully take account of the potential impacts on fisheries, navigation, forests, domestic and industrial water supply, biodiversity, and salinity management.

4.2.5.1 Hydraulic Impacts

The increase in year-to-year variability in flooded area and the increase in the frequency of severe flood events in terms of flooded area despite the growth of flood control projects, as illustrated in Fig. 4.3, can be attributed to the hydraulic impact the projects have had on the river-floodplain systems. Ill-planned growth of flood control projects together with ill-designed transportation and drainage networks make the system unstable during extreme floods. The flood control projects provide protection against normal floods. However, during moderate to major floods the damages to infrastructure including embankments are very pronounced thus causing increased flooded area. Flood control embankments suffer substantial damage during large and moderate floods, while damage during smaller floods is also quite high (Chowdhury 2000). The resulting damage due to embankment failure is very high owing to the accelerated economic activities compared to that in "without embankment" situation.

Floodplains moderate the flood flow by acting as storage during monsoon. Floodplains also augment the post-monsoon river flow by gradually releasing water from storage during the recession phase of floods. The river flow during late October and November is thus increased. Floodplains are also a useful source of groundwater recharge. Rainfall and flood water over the floodplains infiltrate and percolate vertically through the pervious soil to reach the relatively shallow groundwater table and recharge the unconfined aquifer. Flood control embankments have disrupted these hydrologic functions of floodplains. The rising trend in the annual maximum water level series and a declining trend in annual minimum water level series for the Atrai river in the NW region is the consequence of preventing storage of flood water in the floodplain depressions (*beels*) by constructing polders (Chowdhury et al. 1997a, b). Reduced floodplain storage due to embankment leads to an increase in water levels and discharges in adjacent areas. Such transfer of flood risk generates social tension, often leading to upstream–downstream conflicts and consequently forced cutting of flood control embankments by the affected people.

The coastal polders have led to deterioration of river morphology and waterways in the southwest and south-central coastal region. Without the embankment, enormous volume of tidal water would have been stored in the floodplain during rising tide which would have allowed sediment in suspension to be deposited over the floodplain. During ebb tide, the stored water would have drained through the tidal rivers providing a flushing action. As a consequence of the decrease in tidal volume due to polders, the water level and tide velocity dropped during ebb tide, resulting in siltation in the river causing rise in the river bed (Halcrow et al. 1993). This has, in turn, caused severe water logging inside the polders, leading to serious damage to agriculture, forestry, fisheries, livestock, homestead and physical infrastructures. Many people had to leave their ancestral homestead by abandoning traditional livelihood activities.

4.2.5.2 Impact on Crop Production

While many projects have been successful in raising agricultural output, as found in project based evaluations such as the study of HTSL (1992) of 17 FCDI projects, the impact of these projects on the overall food grain production is less clear (Chowdhury et al. 1997a, b). Flood control projects with irrigation projects have been more successful.

4.2.5.3 Impact on Floodplain Ecosystem and Water Transport Function

The objective of increasing rice production by providing flood protection to agricultural land has got so much priority that the consequential stress on the floodplain ecosystem received little attention. Livelihood activities of different groups, especially the marginalized, have been compromised. There has been substantial damage to capture fisheries because of disruption of hydraulic connection and hence fish movement between river and floodplain. There was a decline of 81 % in catch per unit area and a reduction of 33 % in the total number of species recorded annually in the case of Brahmaputra Right Embankment project (ODA 1995), and there was a 91 % decline in capture fisheries inside full flood control projects in the northeast region (Shawinigan Lavalin Inc. et al. 1992). Closure of the outlet of floodplain *khals* by flood control embankments created obstruction to country boats. The study by HTSL (1992) showed that FCDI structures had severely impeded boat transport in half of the 17 projects investigated. Out of the 66 projects studied by Shawinigan Lavalin Inc. (1993), 19 had major and 14 had medium level negative impacts on boat transport. The depletion of fisheries and impediment to floodplain water transport have caused conflicts between different users of floodplain such as farmers, fishermen and boatmen.

4.2.6 Lessons Learned and Paradigm Shift

The experiences of flood control interventions in a floodplain country have proved that it is difficult to attain stated objectives of interventions without giving due consideration to the hydro-morphologic features of the floodplain and the socio-economic conditions and cultural heritage of its inhabitants. It is seen that prevention of flood in floodplain has many negative environmental consequences. On the other hand, population, in general, do demand protection against flood damage. So an important issue is how to mitigate flood damage without causing degradation of floodplain environment. What it requires is a shifting of policy from flood control to flood management, with preservation of floodplain resources like soils, wetlands, capture fisheries, wild life, flora and fauna, and with indigenous production systems receiving their due priority. Floodplain zoning can be developed with the aim of preserving floodplain resources.

The nature of a water resources project is such that while it brings benefits to certain section of population, it may bring disbenefits to another section of the population. While resolution of conflict arising out of inequitable distribution of project benefits and disbenefits is difficult in any country, it takes on special poignancy in Bangladesh where the majority of the population living in the floodplains is overwhelming poor, holds small landholding and depends on many free resources of the floodplain. Hence, calculation of economic return should not be the only guide in selecting water resources projects. Decisions should be based on a multi-criteria framework of economic costs and benefits, and social and environmental impacts. Vulnerability of the society to large floods can be reduced by making communications lines, especially utility services and other infrastructures flood proofed. A floodplain land use regulation needs to be formulated so that planning, design, construction and maintenance of infrastructures account for flood factor and preservation of floodplain resources and environment.

The shifting of priority of the Government from "flood control" to "flood management" started towards the later stages of Flood Action Plan (FAP), initiated by the government after the severe floods of 1987 and 1988. At the initial stage of FAP, the focus was on flood mitigation. After a series of public debates followed by documentation of the concerns of flood mitigation in a number of published books, gradually it was recognized that FAP should pay attention to the complete hydrologic cycle and develop an integrated water management plan covering issues relevant to not only flood but also drainage, irrigation, navigation, environment and socio-economy. A real start in the paradigm shift in water resources management policy and practice towards "integrated" management took place with the preparation of the National Water Policy (NWP) in 1999 (MoWR 1999) and subsequently the National Water Management Plan (NWMP) in 2001 (WARPO 2001b). The NWP envisages full structural protection against floods in regions of economic importance (e.g. metropolitan areas) and reasonable degree of protection in other critical areas (e.g. district town), motivating the people to develop different flood proofing measures such as raising platform for homesteads and community facilities in the rural areas, with the exception of those already covered by existing FC infrastructure, and construction of all highways, railway tracks, and public buildings and facilities above the highest ever recorded flood level in future. The policy also stressed the importance of designating flood risk zones and taking appropriate measures to provide desired levels of protection for life, property, vital infrastructure, agriculture and wetlands. The NWMP includes a total of 84 programs grouped into eight clusters. One of the programs in the "agriculture and water management" cluster is to rationalize flood control and drainage projects considering the adverse impacts on the society and the environment and the institutional requirements for participatory management. The "disaster management" cluster includes programs related to improved warning and preparedness systems, social measures based on improved or more appropriate dissemination and response procedures, physical and social mitigation measures such as elevated platforms, cyclone shelters, raised highways and the like, and multiple use of infrastructure.

4.3 Innovative Flood Risk Reduction Approaches

In the light of the experiences with the flood control projects, the necessity to harness the beneficial impacts of floods especially in maintaining soil fertility and sustaining fisheries resources became apparent. All the socio-economic and environmental concerns of earlier practices gave rise to exploration of alternative management strategies which are more resilient and environment friendly. Three such strategies are tidal river management in the coast, compartmentalization in the central Brahmaputra floodplain and partial flood control in the north-eastern part of Bangladesh.

4.3.1 Tidal River Management

Polders in the southwest and south-central region implemented under the CEP had initially created good scope for growing agricultural crops by preventing the intrusion of saline water. However, as discussed previously in Sect. 4.2.4, obstruction of flood tide by CEP embankments and sluice gates caused reduction of tidal volume, leading to gradual silting up of rivers. As the land development of the *beel* stopped because of lack of sediment deposition, this land became lower than river beds.

After more than a decade of good productivity, drainage congestion began to increasingly affect the northernmost polders from the 1980s when the rivers and creeks silted up to such an extent that most of them became inoperative. This resulted in vast tracts of land remaining waterlogged round the year. In 1995, BWDB undertook the Khulna–Jessore Drainage Rehabilitation Project (KJDRP) (Fig. 4.5) to relieve drainage congestion in the Khulna and Jessore districts. The project embarked on the design of large regulators at the mouth of rivers in order to control tidal inflow into the project area. Such controlled inflow would reduce the possibility of sedimentation of the rivers as the sediment load is tide borne. Such regulator option was opposed by the local people as they apprehended sedimentation in front of the regulators and therefore feared that regulator based option will not be a permanent and sustainable solution. Instead, the local people came up with the concept of tidal river management.

According to the Tidal River Management (TRM) concept, *beels* are to act as tidal storage basins which allow natural tidal flows up and down in the river system.

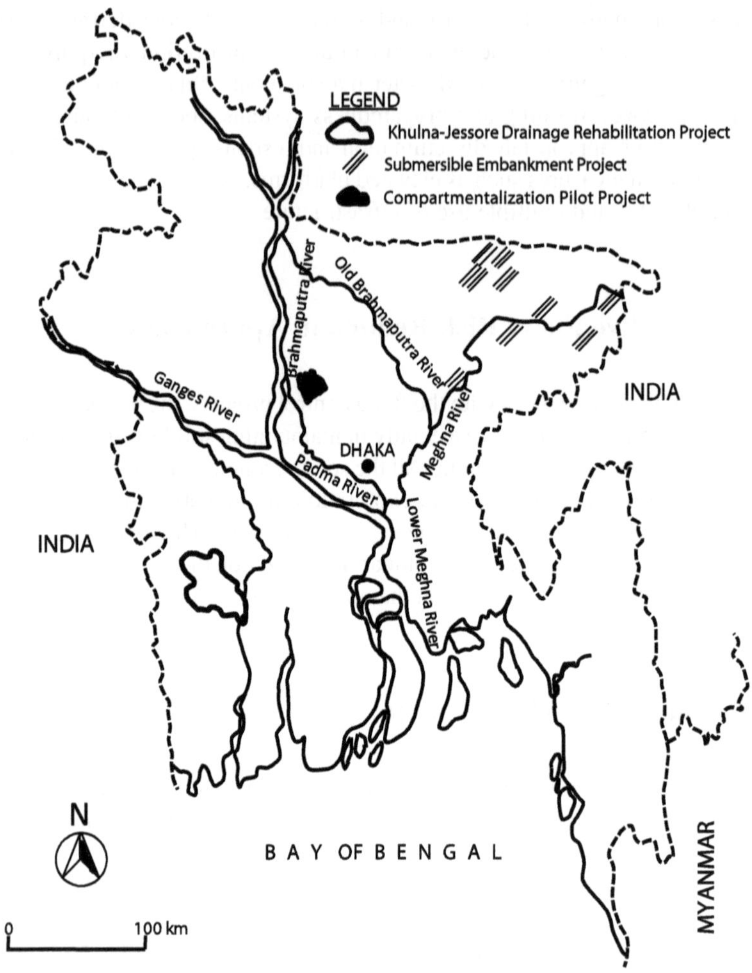

Fig. 4.5 Location of innovative flood management projects in Bangladesh

During high tides, a large volume of water along with sediments flows into the *beels*. These sediments settle down into the basin due to reduction of velocity and flocculation due to high salinity. This sedimentation would occur into the riverbed if the *beels* are not utilized for storage. This sedimentation in the basin lowers the river sediment capacity below the equilibrium condition and as a result, during ebb tide the river erodes the banks and scours the river bed. This continuous process gradually widens and deepens the river increasing the tidal flow volume of the river and thereby keeping the river alive. Tidal river management is in fact a natural water management process with very little human interventions but it needs strong participation and consensus with a great deal of sacrifice by the stakeholders for a specific period (3–5 years or even more depending on the tidal volume and the area of the *beel*).

Fig. 4.6 Location of proposed tidal basins in KJDRP

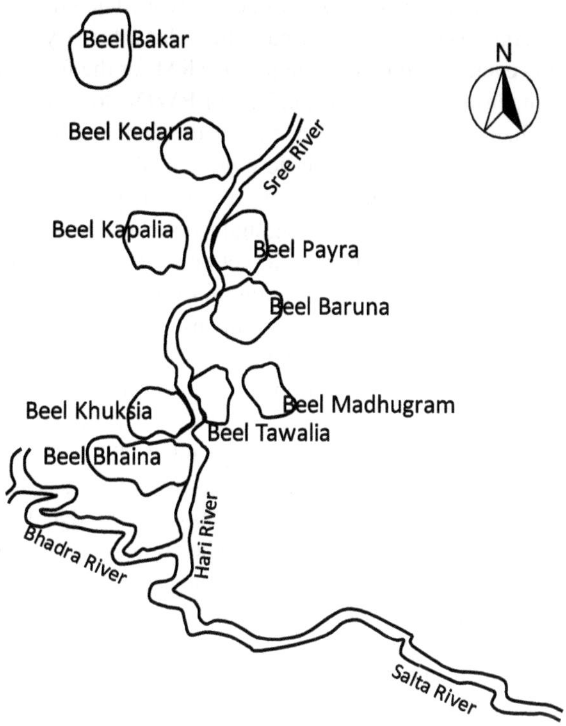

The principle of a rotational tidal basin was proposed to share both the inconveniences and the benefits with adjacent *beels*. In this regard nine *beels* (Fig. 4.6) with average size in the range of 1,000 ha were selected to operate as tidal basin one after another. It was planned that a new tidal basin would be opened after every 2–3 years.

The Environmental Impact Assessment (EGIS 1998) carried out for the KJDRP showed the viability of TRM for the project. The approach was adopted for the Hari river system, with a benefitted area of about 27,000 ha. The KJDRP started its activities in 1998 by first dredging the silted up rivers to restore their natural conveyance capacities. In the mean time, the first tidal basin was implemented and operated in an unplanned way by forced-cutting of the embankment by the local people at Beel Bhaina. This tidal basin operation continued for 4 years starting from 1998 to 2001. The high tidal volume generated by the Beel Bhaina tidal basin increased the flow area of the Hari River downstream of the basin. The riverbed deepened by about 10 m more than the design level, and the river widened as well by eroding both of its banks (CEGIS 2002). A huge amount of sediments deposited within the *beel* bringing the entire *beel* area under cultivation and navigability of the Hari river was restored.

The second tidal basin was implemented during the period from 2002 to 2005 in Beel Kedaria in a planned way without any obstruction from the local people. While deposition of silt in Beel Kedaria was not adequate as per expectation of the landowners, the drainage was perfectly well and the project was free from any water logging. The landowners gave their land for 3 years without any compensation. However, they did not want to run TRM for the fourth year in their *beel* without any compensation despite the efforts of BWDB to convince them. As BWDB could not finalize a new *beel* and complete the construction of peripheral embankment for operation of TRM in the new *beel*, it ultimately resulted in the closing of TRM operation in Beel Kedaria with resultant deposition of sediments in the river system and recurrence of water logging in vast areas in the KJDRP.

The third tidal basin could not be started in Beel Khukshia in the year 2005 due to opposition and non co-operation from the landowners as they were not willing to give their land without any compensation. On the other hand, attempt to continue the on-going tidal basin operation in Beel Kedaria for another 1 year was not possible due to non-cooperation and opposition from a section of people who cultivated crops in the high land of Beel Kedaria. As such, about 17.00 km stretch of the Hari river rapidly silted up within September, 2005; drainage route of the area became blocked and heavy drainage congestion resurfaced. In response, BWDB engaged dredgers, excavators and manual labours to dredge the river again. In the mean time, BWDB was able to prepare another tidal basin in Beel Khukshia, which started operating since 2006. Drainage congestion situation gradually started to improve and in year 2008, water logging in the project area virtually disappeared again.

In the year 2012, Beel Kapalia was due to be the next tidal basin in operation. But again the land owners of Beel Kapalia refused and resisted giving their land for TRM without compensation. Violent clash between local people and law enforcing agency occurred and TRM operation in the KJDRP again started facing uncertainty.

The biggest responsibility in proper operation of tidal basin lies with the beneficiaries. This is for two reasons: firstly, TRM was suggested by the project inhabitants themselves; therefore they are expected to own it and work accordingly, and secondly, if TRM does not work they have to face severe consequences. When a tidal basin operates, the inhabitants of other *beels* enjoy the benefits of the project and nobody wants to hand over their *beel* for next round of TRM. Attitude like TRM is necessary but "not in my *beel*" prevails everywhere. The spirit of TRM is to share long-term benefits and short-term dis-benefits among the beneficiaries. In this regard the vast number of beneficiaries should be able to arrange compensation for the small number of inhabitants of the running tidal basin. On the other hand, BWDB can arrange compensation from the Government fund which will be easier to implement and disburse. This is now being tried in the case of Beel Kapalia. If such mechanism works then recurrent problem of preparing next round of tidal basin can be solved and TRM can become a model of innovative approach to solving tidal flood problem in the coast.

4.3.2 Compartmentalization

From a historical point of view, floods have devastated large parts of Bangladesh, especially during the 1987 and 1988 floods. In the aftermath, various studies were conducted in the assessment of confining these floods or, at the other extreme, apply a strategy of "living with the floods", which is the more traditional way of dealing with these floods in these circumstances. A compromise was made out of these two extreme points of view, and hence the concept of compartmentalization was developed.

Compartmentalization is an established technique in watershed management. The technique is applied when it is necessary to slow down the runoff of river systems or watersheds or to create independent systems to which different drainage criteria can be applied. By applying the technique, the risk of flooding and/or the size and cost of interventions can be reduced. In the compartmentalization approach it is foreseen that floodwater could be regulated in such a way that it gives maximum protection, minimum flood damages and maximum benefits to agriculture, fisheries and navigation. When the system is in full operation (entire floodplain covered with multiple compartments), decisions are needed to determine how the flood (and rain) water has to be distributed among the compartments, given the hydrological conditions (flood levels, flood duration, flood predictions and actual rainfall) and given the degree of liberty as determined by the technical features of the system.

The entire scenario of possible choices regarding the management of these compartments is heavily determined by the prevailing hydrological conditions in a particular year. In hydrologically "wet" years, the choice of flood protection and minimizing flood damages will prevail, while in "dry" years the choice of making use of the beneficial effect of controlled flooding will dominate. Risk management is an essential part of the concept and therefore an analysis of possible scenarios need to be worked out which would provide guidelines for management at (floodplain) regional scale.

In order to test the compartmentalization concept in Bangladesh, the Compartmentalization Pilot Project (CPP) at Tangail was constructed during 1995–1998. The project is located (Fig. 4.5) in the Brahmaputra floodplain and is surrounded by embankments that can withstand a flood with a return period of 20 years. The CPP has established a single pilot compartment covering an area of 13,200 ha, divided by a seasonal river, the Lohajang, and bordered by two other rivers, the Dhaleshwari to the west and the Pungli to the East (Fig. 4.7). The Lohajang River is allowed into the project through a gated regulator at the northern side, not for the supply of water but for drainage, although the river also has a flushing function for the drains of Tangail Town. By lowering the water level of the Lohajang, the river will act as a drain for the numerous outlets that discharge into it. In the peripheral embankment on the northern, western and eastern sides of the project area, gated inlets are built to allow water into the compartment. The southern "embankment" is not an embankment: the road embankment is open and the Lohajang River and a number of other *khals* exit the Tangail Compartment without any control. In the case of long lasting high floods, flood water will enter the project from the southern side as back flow. Water control structures control water levels between sub-compartments and systems.

Early in the monsoon season water levels in the rivers are still low and only the occasional rain shower has to be drained off. Local drainage is thus the objective during this period. All gates are to be kept open, but can be closed if local

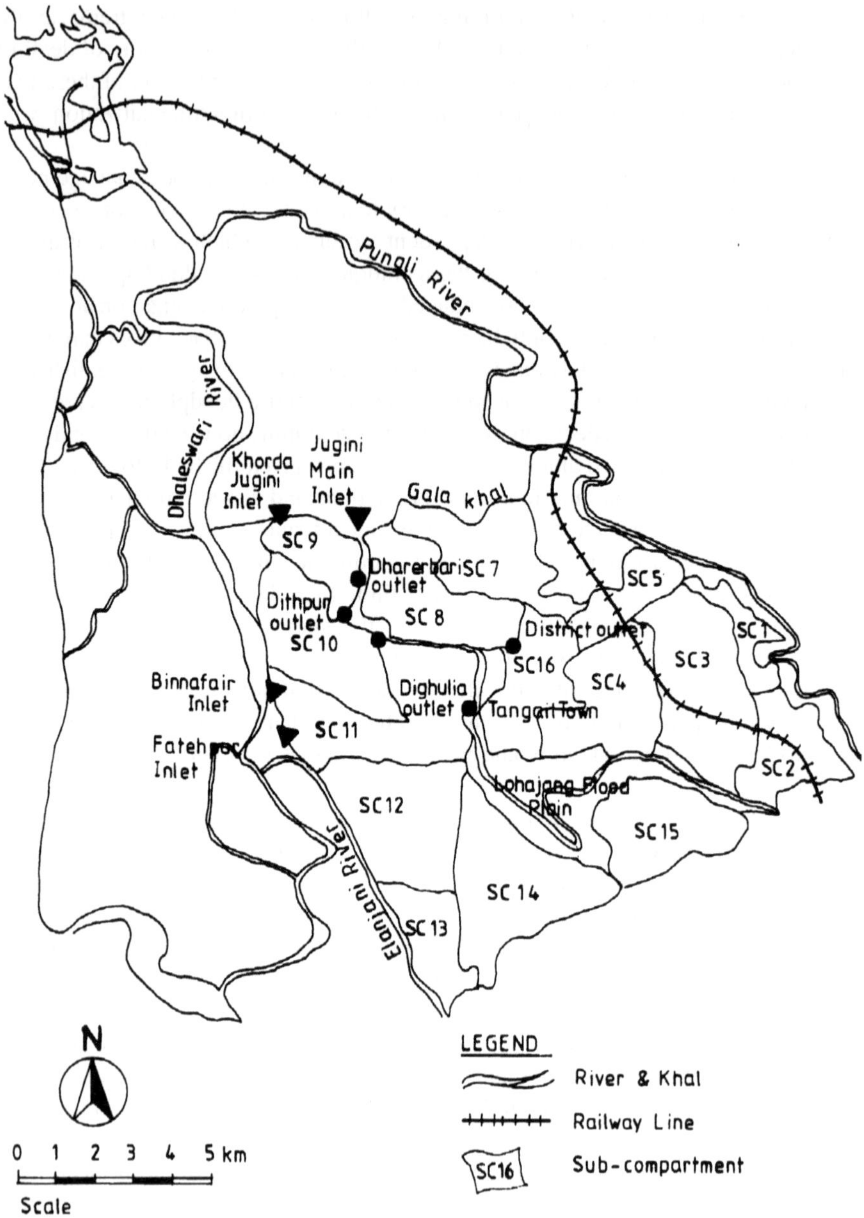

Fig. 4.7 Principal features of Compartmentalization Pilot Project

considerations require. Once the river water levels start to increase, inlets and some outlets are opened or remain open and water enters the compartment. Controlled flooding is thus the objective during this period. Siltation is allowed to take place on the field, and the entry of fish into the compartment is facilitated. The water control structures between sub-compartments and systems are operated as required, to facilitate an even distribution of the silt laden river floods over the compartment, and to deal with rain showers. The key objective is thus to maintain a level of flooding which will give maximum benefits to both agriculture and fisheries.

At a later stage, the emphasis shifts from controlled flooding to water level control. The main regulator will be (partly) closed, lowering the water level in Lohajang River, inlets will be closed, and outlets will generally be open when drainage is possible. The water level will be lowered to such an extent that land preparation and planting can take place. The objective is to keep the water level close to an optimum level. These levels are dictated by crop requirements in compromise with the physical possibilities of the water management infrastructure and fisheries and environment aspects.

When flood levels in the main river system exceed certain levels, then flood control is called for, meaning that the main regulator is completely closed, the inlets are closed, the water control structures are operated as required to limit flooding in sub-compartments and systems, and the outlets are open, downstream water levels permitting. Some local inundation, however, may occur, particularly as a result of rainfall, which cannot be drained off.

Towards the end of the monsoon, once the water levels start to drop, early drainage of the compartment becomes possible. The improved drainage in the compartment will allow early land preparation for the dry season crops, and early harvesting of *aman* rice crops.

Due to high density of required structures in the CPP, the cost was relatively high. On the other hand, the incremental agricultural benefit was relatively low as the cropping intensity at 191 % was already very high in pre-project condition and the CPP attained a cropping intensity of 214 % in 1999. As a result, an EIRR of only 7.3 % and a negative net present value of $4 million were achieved as per the CPP Final Report (Lahmeyer International et al. 2000).

In a later study Rahman (2008) found that the cropping intensity improved further, and that the controlled flood component of the CPP has been highly beneficial for agricultural purposes and is in much demand. The Dhaleswari river is, however, almost dry now and carries very little water. As a result, the project is suffering from water shortage. The innovativeness of controlled flooding lies in the main purpose of the CPP, which was to retain beneficial effects of flood. Among many beneficial effects, two major purposes were under consideration in the CPP; one was to reinvigorate soil fertility each year by allowing flood water in agricultural land to deposit silt and the other was to retain the project area as a suitable ground for capture fisheries by allowing fingerlings migration through fish passes in the main inlet during rainy season. Although the project allows flood water to enter into the project area to a certain extent to take advantage of floods' beneficial effect of increased fertility, but in reality water passing through the narrow inlets reduces sediment load to a great extent. Indiscriminate catching of fish near the main regulator is leading to a decrease in capture fisheries in the project area. Therefore, these two purposes are being served only to a limited extent in CPP area. However, if properly designed and adequately monitored and maintained, controlled flooding through compartmentalization remains an innovative flood management approach.

4.4 Partial Flood Control

The Haor Basin in the North east region of Bangladesh has an area of 6,000 km^2 which is deeply flooded region and is an important fishing ground of the country. During monsoon (June to October) the entire basin is flooded to a depth of 6–8 m. During dry season (November to May), the basin dries up with few pockets of water remaining. The basin is cultivated during dry season and the standing crop is harvested during late April–early May period. During this period the haor basin frequently experiences flash flood. In order to safeguard the only standing crop of the region, about 47 submersible embankment projects (SEP) were constructed during 1975–1990. The projects are small in size usually in the range of 5,000–6,000 ha. The concept of submersible embankments is illustrated in Fig. 4.8. While the typical FCD projects provide full protection throughout

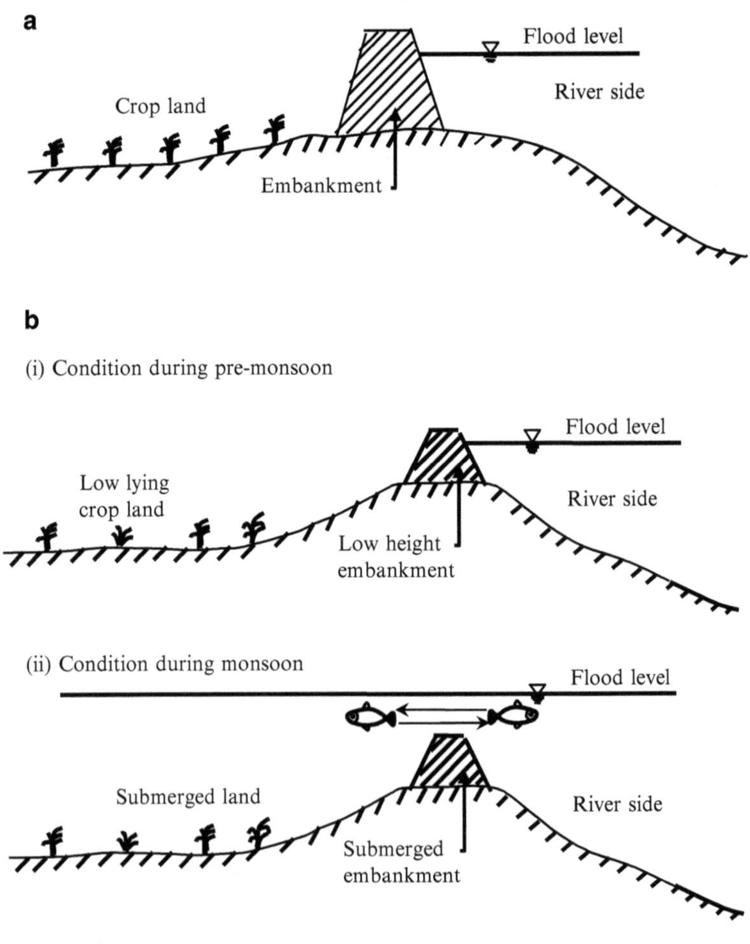

Fig. 4.8 Illustration of protection from river floods by embankments. (**a**) Protection from monsoon floods; (**b**) protection from pre-monsoon floods

the year, the submersible embankments provide partial flood control as it provides protection with low height embankment only during pre-monsoon. During monsoon, the embankments remain submerged, and therefore do not disrupt the lateral migration of fish between river and floodplain.

Saleh and Mondal (2007) found that the SEPs have achieved their desired objective of protecting the *boro* crop from pre-monsoon flood, especially after 1990s. The impact on fisheries production due to these types of projects has been minimal as found out by FAP 17 (1995). This is in contrast to traditional flood control projects where embankments obstruct monsoon flood and consequently the impact on fisheries has been very high. The submersible embankments result in a higher economic return; for example, HTSL (1992) found that in Halir Haor the average yield of *boro* was 19 % higher with the project than pre-project, and the estimated economic rate of return was on the higher side (about 65 %). On the other hand, the Brahmaputra Right Embankment (BRE) project (a full flood control project in nature), despite achieving a moderate increase in incremental value of monsoon paddy production, yielded a zero internal economic rate of return because of huge loss of capture fishery.

References

Akbaruddin (1974) History of Bengal (in Bengali). Bangla Academy, Dhaka, pp 16–17 [translated from Moulvi Abdus Salam's English translation (1904) of 'Reaz-Us-Salatin' (in Persian) by Gholam Hossain Salim]

BIDS (2006) Natural disasters, risks, vulnerability and persistence of poverty: an analysis of household level data, PRCPB Working Paper No. 15, Programme for Research on Chronic Poverty in Bangladesh (PRCPB), Bangladesh Institute of Development Studies (BIDS), Dhaka

Brammer H (1999) Agricultural Disaster Management in Bangladesh. University Press Ltd, Dhaka

Brammer H, Khan HR (1991) Bangladesh country study. In Disaster Mitigation in Asia and the Pacific. Asian Development Bank, Manila

CCC (2009) Characterizing country settings: development of a base document in the backdrop of climate change impacts. Climate Change Cell, Department of Environment, Ministry of Environment and Forests, Component 4b, Comprehensive Disaster Management Programme (CDMP), Ministry of Food and Disaster Management, Bangladesh

CEGIS (2002) Annual Monitoring Report 2001, Khulna-Jessore Drainage Rehabilitation Project (KJDRP), Bangladesh Water Development Board

CEGIS (2009) Cyclone Shelter Information for Management of Tsunami and Cyclone Preparedness. Annex A: Spatial Distribution Maps and Attributes. Center for Environmental and Geographic Information Services in cooperation with Comprehensive Disaster Management Programme (CDMP), Ministry of Food and Disaster Management, Government of Bangladesh

Chowdhury JU (2000) Bangladesh, a state at permanent flood risk. In: Proceedings of the international symposium on River Flood Defence, Kassel

Chowdhury JU, Rahman MR, Salehin M (1997a) Flood control in a floodplain country: experiences of Bangladesh. Islamic Educational Scientific and Cultural organization (ISESCO), Rabat

Chowdhury JU, Haque MM, Rahman MR, Salehin M (1997b) Flood loss reduction in Bangladesh: integrating environmental considerations into economic policy making processes. Prepared for UN ESCAP. IWFM, BUET

EGIS (1998) Environmental and social impact assessment of the Khulna-Jessore Drainage Rehabilitation Project, BWDB

FAP-17 (1995) Fisheries Studies and Pilot Project. Flood Action Plan, Government of People's Republic of Bangladesh
French Engineering Consortium and BWDB (1989) Prefeasibility study for flood control in Bangladesh, vol 2: Present conditions. Bangladesh Water Development Board, Dhaka
Halcrow SW, Partners Ltd. et al (1993) Master Plan Report of River Training Studies of the Brahmaputra River, FAP-1. Bangladesh Water Development Board, Dhaka
Harza Engineering Co. Int. et al. (1991) National Water Plan, Phase II. Master Plan Organization (presently WARPO), Dhaka
Hofer T, Messerli B (1997) Floods in Bangladesh: process understanding and development strategies. Institute of Geography, University of Berne, Berne
Hossain M, Islam ATMA, Saha SK (1987) Floods in Bangladesh: recurrent disaster and people's survival. Universities Research Center, Dhaka University, Dhaka
HTSL (1992) FCD/I agricultural study, FAP-12, Main Report, vol 1. Hunting Technological Services Ltd. FPCO, Dhaka
Islam KMN (1996) Linkages and flood impacts at macro level – a case study of Bangladesh. Research Report No. 146. Bangladesh Institute of Development Studies, Dhaka
Islam KMN (2006) Impacts of flood in urban Bangladesh, micro and macro level analysis. A H Development Publishing House and Community Development Library, Dhaka
Lahmeyer International et al (2000) Compartmentalization Pilot Project – Tangail, Final Report. Ministry of Water Resources, Govt. of Bangladesh
Mahalanobis PC (1927) Report on Rainfall and Floods in North Bengal 1870–1922. Irrigation Department, Government of Bengal, Bengal Secretariat Book Depot, Calcutta
Messerli B, Hofer T (2007) Floods in Bangladesh: history, dynamics and rethinking the role of the Himalayas. United Nations University Press, New York
Ministry of Water Resources (1999) National Water Policy, Government of the People's Republic of Bangladesh
ODA (1995) FAP-17 fisheries studies and pilot project. Final Report, Overseas Development Administration, UK
Rahman R (2004) Formulation of a pro-poor Water Sector Strategy, Final Report No. 01/2004, Institute of Water and Flood Management, Bangladesh University of Engineering and Technology, Dhaka
Rahman R (2008) Investigation into replicability of good practices in flood management. IWFM, BUET, Dhaka
Rahman A, Huq S, Conway G (1990) Environmental aspects of surface water systems of Bangladesh: an introduction, environmental aspects of surface water systems of Bangladesh. University Press Limited, Dhaka
Rahman R, Haque A, Khan MSA, Salehin M, Bala SK (2006) Investigation of hydrological aspects of flood-2004 with special emphasis on Dhaka city. Final Report, Institute of Water and Flood Management, Bangladesh University of Engineering and Technology, Dhaka
Rashid HE (1991) Geography of Bangladesh. University Press Limited, Dhaka
Saleh AFM, Mondal MS (2007) Impact of Submersible Embankment Projects on Boro Production in Haor Areas. IWFM, BUET, Dhaka
Salehin M, Haque A, Rahman MR, Khan MSA, Bala SK (2007) Hydrological aspects of 2004 floods in Bangladesh. J Hydrol Meteorol Soc Hydrologists Meteorologists Nepal 4(1):33–44
Shawinigan Lavalin Inc (1993) Water Transport Study, Northeast Regional Water Management, FAP-6. FPCO, Dhaka
Shawinigan Lavalin Inc. et al (1992) Thematic Study Regional Water Resources Development Status, Northeast Regional Water Management Project, FAP-6. FPCO, Dhaka
WARPO (2001a) National Water Management Plan (NWMP) Development Strategy, Water Resources Planning Organization (WARPO), Ministry of Water Resources, Dhaka
WARPO (2001b) National Water Management Plan (NWMP) Main Report, Water Resources Planning Organization (WARPO), Ministry of Water Resources, Dhaka
Wilcocks W (1930) Lectures on the Ancient System of Irrigation in Bengal. University of Calcutta, Calcutta

Chapter 5
Cyclone and Tornado Risk and Reduction Approaches in Bangladesh

Fuad Mallick and Aminur Rahman

Abstract The physiographical features of Bangladesh coast make it susceptible to cyclones and associated surges. According to the Multipurpose Cyclone Shelter Programme report, 6.4 % of the country is considered High Risk Area where the surge height may exceed 1 m. The country has been devastated by a number of cyclones resulting hundreds of thousands of human deaths. Besides, loss and damage of people's assets and properties have been extensive. With time, Bangladesh has been successful in significantly reducing human casualties from cyclone and the multipurpose cyclone shelters have been playing a great role in this regard. Another highly effective risk reduction initiative with regards to cyclone is the Cyclone Preparedness Programme (CPP) jointly run by the Bangladesh Red Crescent Societies (BDRCS) and the Government. The CPP can be considered as the precondition for the successful working of the cyclone shelters as one of its major activities is to disseminate cyclone warnings and mobilize the people at risk to cyclone shelters. Mangrove forests along the coast have an important role in reducing the wind speed and the surge impact, which led to taking coastal afforestation program in Bangladesh. The earthen coastal embankments in Bangladesh are primarily meant for protecting agricultural land from regular tidal phenomenon. These can be constructed strong enough to safeguard against cyclone-associated surge the cost of which can be justified only in the case of large investment. Disaster resilient habitat is a new concept which advocates for individual houses made as strong as to withstand cyclones or, at least, to enhance reconstruction and recovery. A pilot programme of the same executed in Bangladesh has been discussed in this chapter. Tornado is an under-reported disaster in Bangladesh the reason of which can be attributed to its impact being localized. Though Bangladesh is considered the only other part of the world outside the United States where strong and violent tornadoes are prevalent, little attention has been given to mitigate the risks. Absences of appropriate forecasting

F. Mallick (✉) • M. Rahman
Post Graduate Programs in Disaster Management (PPDM),
BRAC University, Dhaka, Bangladesh
e-mail: fuad@bracu.ac.bd

R. Shaw et al. (eds.), *Disaster Risk Reduction Approaches in Bangladesh*,
Disaster Risk Reduction, DOI 10.1007/978-4-431-54252-0_5, © Springer Japan 2013

and early warning system as well as lack of shelter provision make Bangladesh the country of highest death tolls from Severe Local Convective Storms (SLCS) like tornado and nor-westers.

Keywords Coastal area of Bangladesh • Cyclone and tornado risk • Cyclone preparedness program • Resilience • Severe Local Convective Storms (SLCS)

5.1 Overview of Cyclone Risk

The Bay of Bengal is called a breeding ground for tropical cyclones and Bangladesh is one of the worst victims in terms of fatalities and economic losses. The global distribution of cyclones shows that only 1 % of all cyclones that form every year strike Bangladesh; but unfortunately the fatalities they cause account for 53 % of the global total (Ali 1999). Records show that 16 out of 35 of the tropical cyclones worldwide that caused deaths of more than 5,000, occurred in Bangladesh (SMRC 1998). The funnel shaped coast line and particularly the low topography make the coastal area subject to high surge associated with cyclones (Fig. 5.1).

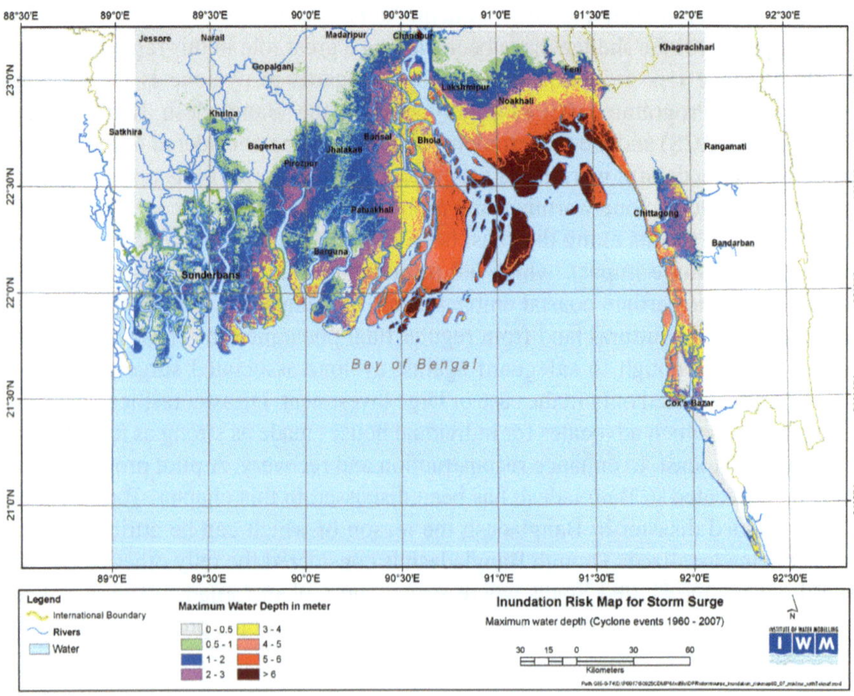

Fig. 5.1 Inundation risk map for storm surge along Bangladesh coast (*Source*: Disaster Management Bureau website: http://www.dmb.gov.bd/reports/stormsurge_inundation_riskmap.pdf; accessed 1 July 2012)

The Multipurpose Cyclone Shelter Programme (MCSP) report (BUET-BIDS 1993) identified a number of factors responsible for differentiated risk along the coastal areas prone to cyclonic storm surge flooding; they are:

- Storm surge height at the coast
- Angle of the storm track with respect to the coast line
- Tidal condition
- Offshore and near-shore bathymetry
- Slope of the land
- Curvature of the coastline
- Width and depth of river-mouth through which the surge will travel
- Presence of islands and chars
- Land topography and
- Landuse and surface resistance including presence of forest.

Based on these factors, the MCSP report identified the area prone to storm surge flooding as the Risk Zone which was further subdivided into High Risk Area (HRA). The HRA, where surge height may exceed 1 m, consisted of an area equalling 9,182 km^2 (6.4 % of the country) (BUET-BIDS 1993) where people are likely to be killed from drowning unless they moved to shelters.

The MCSP report acknowledged that the delineation of the Risk Zones is "an extremely difficult" task given the complexities of the factors listed above and also the associated uncertainties. It was recommended in the report that the Risk Zones are modified/refined as more information is gathered and experiences are gained.

5.2 Historic Cyclone Events and Impacts

Records of cyclone in Bangladesh date back to late sixteenth century. *Ain-i-Akbari* and a few other sources mention a severe cyclonic storm in 1582 that crossed the coast of Backerganj killing approximately 200,000 people (SMRC 1998). Though much detail, including the date of occurrence, is not available, it is clear that the impact of the cyclone was wide-spread and huge casualties took place even when the population density, understandably, was far less than that of today. Records containing partial information for 27 other cyclonic events till the end of the nineteenth century suggest the death of approximately 352,000 people (*ibid*). Among all these, the Great Backerganj Cyclone of 1876 alone caused death of 100,000 people from drowning and another 100,000 from hunger and epidemic which was a consequence of it (*ibid*).

Though still not comprehensive, more detailed information on cyclones in the twentieth century is available. According to the SMRC (1998) report, at least 44 severe cyclonic storms made landfall on Bangladesh coast throughout the century causing half a million deaths. Among all these, the Killer November Cyclone of 1970 alone caused 300,000 deaths and affected 4.7 million officially. The April

1991 cyclone in Bangladesh is considered as the Worst Killer Cyclone in Human History (SMRC 1998); it killed 138,882 people and injured another 139,054.[1]

The cyclones *Sidr* of 2007 and *Aila* of 2009 are the two noteworthy events of recent time. While cyclone *Sidr* is climatologically comparable with the 1970 and 1991 events, the consequence in terms of human casualties have been much less. Around 5,000 people died in cyclone Sidr which is attributed to the enhanced disaster management practices imparted institutionally over last few decades in Bangladesh. Unlike cyclone *Sidr*, Cyclone Aila was an under-quoted event as it had its hard impact on a relatively small area (part of Khulna and Satkhira); but the affected community have been suffering for prolonged periods and in some cases, in unprecedented ways.

Apart from the human casualties that resulted from cyclone events, the damage to properties, which include physical infrastructure, livelihoods means and various direct tangible/intangible and indirect tangible/intangible items, is substantial. The continuum of loss and damage caused by these disaster events is very difficult to frame; it transcends the spatial and temporal boundaries. The population exposed to the risk of cyclones cannot afford shifting their settlements to a safer location since their livelihood activities are localized. Thus, after every cyclone the people, as if, "rise from the ashes" to get ready to face another.

5.3 Cyclone Risk Reduction Approaches

5.3.1 Multipurpose Cyclone Shelter

The construction of designated cyclone shelters started in Bangladesh after the devastating effect of the 1970 cyclone. Over the period 1972–1979, 238 cyclone shelters were constructed in various locations along the Bangladesh coast. Following the 1985 cyclone, BDRCS decided to construct 500 cyclone shelters out of which only 62 were constructed by the year 1990 because of financial constraints.

That the number of cyclone shelters was inadequate was demonstrated by the casualties arising from the cyclone on 29 April, 1991. This prompted the government of Bangladesh, in collaboration with the UNDP and the World Bank, to undertake a major approach to study the need for cyclone shelters in the coastal areas of Bangladesh. The Inter-ministerial Task Force on Cyclone Shelter Construction guided the initiative titled Multipurpose Cyclone Shelter Program (MCSP). The major objective of the study was to formulate a framework for the establishment of a cyclone shelter network in the coastal areas which would define the basic concept, strategy and location pattern to be followed in all future construction (BUET-BIDS 1993). As part of the study, review of all types of existing shelters was made with a view to find suitability for specific locations.

[1]Though the casualties in 1991 cyclone was less compared to the 1970 cyclone, may be because of the widespread impact it caused on the overall economy of the country mark it so.

Fig. 5.2 Traditional cyclone shelters in Bangladesh (*Source*: Postgraduate Programs in Disaster Management, BRAC University)

After an extensive study, the MCSP estimated the need of a total of 2,500 cyclone shelters in the High Risk Area (HRA) along the coast considering the demand in the year 2002 for a projected population of 4.7 million. Out of the 2,500 shelters, 60 % were proposed to be located in existing primary school sites and the rest in new primary schools, *madrasas* and secondary schools.

At the time MCSP was being carried out, the Government of Bangladesh had enacted a legislation to make attendance in primary schools compulsory which required more than 3,000 primary schools to be constructed in the HRA. This requirement of educational facilities matched very well with the requirement of additional shelters (2,500) (BUET-BIDS 1993). Thus the "multipurpose" notion of the shelters was justified and the regular maintenance of the shelters was considered viable (Fig. 5.2).

In total, there are some 2,200 cyclone shelters in the coastal areas of Bangladesh at present. Another 400 are under construction with assistance from the Islamic Development Bank (IDB) and the World Bank while a total of 4,000 shelters are considered as required to ensure the safety of the coastal population.

5.3.2 Cyclone Preparedness Program (CPP)[2]

Organized disaster preparedness in Bangladesh can be said to have started with the formation of the Cyclone Preparedness Program (CPP). The ideas of CPP started in 1965 when the National Red Cross Society, now the Bangladesh Red Crescent Society (BDRCS), requested the International Federation of Red Cross and Red Crescent Societies (IFRCS), then the League of Red Cross and Red Crescent Societies, to support the establishment of a warning system for the population living in the coastal belt. In 1966 the International Federation and the Swedish Red Cross began the implementation of a pilot scheme for Cyclone Preparedness which

[2] From "CPP at a Glance" by the Bangladesh Red Crescent Society (BDRCS).

consisted of warning equipment such as transistor radios, sirens etc. and training of the local militia (*Ansars*), as the backbone of warning and dissemination activities. In 1966 the Cyclone Warning System was operational in 299 locations and supported by 473 team leaders. In May 1969 the first weather radar station was installed in Cox's Bazar by the Swedish Save the Children Fund in collaboration with the National Red Cross, the International Federation and the Swedish Red Cross Society.

Following the 1970 cyclone, the United Nations General Assembly requested the International Federation to take the leading role in establishing and improving the pre-disaster planning program for Bangladesh. The International Federation and the Bangladesh Red Crescent Society undertook an extensive evaluation of the programme and drew up a new strategy, which from February 1972 led to a new program encompassing 20,310 volunteers in 204 Unions of 24 Upazilas (subdistricts) and a transceiver telecommunication network linking 22 coastal stations.

In June 1973 the Government of Bangladesh approved the new program, accepted financial responsibility for recurring expenses and established joint program management through the creation of a Program Policy Committee and a Program Implementation Board.

Since then CPP has continued to operate with the goal of minimizing loss of life and property in cyclonic disasters by strengthening the disaster management capacity of Bangladesh's coastal people. CPP Presently covers 11 districts in the coastal area comprising 32 Upazilas consisting of 274 Unions. There are 2,845 units spread across this area each of which covers one or two villages.

5.3.3 Coastal Afforestation

Mangrove forests act as a shield against cyclone by lowering wind speed as well as reducing the surge impacts; realizing this, coastal afforestation with mangrove species was initiated along the Bangladesh coastal belt in the 1960s. The benefit of coastal afforestation programs then extended to a number of development agenda including protection of agricultural land against salt intrusion, conservation of coastal ecosystem and environment, protection of aquatic resources and wildlife and enhancing land accretion, inter alia. Over the last four decades the Department of Forest has brought about 148,000 ha of land under mangrove plantations scattered over on- and offshore areas mostly along the central part of the coast (Islam n.d.).

5.3.4 Coastal Embankment

Coastal embankments can be constructed to prevent cyclonic surges; but the cost of construction and also the cost of maintenance cannot be justified except for protecting some large investment (Choudhury 1994), as in the case of Chittagong Export Processing Zone.

The coastal embankments along Bangladesh cost have been constructed with a view to protecting the agricultural land from salinity intrusion associated with tidal flooding. Constructed under the Coastal Embankment Project in the 1960s, these earthen embankments are not capable of withstanding the forceful surge nor are they high enough to guard against the surge. Rather the presence of these embankments created a sort of false sense of security among people; in 1991 cyclone, a large number of people were reported to have been killed as a consequence of taking shelter on such embankments (Choudhury 1996). The age-old embankments are also suffering from improper maintenance. While they are not meant to be safeguards against cyclonic surges, their damage/destruction is causing unprecedented havoc to the people.

5.3.5 *Cyclone-Resilient Housing*

Cyclone shelters save lives; this is evidenced by the decline in casualties since the Multipurpose Cyclone Shelter Programme has been implemented. But destruction of houses and infrastructure continue. People, whose lives have been saved by cyclone shelters have to go back to destruction, often total, shortly after the cyclone stops. Besides there are also some who do not go to shelters leaving their assets and livestock behind (RED-BRAC, 2007, Measuring the impact of false tsunami warnings in Bangladesh. Unpublished Research Report, Research and Evaluation Division). Many shelters suffer from lack of maintenance, partly because they are of relatively sophisticated construction given their rural locations.

Idcally, loss of lives and destruction could be greatly avoided if houses and infrastructure in coastal areas were strengthened to withstand cyclones and storm surges. This is not possible because of economic constraints, but not necessarily technical ones. The idea of creating disaster resilient communities, who are quickly able to return to normal lives after a disaster, has resulted in the concept of the Disaster Resilient Habitat (DRH) (Mallick and Rahman 2007). This uses community skills and knowledge of technical personnel like engineers and architects, to develop on the "Building for Safety" options (Seraj and Ahmed 2004) to design and construct houses and infrastructure that are resilient to cyclones and storm surges. An experiment to build stronger houses on a small scale was carried out in Noakhali, using local knowledge and a participatory process (Mallick et al. 2008). This led to testing the idea further with two schools in the same location.

Recently, an entire DRH has been designed and built in Shyamnagar, in Satkhira for the victims of cyclone Aila. This UNDP funded project was executed by BRAC and BRAC University's Department of Architecture and the Postgraduate Programs in Disaster Management and combined the building skills of the community with technical knowledge from architects and engineers. The house owners voluntarily participated in the design and construction management process. Forty-three houses and a school were built using a traditional construction process but on top of a reinforced concrete frame. The homeowners are able to add to the basic structure as needed and on their own. A very strong cyclone may blow away the weaker parts of the house but the basic structure would remain allowing quick reconstruction. The cost for this entire project was US$150,000. A cyclone shelter to hold 1,800 people would cost around US$200,000.

5.4 Overview of Tornado Risk

The geographic location of Bangladesh, with the Bay of Bengal to the south and the Himalayan Mountains to the north, makes tornado a natural phenomenon for Bangladesh (Islam 2007). During the hot weather period, predominantly during the pre-monsoon season from March to May (Yamane et al. 2010) thunderstorms with great destructive potential occur. These thunderstorms are generically called Severe Local Convective Storms (SLCS) (Yamane et al. 2010). Yamane et al. also classified SLCS based on the magnitude of wind speed; wind gusts ranging from 11 m/s are termed as nor'westers (Kal Boishakhi in local term) referring to the direction (north–west) they generate from while wind gusts above 42 m/s are defined as tornadoes.

Bangladesh is considered the only other part of the world outside the United States where strong and violent tornadoes are prevalent (SMRC 2006). Of the ten worst tornadoes in the world in terms of casualties, six were reported to have struck Bangladesh (Discovery UK website). Jonathan D. Finch in his *Bangladesh and East India Tornado Prediction Site* (http://bangladeshtornadoes.org) listed 67 tornadoes that struck Bangladesh between 1838 and 1998. Of the 67 events, 19 were reported to have caused deaths of 100 or more each while at least six incidents caused death of more than 500 people each. The deadliest tornado in recorded history of Bangladesh struck Daulatpur and Saturia in Manikganj district (near Dhaka) in the evening of 26 April, 1989 killing around 1,300 people (Finch website).

The distribution of devastating tornadoes in Bangladesh indicate that about 8,000 mile^2 of the country across the central, south central and southeast region are particularly prone to this atmospheric disturbance; almost 75 % of the incidents have been reported to have occurred in these parts of the country (Finch website) (Fig. 5.3).

5.5 Tornado Risk Reduction Approaches

Despite a much appreciated institutional and functional set-up for disaster risk management in Bangladesh, tornado is still a neglected issue compared to flood or cyclone. There is no such tornado-specific program that addresses the risk reduction aspects associated with tornado while the geographic location makes tornado a recurrent natural phenomenon for Bangladesh.

Ono and Schmidlin (2011) state that absence of storm warnings, poor communication, weak housing and dearth of shelters from strong winds are the reasons why Bangladesh has the highest death tolls in the world from tornadoes and nor-westers.

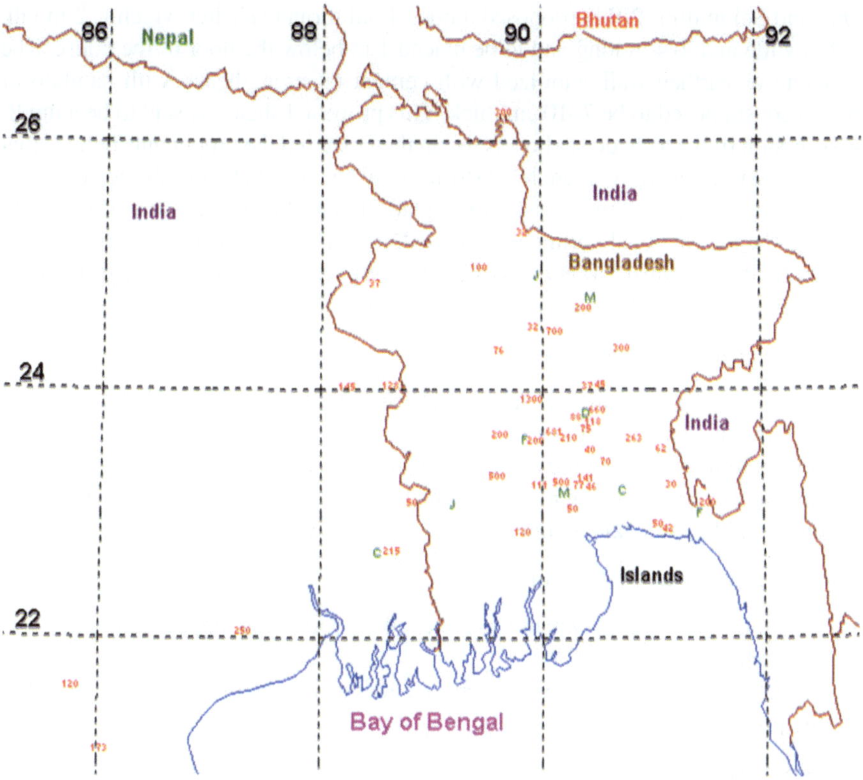

Fig. 5.3 Map showing locations of all tornados that killed at least 30 people (*Source*: http://bangladeshtornadoes.org; accessed on 17 July 2012)

Paul and Bhuiyan (2004) address the absence of an appropriate forecasting and early warning system as one of the major reasons contributing to the high casualties arising from tornadoes. They argue that 28 weather stations are far from adequate a number for an area of 56,000 mile^2 for forecasting tornadoes. The Bangladesh cyclone early warning dissemination system is a much lauded one across the world which is run through the Cyclone Preparedness Program (details in Sect. 6.3.2), a community based approach. For tornadoes, such dissemination model can be used provided a state of the art early detection system is in place. Hammer and Schmidlin (2002) discuss that early warning messages are more effective when they come from multiple sources which, for Bangladesh, can be ensured by the use of mobile phones as means for warning families. Beside all these, year round community awareness raising programs can help reduce the risk of damage and loss from tornado.

The construction materials (bamboo thatch, mud, CI sheet) for typical Bangladeshi houses result in damage of larger extent arising from tornadoes. People, if outdoor at times of tornado, tend to run to get home which is not strong enough to withstand the impact of a tornado. Moreover, the flying corrugated iron (CI) sheets become lethal during tornadoes and are reported to have killed many (Islam 2007).

Ono and Schmidlin (2011) proposed a household tornado shelter which is 2 m tall, 1.2 m wide and 2–4 m long and to be placed 1 m below the floor of the house. The concrete or earthen wall stabilized with cement or strengthened with bamboo or bricks are supposed to be 7–10 cm thick. This proposed shelter is said to be a modified version of the tornado shelters is use in the USA and the proponents believe that deaths and injuries from severe local storms may be reduced by installation of those.

Given the state of tornado risk in Bangladesh and the vulnerability of the community, the International Forum on Tornado Disaster Risk Reduction for Bangladesh in a meet in December, 2009 made the following ten-point recommendation under four thematic areas (IAWE website) to mitigate the effects of tornado in Bangladesh:

Governance and Policymaking

- Prioritize tornado disaster risk reduction and integrate it into national policy of the Government of Bangladesh.
- Obtain commitments of the Government at the local and national levels to implement an integrated action plan for tornado disaster risk reduction in Bangladesh supported by national and international stakeholders.

Public awareness and education

- Raise public awareness about tornadoes, their formation, appearance, behavior, and impacts.
- Educate the public about the appropriate actions to take in order to preserve life and property.

Technological improvements

- Develop a tornado warning system with improved capacity to identify weather conditions that pose a high risk of tornadoes for early detection.
- Improve capacity to disseminate early warnings to threatened local communities in an innovative way.
- Identify buildings in the community that would serve as safe shelter from a tornado and develop and test a household storm shelter for use in Bangladesh.

Recovery and Monitoring

- Strengthen the procedures for response preparedness during and after a tornado.
- Enhance climatological and meteorological research on tornadoes and develop and maintain a national database of tornado occurrences and their impacts.
- Conduct a detailed field survey after a major tornado for a better understanding of risk factors.

5.6 Conclusion

Bangladesh has made significant progress in dealing with cyclones and floods. Innovative and original ideas have been transformed to reality. Loss of life has reduced significantly. The MPCS project is a much talked about one and other countries look at it as an example to follow or develop their own schemes on its basis. The CPP volunteer program is much lauded one and a lot has been written and said about it. The world looks at Bangladesh not only as a disaster prone country but also one that has a well-organized disaster management system. The idea of the DRH is also a new concept and also gaining attention from various levels.

Unfortunately tornadoes have not received the attention they deserve; this is probably because casualties in each instance are not much, but when added up they are significant. The approach to tornado mitigation is quite simple. Putting together an early warning system, structural strengthening of houses and the introduction of shelters on a domestic level. Given Bangladesh's recognition as a disaster resilient country there is no reason why tornadoes cannot be tackled with proper research and action.

References

Ali A (1999) Ghurnijhar (cyclone). Bangla Academy, Dhaka

BUET-BIDS (1993) Multipurpose Cyclone Shelter Programme. Government of Bangldesh and UNDP/World Bank

Choudhury JR (1994) Cyclone Shelter and its Multipurpose Use. In Disaster Management in Metropolitan Areas for the 21st Century. Proceedings of the IDNDR Aichi/Nagoya International Conference, 1–4 November, Nagoya, Japan, 369–376. UNCRD Proceedings Series, No. 1. Nagoya, Japan: United Nations Centre for Regional Development, 1994

Choudhury JR (1996) Protection of Society from Storm Surge Hazard, Proceedings of the 50th Anniversary Conference of Commonwealth Engineering Council held in London, 18–20 March, 1996

Finch JD (n.d.) Bangladesh and East India Tornado Prediction Site. Retrieved July 3, 2012, from http://bangladeshtornadoes.org

Hammer B, Schmidlin T (2002) Response to warnings during the 3 May 1999 Oklahoma City tornado: reasons and relative injury rates. Weather Forecasting 17:577–581

International Association for Wind Engineering (IAWE) (2009) Proceedings of the International Forum on Tornado Disaster Risk Reduction for Bangladesh, held in Dhaka on 13–14 December, 2009. Website: (http://www.iawe.org/about/BangladeshFinalReport.pdf). Accessed 30 Dec 2012

Islam T (2007, May 18) Tornadoes: can we reduce vulnerability? Retrieved July 8, 2012, from The Daily Star: http://www.thedailystar.net/2007/05/18/d705181803127.htm

Islam MM (n.d.) Coastal forest rehabilitation and management in Bangladesh. Retrieved July 11, 2012, from FAO: http://www.fao.org/forestry/12669-09d3e208c72f517f191a02fed14d9bb8a.pdf

Mallick FH, Rahman MA (2007) Cyclone shelters and alternatives for sustained development in Bangladesh. J South Asia Disaster Stud 1(1):59–67

Mallick FH, Kabir KH, Kabir MH (2008) Improved design and construction of rural housing in Noakhali. IUCN, Dhaka

Ono Y, Schmidlin TW (2011) Design and adoption of household tornado shelters for Bangladesh. Nat Hazards 56:321–330

Paul B, Bhuiyan R (2004) The April 2004 tornado in north-central Bangladesh: a case for introducing tornado forecasting and warning systems. Quick Response Research Report 169, Natural Hazards Center, Boulder

Seraj SM, Ahmed KI (2004) Building safer houses in rural Bangladesh. BUET, Dhaka

SMRC (1998) The impact of tropical cycones on the coastal regions of SAARC countries and their influence in the region. SAARC Meteorological Research Centre (SMRC), Dhaka
SMRC (2006) Environmental and synoptic conditions associated with nor'westers and tornadoes in Bangladesh – an appraisal based on numerical weather prediction (NWP) guidance products. SAARC Meteorological Research Centre, Dhaka
Yamane Y, Hayashi T, Dewan AM, Akter F (2010) Severe local convective storms in Bangladesh: Part I. Climatology. Atmos Res 95:400–406

Chapter 6
Earthquake Risk and Reduction Approaches in Bangladesh

A.S.M. Maksud Kamal

Abstract Bangladesh is a disaster prone country. Though cyclone, flood, landslides, river erosion as well as climate induced hazards are the frequent disastrous phenomena in the country, in the last 150 years, the country experienced damages of five earthquakes having magnitude over 7.0 (Richter Scale). There are subduction zones in the east and north of the country while three major active faults have been studied so far having potential to generate 7.0–8.5 magnitude earthquake. Geophysical and paleoseismic investigations also conducted to estimate the slip-rate as well as recurrence period of the historical earthquakes. Considering the potential risks, the Government has taken initiatives through Comprehensive Disaster Management Programme (CDMP) to assess the seismic risk following the deterministic and probabilistic procedures mainly to address the scenario based earthquake risk management in the country. Following the estimated scenarios, spatial emergency risk management plan in the name of contingency plan developed for the main cities, viz., Dhaka. Chittagong and Sylhet as well as other ten major responding agencies. Numbers of simulation exercises were conducted to implement the contingency plan. In order to rightly address the earthquake risk reduction, government has taken number of initiatives onwards 2006 under the umbrella of CDMP. These are development of 62,000 Urban Community Volunteer mainly in the earthquake prone cities, training for masons and bar binders as well as engineers, preparation of risk sensitive landuse planning, building awareness to the public representatives and city dwellers, increasing the research activities in the government agencies and academic institutions as well as gradual procurement of search and rescue equipments. It can be said that a comprehensive approach has been launched for earthquake risk management in Bangladesh, which need to be nourished by all relevant quarters to develop a sustainable urban risk management environment in the country.

Keywords Building vulnerability • Earthquake risk management • Microzonation

A.S.M.M. Kamal (✉)
Department of Disaster Science and Management, University of Dhaka, Dhaka, Bangladesh
e-mail: maksudkamal@yahoo.com

R. Shaw et al. (eds.), *Disaster Risk Reduction Approaches in Bangladesh*,
Disaster Risk Reduction, DOI 10.1007/978-4-431-54252-0_6, © Springer Japan 2013

6.1 Earthquake Hazard of Bangladesh and Brief Description of the Historical Earthquakes

Half of the world population already lives in the urban area having urban growth rates highest in the developing world. By 2050, the urban population of the developing world will be 5.3 billion; Asia alone will host 63 % urban population, or 3.3 billion people. With an urban growth rate more than 4 % annually, Dhaka, which already hosts more than 13 million people, is one of the fastest but unplanned growing cities in Southern Asia, and is projected to accommodate more than 20 million by 2025. Recent earthquake in India (Bhuj 2001), Pakistan (2005), China (Sichuan 2008) and Haiti (2010) implied that earthquake disaster are mostly human-induced due to poor constructions because of unavailability or noncompliance of building code and unplanned urbanization practices. Moreover, earthquake impact on Gross Domestic Product (GDP) implies that poor nations are most suffered than the rich nations. For instances, almost same magnitude earthquake cost the nation's economy in Haiti at least 15 % of its GDP while it is calculated less than 1 % in China. The seismic potential of Bangladesh is very high, because there are several active structures capable of generating potential earthquake along Himalayan Main Frontal Thrust (MFT), Dauki Fault (DF), Ramree and Chittagong sections of Arakan megathrust (AMT); PBF-1 and PBF-2 of CDMP-I, and Naga Thrust Fault (NTF); PBF-3, and other faults in Ganges basin such as Madhupur Fault (MF). They are estimated to generate magnitude Mw 7–8.5 earthquakes in future. The secondary active structure in Chittagong–Tripura Fold Belt (CTFB) is also investigated, and Mw 6.3–7.5 earthquakes will be averagely generated once in 30–50 years, though with large uncertainties. Figures 6.1 and 6.2 represent the positions of active morphotectonic features and faults in and around Bangladesh.

In the last 250 years history, Bangladesh has suffered from severe large earthquakes, such as the 1762 Arakan earthquake, the 1869 Cacher earthquake, the 1885 Bengal earthquake, the 1897 Great Assam earthquake, and the 1918 Srimangal earthquake (Earthquake in website Banglapedia; Oldham 1883; Ambraseys and Douglas 2004; Bilham and Hough 2006, etc.). The 1885 Bengal earthquake and 1918 Srimangal earthquake had their epicenter in Bangladesh (Fig. 6.3). CDMP study (2009a, b, c, d, e) mapped out several major active faults, e.g. the plate boundary faults, Madhupur Fault, and the Dauki Fault illustrating their characteristics like magnitude, detailed location, faulting history, recurrence period, lengths, types etc. All theses faults have the potential to generate large earthquakes with magnitude more than 7.5 or 8.5. Table 6.1, illustrates the characteristics of the inferred active faults as mentioned above based on available information of the CDMP study and other investigations. However, the active fault mapping, namely to clarify the nature, detailed location, faulting history and recurrence period etc. are not well identified yet.

Historical earthquakes caused significant damages in Bangladesh are not well documented, in this section; a short description is demonstrated highlighting the known records of some major historical to recent damaging earthquakes in and its surroundings Bangladesh (Fig. 6.4).

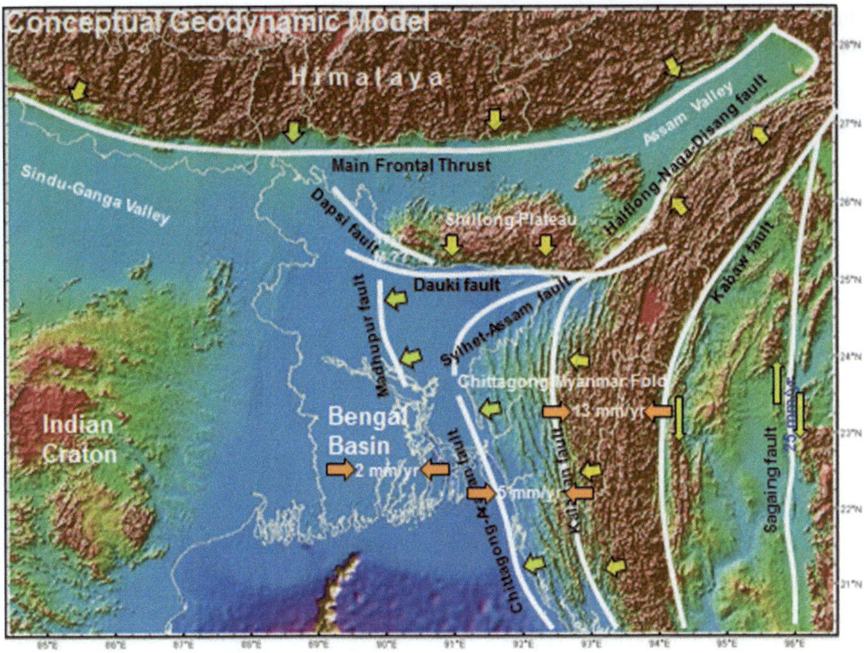

Fig. 6.1 Active structure in and around Bangladesh and shortening rates by GPS observation (Kaneko 2012)

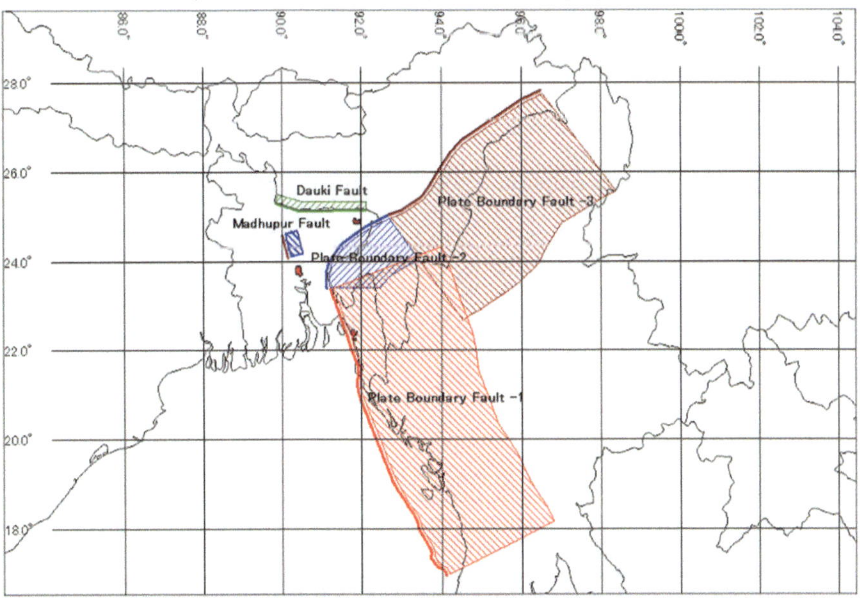

Fig. 6.2 Major active (CDMP-I 2009) faults in Bangladesh

Fig. 6.3 Earthquakes having epicenter within Bangladesh

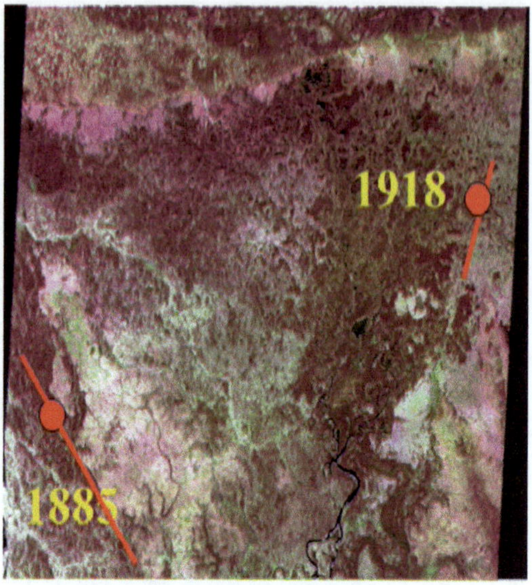

Table 6.1 Characteristics of the major active faults based on available information

Segment		Event	Observed recurrence period (years)	Elapsed since last event (years)	50 Year prob. time-dependent (%)	Estimated magnitude (Mw)	Length (km)
Plate Boundary Fault	PBF-1	*AD 1762*[a] AD 680 to 980[a] BC 150 to AD 60[a,b] BC 1395 to 740[a]	900	246	1.1	8.5	795
	PBF-2	Before sixteenth century	>900	>508	>6.7	8.0	270
	PBF-3	Before sixteenth century	>900	>508	>6.7	8.0	504
Dauki Fault (DF)		*AD 1897*[c] AD 1500 to 1630[c] (AD 1548)	349	111	7.0	8.0	233
Madhupur Blind Fault (MF)		*AD 1885*	350	123	8.7	7.5[d]	60
Non characteristic but relating to fault[c] (PBF-2, PBF-3, DF)		AD 1918 (PBF-2) AD 1869, 1954, 1988 (PBF-3) AD 1664, 1923, 1930 (DF)	20	–	–	7.0–7.4[e]	–

[a]Shishikura et al. (2009)
[b]Based on the trench investigation of this project at Feni
[c]Based on the trench investigation of this project at Northeast of Haluaghat
[d]AD 1885 earthquake was M = 6.9–7.0 but possibly M = 7.5 in the future
[e]M = 7.0–7.4 earthquake have often occurred in the interval of Characteristic earthquake, and may not repeat on a specific fault and occurred randomly

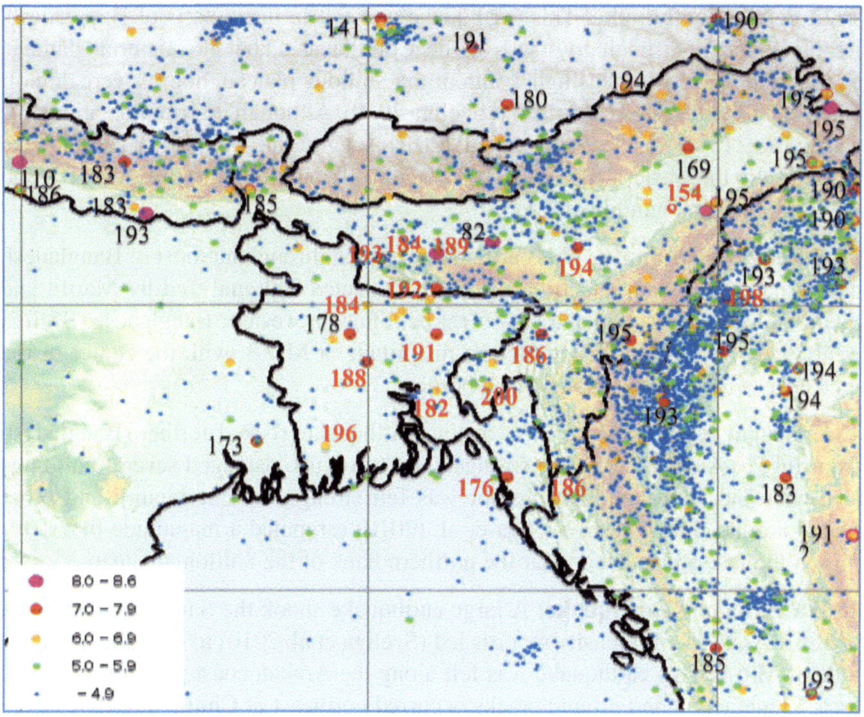

Fig. 6.4 Significant historical earthquakes in and around Bangladesh

1548 Assam earthquake: The 1548 earthquake is one of the earliest damaging earthquakes recorded in the region. Detailed information about ground shaking and damage is lacking, but the earthquake is known to have damaged both Sylhet and Chittagong and to have caused significant liquefaction (Rizvi 1970; Bilham and Hough 2006; Steckler et al. 2008). Iyengar et al. (1999) suggested that the intensities might have been as high as IX in the southern Assam Valley. Steckler et al. (2008) suggested the earthquake resulted from slip on the megathrust beneath the Chittagong–Tripura Folding Belt (CTFB), because of the severity of shaking and damage in the area. Morino et al. (2011) suggested that rupture of the Dauki Fault (DF) produced the earthquake, based on paleoseismologic investigation and geochronological analysis of the Dauki Fault.

Great Arakan earthquake of 1762: The Aarakan earthquake in 1762 is best-known historical earthquake along the eastern side of Bay of Bengal. The earthquake was strongly felt from Cheduba and Ramree Islands of western Myanmar to the area near Dhaka and caused heavy damage in the Chittagong region. A British officer who surveyed the Myanmar coast about 80 years later documented several meters of uplift associated with the earthquake (Halsted 1841). It was reported by Oldham (1883) that Chittagong and Dhaka suffered severe damage by the 1762 earthquake and tsunami. The subduction fault of 1,200 km in length located from off Sumatra to Andaman Islands was ruptured by the 2004 off Sumatra earthquake (Mw 9.2).

1822 Bengal earthquake: This earthquake was felt in many parts of Bangladesh. People found it difficult to stand in the Comilla area, but no material damage occurred there. At Mymensingh, south of the Shillong plateau, houses were demolished or badly fractured (Martin and Szeliga 2010). Although information about this earthquake is very sparse, Szeliga et al. (2010) suggest the source of the earthquake was close to the western edge of the Chittagong–Tripura Folding Belt (CTFB), and they estimate a magnitude of 7.1.

1842 Bengal earthquake: This earthquake was felt throughout most of Bangladesh and in parts of Assam and Bihar. Records collected and analyzed by Martin and Szeliga (2010) report damage of several buildings in western Bangladesh. Szeliga et al. (2010) estimates the earthquake magnitude at M 7.3, with the center of the quake close to the India–Bangladesh border.

1845 Siraganj earthquakes: Three distinct earthquakes rocked northern Bangladesh from July to August 1845. The strongest, on 6 August, damaged several buildings at Sylhet and Guwahati. The tremor was felt strongly at Cherrapunji and other places around Bangladesh. Szeliga et al. (2010) estimated a magnitude of 7.1 for this earthquake and a center near the northern flank of the Shillong plateau.

1865 Chittagong earthquake: A large earthquake shook the Sandwip Island area in 1865. Analysis of intensity records led (Szeliga et al. 2010) to estimate a magnitude of M 6.8. This earthquake was felt along the Arakan coast and in the Bengal area. Liquefaction and ground cracks occurred northeast of Chittagong city, where the intensity was strongest (Martin and Szeliga 2010). The location of the strongest shaking suggests a source within the Chittagong–Tripura Fold Belt (CTFB) or on the Arakan Megathrust (AMT).

1869 Cacher earthquake: The Cachar earthquake of 1869 occurred in the Silchar area, in the Indo Burman range east of Bangladesh. Damage extended from Silchar to the Manipur area and included extensive liquefaction. The quake was felt in both northeast and eastern India as well as the in adjacent parts of Myanmar. This was followed by three minor tremors within a period of 3 h (Rizvi 1970). On the Brahmaputra River, people observed energetic seiches (Martin and Szeliga 2010). Based upon intensity records, Szeliga et al. (2010) estimated the center of this earthquake to have been near the India–Myanmar boarder and that its magnitude was M 8.3.

1885 Bengal earthquake: This earthquake, known as the Bengal Earthquake, occurred on 14 July 1885, and is one of the most seven earthquakes. The earthquake was followed by 11 aftershocks during the period of 21 July to 5 September 1885 (Middlemiss 1885). Though no exact figures for the casualties in Dhaka caused by this earthquake are available, it was considered one of the major earthquakes in Bangladesh. Martin and Szeliga (2010) noted: "At least 75 deaths reported, including 40 at Sherpur, 11 at Azimgunj, 11 at Rajbari, 5 at Bogra, 4 at Govindpur and 4 at Dum Dum. Serious damage occurred at Mymensinh, Nattore, Serajganj and neighboring towns in north Bangladesh". The earthquake was widely felt with reports from as far as Chittagong, Chunar, Darjeeling, Giridh and Imphal.

Great Assam earthquake of 1897: The 1897 earthquake is the first Indian earthquake for which levels of shaking were documented in a contemporary earthquake report (Oldham 1899). The region of very strong shaking (EMS intensity IX) includes the Attrabari, Rambrai and Shillong areas (Martin and Szeliga 2010). The Sylhet and Mymensingh regions experienced severe damage. The tremor was felt less strongly over much of South Asia, including lower Myanmar, the Assam valley and much of the Indian subcontinent. Richter (1958) calculated Ms 8.7. Abe (1994) recalculated the magnitude as Ms 8.0 from instrumental records. Szeliga et al. (2010) suggested a magnitude of M 8.4 based on seismic intensity records. All of these magnitude estimates suggest that surface rupture would have been at least 200 km long, based on magnitude-rupture empirical relationships (Wells and Coppersmith 1994). Since no evidence of surface rupture has been found, the source of the earthquake currently is still debated.

Srimongal earthquake of 1918: The Sirmongal earthquake of 1918 caused extensive damage along the eastern Bangladesh border, particularly in the Balisera Valley near Srimongal, where seismic intensity reached EMS VIII to IX (Martin and Szeliga 2010). Buildings were also damaged in Dhaka and Sylhet, as well as some adjacent parts of in India. The tremor was throughout Bangladesh and adjacent parts of India and Myanmar (Stuart 1920). Pacheco and Sykes (1992) estimated a magnitude Ms 7.4 centered close to the western front of the Chittagong–Tripura Fold Belt (CTFB). The highest intensities of this earthquake occurred in the northernmost part of the CTFB. This suggests that the earthquake was generated by a structure within the CTFB or the Arakan Megathrust (AMT) beneath it.

1923 Shillong earthquake: A major earthquake with magnitude Ms 7.1 occurred on September 1923 at 22:33:42 BST. The epicenter was located 180 km north-northeast of Dhaka in southern Meghalaya near the Bangladesh–India border. The earthquake causes heavy damage at Mymensingh and was felt all over Bangladesh including Dhaka. Martin and Szeliga (2010) noted, "At least 50 people were killed in the Mymensingh district in northern Bangladesh (Anonymous 1923). Damage occurred in Mymensingh and to a lesser extent at Agartala, Guwahati and Kolkata".

1930 Dhubri earthquake: This earthquake is known as the Dhubri earthquake. It was triggered on the early morning of 2 July 1930 at 03:53:34.4 BST with magnitude Ms 7.1. This major quake was followed by six strong aftershocks of magnitude 6.0. The epicenter was located 230 km north of Dhaka in northwestern Shillong Plateau near Dabigiri, Meghalaya. The earthquake was felt widely in Bangladesh. Martin and Szeliga (2010) note: "One person was killed at Bishalpur in Bangladesh. Damage was not widespread though serious at places such as Cooch Behar, Dhubri, Lalmonirhat, Nilphamari and Tura near the epicenter".

1943 Assam earthquake: This earthquake was widely felt in Bangladesh and northeast India including at Darjeeling, Hazaribagh, Patna, Kolkata and Shillong. It was also felt in much of Bengal, at Kathmandu in Nepal, in southern Tibet and in parts of Burma (Ambraseys and Douglas 2004). The epicenter was centered in Hojai, Assam, and magnitude would be 7.2 around.

1964 Saugor Island earthquake: One death occurred at Debipur and three at Abasbadi near Nandigram. Old and dilapidated buildings collapsed on Fraserganj Island while at Namkhana, tiles were dislodged from buildings. Minor damage extended to other places in the Sunderbans such as Diamond Harbour, Digha and even as far north as Kolkata. Seismic seiches were observed in ponds and tanks in Diamond Harbour, Mahishadal and Saugor Island following the quake and lasted for more than 30 min. Variations in tide heights following the earthquake was also recorded at Beguakhali, Digha, Gangasagar, Junput and Kachuberia.

1988 Indo-Burma Border earthquake: A major earthquake occurred on the early morning of 6 August 1988 at 06:36:26.29 BST and had a magnitude Ms 7.3. Two people were reported dead and 12 injured in house collapse in Silchar, Assam. The earthquake was located in Tonmalaw, northern Myanmar and hadhypocentral depth of 90 km. The earthquake was widely felt in Bangladesh with minor damage (Martin and Szeliga 2010). Many people awakened and ran out of their houses in Dhaka.

2003 Kolabonia earthquake: Three people were killed and nearly 32 injured in the Barkhol-Rangamati area of Chittagong Division, Bangladesh (Ansary and Sadek 2006). Damage was reported from Aimachara, Barkhol, Boroharina, Langadu and Rangamati while buildings developed cracks a far as Chittagong, Cox's Bazaar, Moheskhali, Kutubdia and Sonadia. In Chittagong, a small fire occurred at the Modunaghat Grid Sub-station temporarily disrupting power supply. Subsidence was reported and liquefaction was observed near Kolabonia. Tremors from this earthquake were felt as far as Dhaka in Bangladesh as well as in parts of Tripura, India.

Due to the presence of light structures in the urbanized area, it is postulated less damagers by the historical earthquakes. Moreover, the loss and damages of the historical earthquakes have not been documented well. But a present study conducted on a part of Dhaka City revealed that unplanned development with the existence of extreme population density, old dilapidated unreinforced masonry buildings, narrow road network, close proximity of adjacent buildings, irregular building shape, lack of open spaces and others have made older part of the Dhaka City more vulnerable to any imminent earthquake (Jahan et al. 2011). In order to figure-out the social vulnerability in this study it was found that most of the surveyed people know more or less about earthquake and feel them aware. But they don't have any practical idea on what should be done in preparatory phase, during and after the occurrence as no earthquake has attached the locality in recent times and so there is doubt on how effectively they can perform in time of earthquake.

It seems that Bangladesh has not been affected by any large earthquake in the last 100 years but there are the existence seismic gaps in the morphotectonic features of the region. The seismic gaps clearly exist in the Dauki Fault and Plate Boundary Fault-3 zones. The presence of seismic gaps in the morphotectonic features reveals that Bangladesh has a high risk for a big earthquake in near future. Considering the potential threat of the earthquake hazards, Disaster Management and Relief Division of the Ministry of Food and Disaster Management has been undertaken systematic approaches in reducing the earthquake vulnerability of the country through

Comprehensive Disaster Management Programme (CDMP) since 2007 onwards. Beside CDMP, other wings of the Government like Disaster Management Bureau (DMB), City Development Authorities (RAJUK), Public Works Department (PWD), Fire Service and Civil Defence (FSCD), Armed Force Division (AFD) and City Corporations are also working in addressing and reducing the earthquake risk of the country. NGOs initiatives should also be noted in this context. CDMP initiatives considered following procedures having onboard the respective organizations for earthquake preparedness and risk reduction: hazard, vulnerability and risk assessment following microzonation techniques, developing risk-based spatial contingency plan for national, city and relevant agency levels, capacity build-up of the major organizations for conducting search and rescue operations and promoting scientific research as well as raising awareness to the city dwellers and decision makers.

6.2 Earthquake Hazard, Vulnerability and Risk Assessment of Some Major Cities

6.2.1 Earthquake Hazard Assessment

Primary earthquake hazards are represented by strong ground motion, soil liquefaction and slope failure. Secondary hazards are fire, subsidence and tsunami. This paper demonstrates the primary hazards of some major cities, like, Dhaka, Chittagong and Sylhet City areas. Fire hazards are also been illustrated here. The approaches of earthquake motion estimation in hazard analysis are roughly classified into two groups. One is called "Deterministic" and the other is called "Probabilistic". Result by the deterministic study is the seismic motion distribution in case a certain scenario earthquake may occur. Output of the probabilistic analysis is expressed as, for example, the seismic motion distribution with 10 % probability of exceedance in 50 years exposure time. The probabilistic study results can be applied to seismic regulation for building or facility construction such as building codes, it cannot be directly used for vulnerability/damage, and on the contrary, the deterministic results are used to develop the scenario of the damage, loss and risk assessment, hence, mostly applicable for earthquake risk management through the development of contingency plan, emergency management plan and risk sensitive land use plan. Considering the earthquake risk management aspects, this paper discussed the findings obtained from deterministic study.

Deterministic seismic hazard has been developed using the geomorphological/geological maps, scenario fault and earthquakes, geophysical and geotechnical information, soil maps, amplification maps and attenuation formula. To characterize the soft soil over seismic baserock, Standard Penetration Test (SPT), profiling of Primary and Secondary waves using PS logging equipments, Shallow Seismic array applying the procedures of Multichannel Analysis of Surface Waves (MACW) and

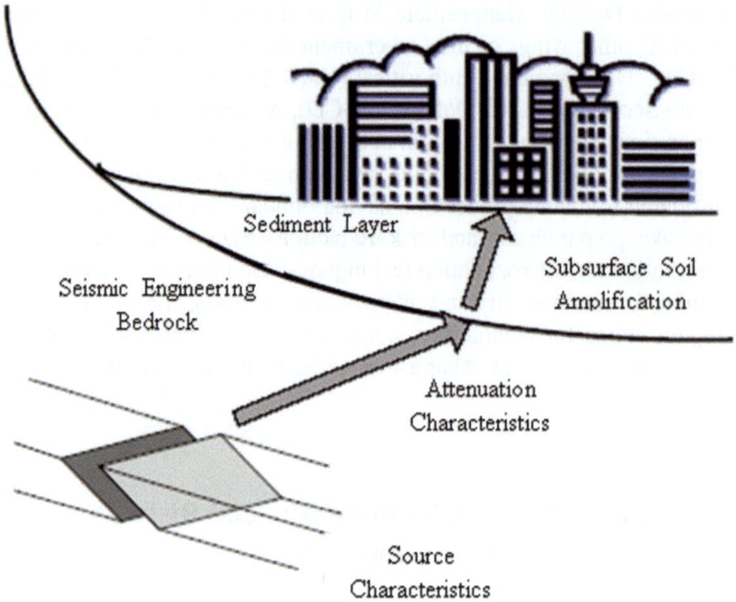

Fig. 6.5 Procedures of determining strong ground motion at surface

microtremor records of single and array techniques. Principle used to determine the ground motion for deterministic seismic hazard mapping is given in Fig. 6.5.

It is to be noted that in order to determine the ground motion *Next Generation of Ground-Motion Attenuation Models* (NGA) was used in this study, because NGA attenuation relations are for shallow crustal earthquakes and Bangladesh with its surroundings belong to shallow crustal zones. The NGA project is a multidisciplinary research program coordinated by the Lifelines Program of the Pacific Earthquake Engineering Research Center (PEER) in US, in partnership with the U.S. Geological Survey and the Southern California Earthquake Center. The scenario earthquakes of MF and DF are shallow crustal earthquakes. The PBF-1, PBF-2 and PBF-3 are interface earthquakes but the three study sites are located on hanging wall block and most of the propagation path is in shallow crust. The similarity of attenuation of interface motion for larger events to those of California crustal events is also shown by Atkinson and Boore (2003).

In order to represent the ground motion, Peak Ground Acceleration (PGA) and Peak Ground Velocity (PGV) with (h = 5 %, T = 0.3 and 1.0 s) at engineering seismic base rock (Vs30 = 760 m/s) were calculated. Then strong ground motion is calculated at the surface integrating the amplification of soft soil over the engineering seismic base rock.

The scenario earthquake can be generated for Dhaka in MF and the PGA for which is calculated in the engineering seismic baserock 130–230 gals. Likely, the scenario earthquake can be generated for Chittagong in PBF-1 and the PGA for which is 650–770 gal. The scenario earthquake can be generated for Sylhet in DF

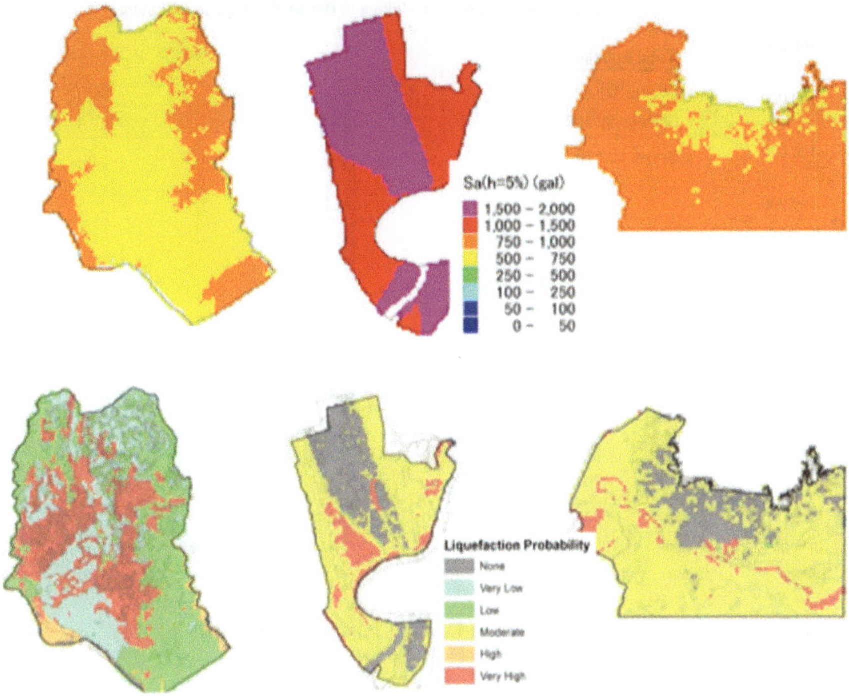

Fig. 6.6 *Top left*: Ground shaking for Madhupur Fault (MF) in Dhaka. *Top middle*: Ground shaking for Plate Boundary Fault (PBF-1) in Chittagong. *Top right*: Ground shaking for Dauki Fault (DF) in Sylhet. *Bottom left*: Soil Liquefaction Potential for Madhupur Fault (MF) in Dhaka. *Bottom middle*: Soil Liquefaction Potential for Plate Boundary Fault (PBF-1) in Chittagong. *Bottom right*: Soil Liquefaction Potential for Dauki Fault (DF) in Sylhet

and PBF-2. PGA by DF is 180–240 gal and 160–240 gal by PBF-2. The PGA, PGV and Soil Amplification (Sa) with h = 5 %, T = 0.3 and 1.0 s at ground surface were calculated. The PGA in Dhaka by MF is 220–410 gal. The PGA in Chittagong by PBF-1 is 600–770 gal and the effect of non-linearity of soils is remarkable. The PGA in Sylhet by DF is 270–420 gal and 230–420 gal by PBF-2. Calculated strong ground motions reflected that Dhaka and Sylhet are susceptible for experiencing intensity VIII, while Chittagong could experiences intensity around X. Intensity "VIII" represents collapse of poorly constructed buildings and damages to well built structures, while intensity "X" represents collapse of even well built structures. Figure 6.6 (top left, top middle, top right) represents the potential ground shaking (PGA) hazard of Dhaka, Chittagong and Sylhet.

As mentioned before, soil liquefaction is the hazard phenomena, which reflect losing of the bearing capacity of the soil due to ground shaking. The liquefaction potential was evaluated based on the procedure given in the software named Hazard United States (HAZUS) with geologic/geomorphic condition, PGA, magnitude (Mw) and groundwater depth. At first, liquefaction susceptibility is evaluated by

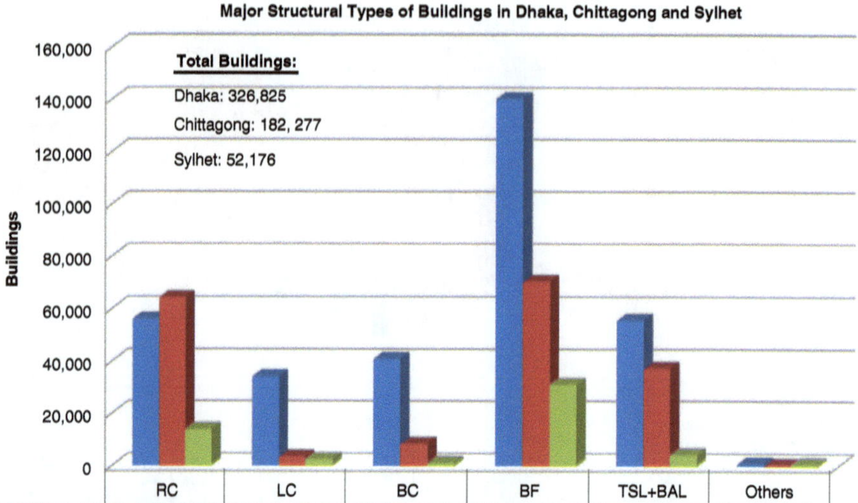

	RC	LC	BC	BF	TSL+BAL	Others
Dhaka	55,923	33,993	40,710	140,078	55,487	634
Chittagong	63,905	3,190	8,047	70,173	36,962	0
Sylhet	13660	2300	827	30896	4000	493

Fig. 6.7 Building inventory information of Dhaka, Chittagong and Sylhet City area

geologic/geomorphic data and information of geological age. Secondary, liquefaction probability is estimated by inputting PGA, Mw and groundwater level into the above evaluated liquefaction susceptibility map. The probability of slope failure was estimated for scenario earthquakes. For Chittagong, the most hazardous scenario is PBF-1 and analysis revealed that 20–40 % slopes are highly potential for landslide. For Sylhet, the most hazardous scenario is DF and PBF-2 it was found that 10–20 % slopes is potential for landslides. Figure 6.6 (bottom left, bottom middle, bottom right) represents the potential soil liquefaction in the city Corporation areas of Dhaka Chittagong and Sylhet City Corporation.

6.2.2 *Building and Lifelines Inventory as Well as Their Vulnerability*

Vulnerability is the degree to which a person, system or unit is likely to experience damage due to exposure to hazard. To assess the vulnerability of the exposures building stocks, lifelines and essential facilities database were developed using high resolution satellite images as well as conducting random field survey. It is estimated that there are 326,000, 182,000, and 52,000 buildings in Dhaka, Chittagong, and Sylhet City Corporation (Fig. 6.7) areas respectively. Afterwards, building stocks were differentiated into structural types. Such as, in Dhaka city corporation area,

Table 6.2 Building types surveyed

RC	Reinforced concrete
LC	Lightly reinforced concrete frames
BC	Brick in cement mortar masonry with concrete floor
BF	Brick in cement mortar masonry with flexible roof
STC	Steel truss with concrete column
STM	Steel truss with masonry wall
TSL	Tin shed
BAL	Bamboo refers to building

Fig. 6.8 Vulnerability factors of buildings frequently observed in Bangladesh

different types of reinforced concrete and masonry buildings are counted around 90,000 and 181,000 respectively. Table 6.2 illustrates the types buildings surveyed.

To assess the vulnerability of the building, three level surveys were conducted in this study. The vulnerability factors considered in level one survey were number of stories, occupancy class, structural type, number of occupants during the day and the night, age of the building, presence of soft story, presence of heavy overhangs, shape of the building in plan view (rectangular, narrow rectangular, irregular), shape of the building in elevation view (regular, setback, and narrow tall), pounding possibility, building in slope land, visible ground settlement, presence of short columns, visible physical condition (poor/average/good). In the Level-2 survey, a sketch of the building plan at the ground story, the dimensions of columns, concrete and masonry walls were measured. Level-3 survey was conducted for dynamic measurement on few selected buildings in three cities using modern equipment. For concrete buildings, five significant vulnerability factors are usually counted, which are (1) soft story, (2) heavy overhang, (3) short column, (4) pounding possibility between adjacent buildings, and (5) topographic effects (buildings constructed on slope ground). Figure 6.8 represents the soft story, heavy overhang and short column buildings observed in Bangladesh. Relationships between structural types and the presence of the above vulnerability factors of the surveyed buildings are shown in Fig. 6.9 for Dhaka, Chittagong and Sylhet City Corporation areas. Note that such relationships for structural types contributing major portion of the building population are only shown here (Table 6.3).

Number of building occupants is an important parameter for earthquake loss estimation of a number of casualties, a number of refugees and etc. it is also

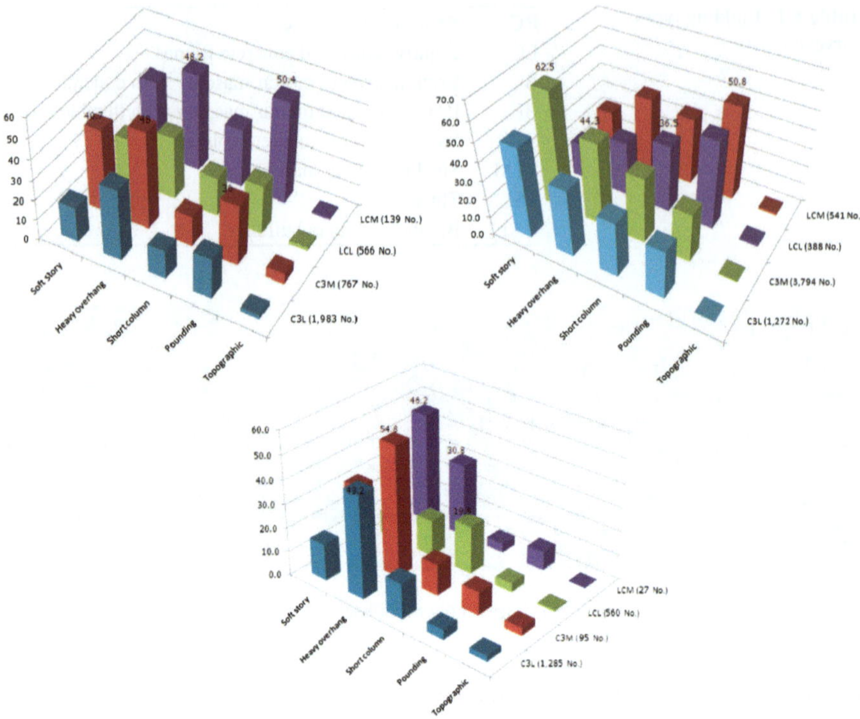

Fig. 6.9 Relationships between structural type and presence of vulnerability factors for Dhaka, Chittagong and Sylhet City Corporation areas

Table 6.3 Concrete framed building surveyed

C3L	Low-rise	Concrete frame with masonry infill walls
C3M	Mid-rise	
LCL	Low-rise	Lightly reinforced concrete frames
LCM	Mid-rise	

Table 6.4 Daytime and nighttime population and density in Dhaka, Chittagong and Sylhet City Corporation areas (*Source*: CDMP 2009a, b, c, d, e)

City	Calculated population		Population density (no./km^2)	
	Daytime	Nighttime	Daytime	Nighttime
Dhaka	6,457,483	7,279,663	53,812	60,664
Chittagong	2,001,691	2,332,599	11,819	13,773
Sylhet	357,079	401,776	13,242	14,900
Total	8,816,253	10,014,038	26,291	29,779

important to know the number of building occupants in the different period of time, mainly in during day and night. Table 6.4 shows population (computed by using a weight arithmetic mean of the number of occupants per floor area) and population density during Daytime and Nighttime.

Table 6.5 Road Length (in km) and Number of Highway Bridge in Dhaka, Chittagong, and Sylhet (*Source*: CDMP 2009a, b, c, d, e)

No.	Component	Dhaka	Chittagong	Sylhet
	Road network			
1.	Major road	*240.49*	*115.88*	*40.39*
2.	Urban road	*1,022.82*	*505.16*	*108.40*
3.	Local road	*1,097.73*	*893.57*	*264.22*
	Total length (km)	*2,361.04*	*1,514.61*	*264.22*
4.	Highway bridge	*10*	*4*	*2*

Table 6.6 Number of Essential Facility in Dhaka, Chittagong and Sylhet City Corporation (*Source*: CDMP 2009a, b, c, d, e)

No.	Essential facility	Dhaka	Chittagong	Sylhet
Medical care				
1.	Large Hospital	*75*	*45*	*7*
2.	Medium Hospital	*59*	*15*	*6*
3.	Small Hospital	*98*	*36*	*7*
4.	Medical Clinic	*368*	*66*	*67*
Emergency response				
5.	Police Station	*10*	*11*	*6*
6.	Fire Station	*62*	*12*	*2*
7.	Emergency Operation Center	*1*	*1*	–
Schools				
8.	Grade School	*2,026*	*906*	*162*
9.	College/University	*711*	*127*	*49*

Before developing the vulnerability of lifelines and essential facilities, there need to develop the database of both the system. The lifeline includes transportation system and utility system. While essential facility include medical care facilities, emergency response facilities, and schools. Lifeline facilities and essential facilities have been incorporated with the base map. Exposed lifeline features have been surveyed. However, underground installations records have been collected from respective organizations and incorporated in the main map. Tables 6.5 and 6.6 represents the statistics of a sector of lifelines and essential facilities.

There are 1,420 km natural gas system in Dhaka, 227 km in Chittagong and 147 km in Sylhet (Titas Gas Transmission and Distribution Company Ltd. 2008). Electric power system consists of electric poles, which are 52,765, 26,703 and 9,040 respectively for above mentioned three cities (Power Development Board, Power Grid Distribution Company 2008). The pipelines consist of portable water system are 1,562 km (DWASA 2008), 559 km (CWASA 2008) and 131 km (Sylhet City Corporation 2008) of the three cities.

The vulnerability of the lifeline system has been estimated overlaying them on the hazard maps. For instances, in potable water system, vulnerability is identified from the pipe ductility type, distribution facility type, and soil liquefaction susceptibility on which the pipe and facility lie on. From the field survey, it is found that most of distribution facilities are not prepared with the seismic design, proper and

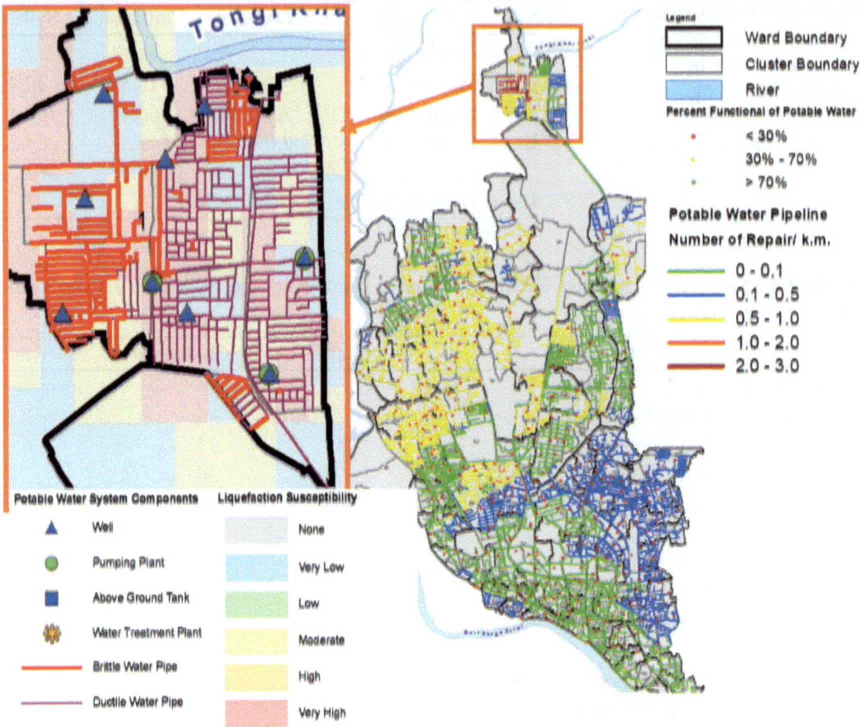

Fig. 6.10 Vulnerability of the portable water system of Dhaka City areas

horage, and backup power system. Based on material type, the pipe is classified into brittle (PWP1) and ductile (PWP2). Brittle pipe material includes asbestos cement (AC), cast iron (CI), and reinforced concrete (RCC), while ductile pipe material includes galvanized iron (GI), ductile iron (DI), mild steel (MS), PVC, and steel. Brittle pipe is more vulnerable than ductile one. Potable water facility include well, pumping plant, above ground tank, and water treatment plant. Pipeline as well as distribution facilities located on higher liquefaction susceptibility soil are more vulnerable than those on lower liquefaction susceptibility level. Figure 6.10 represents the vulnerability of the portable water system of Dhaka City areas.

6.2.3 Risk Assessment

Risk is the conditional probability and magnitude of damage occurs on the exposures to a hazard event. For assessing the seismic risk, the fragility functions (capacity of the buildings to withstand during earthquake) has been developed for the structural types of the buildings and lifeline systems. Combing the fragility functions of the buildings and infrastructures with the hazard maps building damages

Table 6.7 Expected Building Damage of Occupancy class in Dhaka City Corporation Area

If earthquake occurs in Madhupur Fault with magnitude 7.5										
	None		Slight		Moderate		Extensive		Complete	
	Count	(%)	Count	(%)	Count	(%)	Count	(%)	Count	(%)
Agriculture	261	0.21	291	0.67	485	0.91	264	0.80	343	0.47
Commercial	8,199	6.56	5,232	12.10	11,973	22.52	8,841	26.67	12,500	17.29
Education	476	0.38	343	0.79	844	1.59	629	1.90	870	1.20
Government	122	0.10	100	0.23	269	0.51	202	0.61	230	0.32
Industrial	1,245	1.00	748	1.73	1,538	2.89	1,137	3.43	1,708	2.36
Non-SFD Residential	112,110	89.73	35,473	82.05	36,242	68.17	20,891	63.01	54,644	75.56
Religion	407	0.33	307	0.71	567	1.07	372	1.12	526	0.73
Single Family Dwelling (SFD) Residential	2,118	1.70	739	1.71	1,248	2.35	817	2.46	1,495	2.07
Total	124,939		43,232		53,166		33,153		72,316	

and functionality of the lifelines and critical infrastructures were estimated using software known as Hazard United States (HAZUS). To run the assessment, City Corporation wards are divided into smaller area called cluster to provide detail level of data and corresponding results. In Dhaka City Corporation area, there are 552 clusters in 91 wards (90 ward and cantonment area). While 41 wards Chittagong City Corporation are divided into 285 clusters, and 27 wards of Sylhet City Corporation into 82 clusters.

Risk assessment reveals that if an earthquake occurred at Madhupur Fault with magnitude 7.5, about 72,000 buildings (Table 6.7) will be damaged in Dhaka beyond repair. The building related economic losses for this earthquake are estimated more than 6,000 million USD. It is estimated that about 140,000 buildings (Table 6.8) will be damaged potential to generate 8.5 magnitude earthquake from Plate Boundary Fault (PBF-1) located near Chittagong.

During an earthquake of 8.0 magnitude potential to generate from Dauki Fault about 25,000 (Table 6.9) buildings will be damaged beyond repair in Sylhet. HAZUS estimates the amount of debris that will be generated by the earthquake. The model breaks the debris into two general categories: (a) Brick/Wood and (b) Reinforced Concrete/Steel. This distinction is made because of the different types of material handling equipment required to handle the debris. Table 6.10 represents the amount of debris to be garneted by the scenario earthquakes.

There will be severe damage to essential facilities like hospital, schools, and police stations. For instances, Madhupur Fault earthquake will cause moderately to complete damage in Dhaka at about 250 hospital or clinics, 1,300 schools as well as 30 police and 4 fire service stations. In Dhaka, there are about 60,000 hospital beds available for use. On the day after the earthquake, it is estimated that only 24,200 hospital beds (41 %) will be available for use by patients already in the hospital and

Table 6.8 Expected Building Damage of Occupancy class in Chittagong City Corporation Area (*Source*: CDMP 2009a, b, c, d, e)

If earthquake occurs in Plate Boundary Fault (PBF-1) with magnitude 8.5										
	None		Slight		Moderate		Extensive		Complete	
	Count	(%)	Count	(%)	Count	(%)	Count	(%)	Count	(%)
Agriculture	15	0.29	21	0.23	30	0.23	43	0.35	262	0.18
Commercial	751	15.04	1,424	15.60	2,018	15.35	1,786	14.70	22,482	15.74
Education	74	1.48	102	1.11	114	0.87	98	0.80	956	0.67
Government	6	0.12	8	0.08	11	0.09	12	0.10	192	0.13
Industrial	53	1.07	66	0.72	83	0.63	85	0.70	1,793	1.25
Non-SFD Residential	2,946	58.96	5,888	64.46	8,906	67.75	8,167	67.22	67,569	47.30
Religion	49	0.98	84	0.92	102	0.78	86	0.70	360	0.25
Single Family Dwelling (SFD) Residential	1,102	22.05	1,541	16.87	1,882	14.32	1,872	15.41	49,242	34.47
Total	4,996		9,133		13,146		12,149		142,856	

Table 6.9 Expected Building Damage of Occupancy class in Sylhet City Corporation Area (*Source*: CDMP 2009a, b, c, d, e)

If earthquake occurs in Dauki Fault with magnitude 8.0										
	None		Slight		Moderate		Extensive		Complete	
	Count	(%)	Count	(%)	Count	(%)	Count	(%)	Count	(%)
Agriculture	6	0.09	5	0.12	10	0.14	12	0.13	44	0.17
Commercial	982	15.28	624	13.84	796	11.43	964	10.40	2,717	10.89
Education	107	1.67	63	1.40	63	0.91	73	0.79	250	1.00
Government	27	0.42	16	0.35	16	0.23	20	0.22	89	0.36
Industrial	44	0.68	33	0.73	55	0.79	77	0.83	239	0.96
Non-SFD Residential	2,013	31.32	1,294	28.73	1,416	20.35	1,454	15.68	4,247	17.03
Religion	103	1.61	60	1.34	56	0.81	50	0.53	96	0.38
Single Family Dwelling (SFD) Residential	3,146	48.94	2,409	53.48	4,547	65.33	6,620	71.42	17,264	69.21
Total	6,429		4,504		6,960		9,269		24,945	

Table 6.10 Amount of debris to be generated by the scenario earthquakes (*Source*: CDMP 2009a, b, c, d, e)

Debris generation	Scenario Earthquake for Dhaka (MF)	Scenario Earthquake for Chittagong (PBF-1)	Scenario Earthquake for Sylhet (DF)
Debris (millions of tons)	30	17	2
Brick/wood (%)	27.00	26.00	29.00
Concrete/steel (%)	73.00	74.00	71.00
Truck load required (@ 25 tons/truck)	1,200,000	680,000	80,000

those injured by the earthquake. After 1 week, 54 % of the beds will be back in service. By 30 days, 72 % will be operational. During this period there will be around 80 leaks and 270 breaks in water supply system, 100 leaks & 350 breaks in waste water system and 60 leaks & 200 breaks in gas supply network. It has been observed that about 50 % of highway road, 70 % of railway track and 6 major highway bridges in Dhaka are located in moderate to very high liquefaction susceptibility area. Moreover, about 55 % potable water pipeline, 52 % waste water treatment pipeline, 56 % gas pipeline in Dhaka are located on moderate to very high liquefaction potential area.

6.3 Earthquake Risk Management Initiatives

6.3.1 Scenario-Based Earthquake Contingency Plan

Usually many agencies are involved in accomplishing the activities for preparedness, response and early recovery to reduce the risk of any big disasterous event. Experiences have shown that all agencies need to work together *under a well-designed and fully-coordinated plan for optimum and efficient preparedness, response and early recovery, usualy known as Contingency Plan*, in a systematic manner so that their capacities and resources are best utilized to fulfil the need complimenting and supplementing other agencies. From CDMP-I (2009) initiatives, scenario-based contingency plans have been developed for national level, three city corporations (Dhaka, Chittagong and Sylhet) and five key Responder Agencies viz., Armed Forces Division (AFD), Fire Service & Civil Defense (FSCD), Directorate General of Health Services (DGHS), Disaster Management Bureau (DMB) and Directorate of Relief and Rehabilitation (DRR). Moreover, lifelines & utilities agencies, such as Power Distribution Company Limited, Dhaka Water Supply and Sewerage Authority, Titas Gas Transmission and Distribution Company Limited, Bangladesh Telecommunication Company Limited (BTCL) and humanitarian assistance providers have also been addressed in the contingency planning process. National Earthqauke Contingency Plans segregated into operational functional clusters with lead and support agencies, which are: Emergency Operations Cluster 1—Overall Command and Coordination (lead-DMB), Emergency Operations Cluster 2—Search, Rescue and Evacuation (lead-FSCD), Health Cluster- (DGHS), Relief Services (Food, Nutrition and other Relief) Cluster (lead-DRR), Shelter (Including Camp Management) Cluster (lead-DMB), Water Supply, Sanitation and Hygiene Cluster (lead-Local Government Bodies: City Corporations/Pourashava), Restoration of Urban Services Cluster (lead-Local Government Bodies: City Corporations/Pourashava), Transport (Road, Rail, Air, Sea) Cluster (lead-BRTA), BIWTA, Bridge Authority, Security and Welfare Cluster (lead-Police). The role of Armed Force Division is actively associated with almost all the clusters.

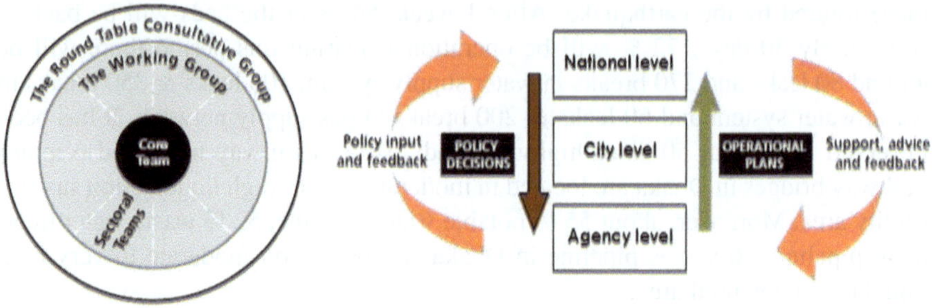

Fig. 6.11 Development procedures of the contingency plan

The contingency plans have been developed with around 30 interactive meetings with the agencies following a certain procedure in national to agency level as given in Fig. 6.11. Agency professionals were also trained on the technical aspects of the plan. Even though the earthquake that occurs may be very different from the one planned for, the plan will still be useful. A good contingency plan ensures better preparedness for any emergency that may occur, even one that is very different from the scenario in the plan. Legal provisions and organizational set up, functional response roles and responsibilities assigned for the agency, operating procedures guideline and readiness checklists are also outlined in this plan.

With regard to each functional cluster, spatial planning requirements in terms of locating facilities at appropriate locations, need assessment based on spatial coverage, identifying alternative locations and routes for emergency supplies, identifying emergency evacuation routes and places for temporary shelters and medical facilities, pre-positioning of facilities etc. have been identified and mapped under GIS environment that would enable the key emergency responders to act in a coordinated as well as cohesive way.

For instances, the total number of displaced population needing temporary shelter has been estimated at around 870,000 within the Dhaka city during Madhupur earthquake. This is estimated as 50 % of the displaced people need immediate shelter after the earthquake and the remaining 50 % will manage their shelter in their relatives and other places. Open spaces within the city corporation area that are more than 2,500 m^2 (area with capacity for sheltering 500 families or more) are mapped (Fig. 6.12) and their shelter capacities are calculated. There are only 13 open spaces within Dhaka City area having only about one fourth capacities of the total population needing shelters. The evacuation routes are also mapped to go to the open spaces under spatial contingency plan.

The population evacuated in immediate shelter requires about 3,800 m^3 of emergency water with the rate of 15 l per capita per day. The populations who are taken outside require about 9,200 m^3 of water per day. So, the total emergency water needed in the emergency shelters is about 13,000 m^3 per day. Pre–positioning of this amount of the water at the above planned spaces is before earthquake is the most appropriate way to provide immediate emergency response of water. To manage toilets in the immediate shelters spaces is the main challenge related to sanitation.

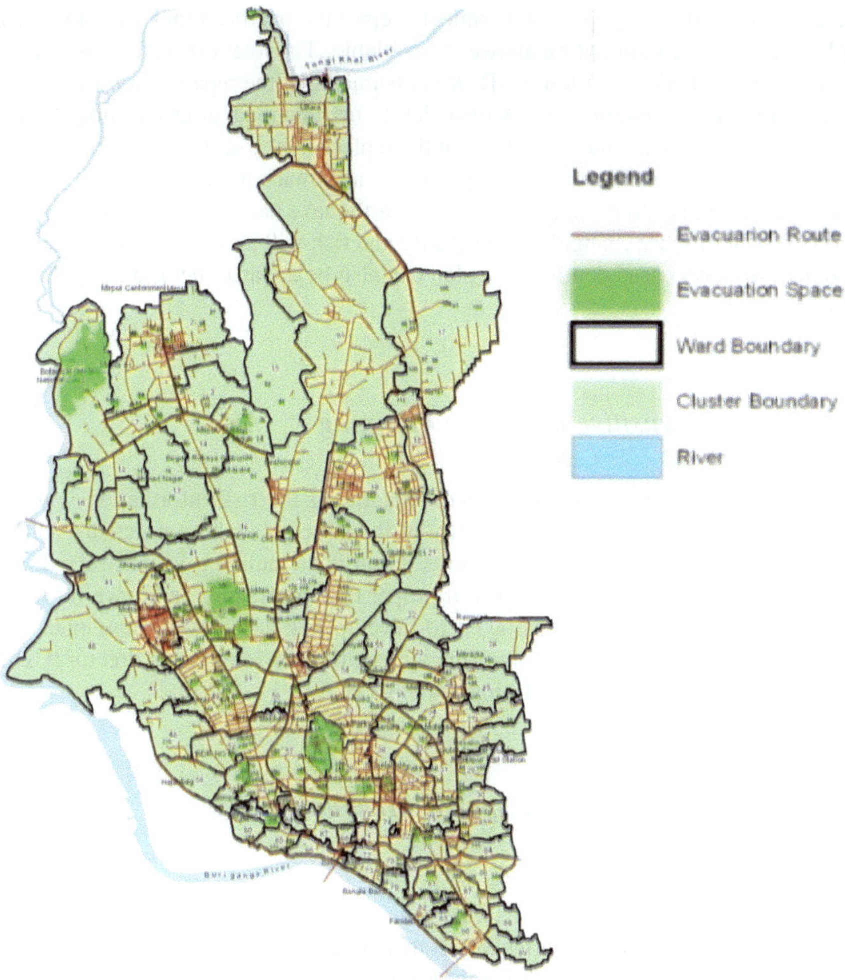

Fig. 6.12 Identified emergency open spaces with evacuation routes

Sphere standard demand one toilet for each 20 people in the emergency shelter. The calculation shows that about 13,000 toilets are needed in the open spaces within Dhaka City and additional 30,000 toilets need to plan for the people of Dhaka who need immediate shelter outside Dhaka city. The total number of skilled/trained workers required for repairing the potable water and waste water system within Dhaka city corporation area after an earthquake is about 1,800 people per day to repair the system within 7 days. If it is planned to repair in 14 days, it require about 900 skilled workers per day and require about 420 people working per day to repair the system in 30 days time. Skilled workers required to repair overhead water tank and waste water treatment plant is not calculated here, as it require thousands of skilled workers if planned to repair within a month, which is practically not possible even if planned. It requires 3–6 months to repair the overhead tanks and treatment plant, even if the required manpower is available. The total estimated cost require for repairs is about 27 million USD. It is assumed that the repair cost for the waste water treatment plant, probability of which to be functional after the earthquake is 40–50 % has been assumed that 25 % of the replacement cost. Priorities for recovery of different components within the system are also made in this plan.

Since, there is still room to improve the earthquake hazard assessment. Once the hazard scenarios are changed, vulnerability and risk will also bring updated results. Hence, scenario based contingency plan can also be updated incorporating the risk information.

6.3.2 Risk Sensitive Land Use Planning

Risk sensitive landuse planning means integrating the risk information with the landuse plan to reduce the vulnerability of the structures taking necessary actions during the construction of the structures. Even if the building is constructed, still risk can be reduced to a certain limit if proper risk information in known for a particular location. Initiatives has taken to develop risk sensitive landuse plan of a municipality known as Mymensingh engaging Urban Development Director (UDD) with the technical and financial supports of CDMP-II.

6.3.3 Improvement of Response Mechanism

Development of urban Community volunteers, a unique initiate of Bangladesh for community involvement in search and rescue operations to support regular forces of the government: Global scenarios reflect that during any disaster community people are the first responders. Considering the model of Cyclone Preparedness Programme (CPP) of the Government of Bangladesh and Red Crescent Society, Government has taken initiatives to develop urban community volunteers through Fire Service and Civil Defence (FSCD). The aim of FSCD is to train 62,000 volunteers in the long run keeping in mind their support mainly preliminary search and rescue operations during major earthquake and fire events. So far, more than 16,000 volunteers received training in Dhaka, Chittagong and Sylhet. It has been planned to develop around 250 volunteers in each ward of city corporations. The volunteers are selected following specific criteria set by joint consultation of FSCD professionals and councilor of the respective wards of City Corporations as well as elite citizens of the community. They received training on search, rescue and first aids to support the regular forces of the government during the period of emergency. Figure 6.13 represents the training on search and rescue operations using light equipment.

Fig. 6.13 Training of the volunteers on search and rescue operations

6.3.4 *Safety and Evacuation Drill to Schools*

Due to irregular shape and poor constructions, schools are most vulnerable. In all earthquake prone countries, school safety and evacuation drills are found as a good practice. It has been observed that even in the most severe earthquakes, buildings rarely collapse completely. Injury and even death are most often caused by the shattering and falling of non-structural elements, such as window glass, ceiling plaster, lighting fixtures, chimneys, roof tiles, and signs. There will be no time to think what to do; therefore, of all earthquake-preparedness measures, earthquake drills are the most important. There is a need to educate the school children and teachers on safety and evacuation measures/drills to make the schools safer, and to know what to do before and after an earthquake occurs in their area. A standard school earthquake safety guidebook and posters have been prepared and distributed to the schools. It is planned to conduct pilot drill in more effective manner at 100 schools in the earthquake risk-prone cities of the country. In order to make safety and evacuation drills sustainable, equipments (fist aid box, stretcher, hand mike, etc.) to conduct the drills have been provided to the schools. To make the programme sustainable, Government has given a directive to primary and secondary school directorate to conduct such drills in regular basis. Figure 6.14 shows the IEC materials and training on school drills.

6.3.5 *Training for Religious Leaders (Imams) for Awareness About Earthquake Dangers*

Natural disasters have had usually some religious meaning for people from ancient times. But, this kind of religious interpretation of natural disasters has declined gradually because of the influence of education in the society. Religious leaders have the access in the society to motivate and educate the people in reducing the

Fig. 6.14 IEC materials and training on school drills

vulnerability of different disasters. Keeping this in mind, religious leaders were educated through this activity on the earthquake dangers and what to do to reduce the non-structural vulnerability in household level. The religious leader were given preliminary knowledge on the earthquake hazard and training on the vulnerability of the non-structural elements, such as, window glass, ceiling plaster, lighting fixtures, chimneys, roof tiles, and signs. After getting training, they were requested to talks on the household non-structural earthquake vulnerability reduction measures during the recitation of "khutba" of Juma prayer on Friday. It is to be noted that CDMP has taken initiative to incorporate an earthquake non-structural vulnerability reduction module with the regular training programe of Islamic Foundation of Bangladesh. Steps have been taken simultaneously to incorporate such training by other religious agencies. NGOs also came forward to conduct such trainings among the religious agencies as well as community group. Figure 6.15 illustrates religious leader training on non-structural vulnerability of Earthquake organized through Islamic Foundation of Bangladesh

6.3.6 *Training for Masons and Bar-Binders for Safe Construction Practices*

Construction of residential buildings in a developing country like Bangladesh is primarily carried out by the informal sector, mostly the owners/builders. The workforce (masons, bar- binders and plumbers) employed in this sector does not have any formal training. Most of them acquire skill either through trial and error or through practical experience. While there have been initiatives by the Governments towards improving seismic performance of new constructions by incorporation of seismic designs in the building construction process, it can be still anticipated that local masons and small contractors will continue to play significant role in the

Fig. 6.15 Training to religious leader on the nonstructural vulnerability of the earthquake

Fig. 6.16 IEC materials and basic training on earthquake safety constructions for masons and bar binders

building construction process. So, any enhancement in their skills in seismic-resistant constructions can significantly help in improving the earthquake resistance of informal, non-engineered buildings and hence considerable reduction in the loss of human lives and properties can be addressed due to earthquake. Government has taken initiative to train masons and bar binders through House Building Research Institute (HBRI) of Bangladesh. A training manual has been developed and presently, 03-days long training is going on over the country to the masons and binders. The aim of the training is to train masons and bar binders with basic knowledge of earthquake resistant building construction technology. Government is planning to launch a short certificate course and systematically engage of the masons and bar binders in the construction industries to ensure the implementation of the building code gradually (Figs. 6.16 and 6.17).

Fig. 6.17 Equipments handover ceremony by Honorable Prime Minister

6.3.7 *Search and Rescue as Well as Research Equipment Procurement*

Governments procured a good number of light and heavy equipment for conducting search and rescue operations and handed over them to Bangladesh Armed Force Division (AFD), FSCD and City Corporations. Research equipment also been procured for Geological Survey of Bangladesh, Public Works Department, Urban Development Directorate, Academic institutions for conducting advanced earthquake risk reduction study in the country.

6.4 Conclusion

Historical evidences reveal that Bangladesh is an earthquake prone country. Hazard, vulnerability and risk assessment implies that major cities of the country are subjected to high vulnerability of the earthquake. Considering the potential earthquake risk of the country, systematic initiatives are taken for earthquake risk management by the relevant Government and non-government agencies. Risk management approach include risk assessment, development of risk mitigation plan through risk sensitive landuse planning, preparation of scenario-based contingency plan for emergency management, introduction of various training for vulnerability reduction and capacity enhancement as well as creation and widely distribution of IEC materials to the relevant professionals and city dwellers for preparedness and awareness build-up. Since in the last 100 years, there was no damaging earthquake in the country, the preparedness and awareness level of the concerns are not within the scale of mapping. After taking various initiatives of CDMP, the Government and Non-government agencies have taken visible initiatives for earthquake risk reductions and management in the country. There need to involve experienced international agencies of the earthquake prone countries to ahead the initiatives and to introduce innovative structural, non-structural and social techniques to set-up the mind of accepting risk management approaches. These efforts need well coordination and consistency among the agencies.

References

Abe K (1994). Instrumental magnitudes of historical earthquakes, 1892 to 1898: Bull Seismol Soc Am 84:415–425

Ambraseys NN, Douglas J (2004) Magnitude calibration of north Indian earthquakes. Geophys J Int 159:165–206

Ansary MA, Sadek A (2006) Assessment of 2003 Rangamati earthquake, Bangladesh, in 8th US National Conference on Earthquake Engineering, San Francisco

Atkinson GM, Boore DM (2003) Empirical Ground-Motion Relations for Subduction-Zone Earthquakes and Their Application to Cascadia and Other Regions. Bull Seism Soc Amer 93(4):1703–1729

Banglapedia (2012) The National Encyclopedia of Bangladesh, http://www.banglapedia.org/HT/E_0002.HTM. Accessed on Dec 15, 2012

Bilham R, Hough S (2006) Future earthquakes on the Indian subcontinent: inevitable hazard, preventable risk. South Asian J 12:1–9

Comprehensive Disaster Management Programme (CDMP) (2009a) Active fault mapping and modeling of Bangladesh

Comprehensive Disaster Management Programme (CDMP) (2009b) Earthquake Hazard Assessment Report of Bangladesh

Comprehensive Disaster Management Programme (CDMP) (2009c) Earthquake Vulnerability Assessment Report of Bangladesh

Comprehensive Disaster Management Programme (CDMP) (2009d) Earthquake risk assessment of Bangladesh

Comprehensive Disaster Management Programme (CDMP) (2009e) National Level Earthquake Contingency Plan of Bangladesh

Halsted EP (1841) Report on the Island of Chedooba

Iyengar RN, Sharma D, Siddiqui JM (1999) Earthquake history of India in medieval times, Indian. J Hist Sci 34:181–237

Jahan I, Ansary M, Ara S, Islam I (2011) Assessing social vulnerability to Earthquake Hazard in Old Dhaka, Bangladesh; Asian Journal of Environment and Disaster Management (AJEDM) 3(3):285–300

Kaneko F (2012) Draft revised report on Regional Morphotectonic Mapping and Crustal Movement Modeling for the Identification of Potential Earthquake Sources of Bangladesh

Martin S, Szeliga W (2010) A catalog of felt intensity data for 589 earthquakes in India, 1636–2008. Bull Seismol Soc Am 100(2):562–569

Middlemiss CS (1885) Report of the Bengal Earthquake, 14th July, 1885. Rec Geol Surv India XVIII(Pt 4):200–221

Morino M, Kamal ASMM, Muslim D, Alid RE, Kamal MA, Rahman Z, Kaneko F (2011) Seismic event of the Dauki Fault in 16th century confirmed by trench investigation at Gabrakhari Village, Haluaghat, Mymensingh, Bangladesh. J Asian Earth Sci 42(3):492–498

Oldham T (1883) Catalogue of Indian Earthquakes. Mem Geol Surv India 19(3):170

Oldham RD (1899) Report on the great earthquake of 12th June 1897. Mem Geol Surv India 29:1379 (Calcutta)
Pacheco JF, Sykes LR (1992) Seismic moment catalog of large shallow earthquakes, 1900 to 1989. Bull Seismol Soc Am 82:1306–1349
Richter CF (1958) Elementary seismology. W. H. Freeman, San Francisco
Rizvi SNH (1970) Bangladesh District Gazetteers, Government of the People's Republic of Bangladesh, Bangladesh Government Press, Dhaka
Shishikura MY, Okamura K et al (2009) Geomorphological evidences of great Holocene Earthquakes off West Myanmar. In: Proceedings of the international workshop of Tsunami and Storm Surge hazard assessment and management in Bangladesh
Steckler MS, Akhter SH, Seeber L (2008) Collision of the Ganges–Brahmaputra Delta with the Burma Arc: implications for earthquake hazard. Earth Planet Sci Lett 273:367–378
Stuart M (1920) The Srimangal earthquake of 8th July 1918. Mem Geol Surv India 46(1)
Szeliga W, Hough S, Martin S, Bilham R (2010) Intensity, magnitude, location and attenuation in India for felt earthquakes since 1762. Bull Seismol Soc Am 100(2):570–584
Wells DL, Coppersmith KJ (1994) New empirical relationships among magnitude, rupture length, rupture width, rupture area, and surface displacement. Bull Seismol Soc Am 84:974–1002

Chapter 7
Drought Risk and Reduction Approaches in Bangladesh

Umma Habiba, Rajib Shaw, and Abu Wali Raghib Hassan

Abstract The insidious disaster drought is common in Bangladesh. Particularly the Northwestern part is severe drought-prone area in Bangladesh due to high rainfall variability along with high temperature. Agriculture as well as farmer's livelihood is badly affected due to the effect of drought. Therefore, different levels such as institution, community and individuals are following a number of adaptation actions to cope with drought. However, to deal better with drought, it is imperative to take effective adaptive practices that will help them to minimize its consequences. In this context, an innovative approach named "SIP approach" has been carried out in two severe drought-prone districts in Northwestern region that helps to measure the socio-economic, institutional and physical aspect of drought-affected area. It also shows the comparison among different drought affected Upazila (sub-district) of these two districts, which is more or less resilient against it. Furthermore, this chapter highlights irrigated and non-irrigated farmer's prioritized drought adaptive practices that derived from the indicators of SIP approach. To carry out these drought adaptive practices successfully, it also discloses various stakeholder' role and mentions the timeframe for each drought adaptive practice. Based on the findings of SIP approach and drought adaptive practices, the drought risk management policy and action framework has developed. This framework brings together both community and national government in identifying timeframe and responsibilities, thus strengthening the link between the drought adaptive practices to national policy.

Keywords Drought • Drought adaptive practice • Drought risk management policy and action framework • Northwestern Bangladesh • SIP approach

---

U. Habiba (✉) • R. Shaw
Graduate School of Global Environmental Studies, Kyoto University, Kyoto, Japan
e-mail: shimuagri@yahoo.com

A.W.R. Hassan
Department of Agricultural Extension, Dhaka, Bangladesh

R. Shaw et al. (eds.), *Disaster Risk Reduction Approaches in Bangladesh*, Disaster Risk Reduction, DOI 10.1007/978-4-431-54252-0_7, © Springer Japan 2013

7.1 Introduction

According to National Drought Mitigation Center (2006), the insidious onset disaster drought takes place in Bangladesh more frequently because of climate change. Every year Bangladesh experiences a dry season for 7 months, from November to May, when rainfall is normally low. Hence, drought mostly occurs in pre-monsoon (March to May) and post-monsoon (October to November) seasons in Bangladesh (Banglapedia 2006) (National encyclopedia of Bangladesh). Beside this, inadequate pre-monsoon showers, a delay in the onset of the rainy season or an early departure of the monsoon may create drought conditions in Bangladesh. ADB (2005) further argued that lower precipitation during the dry season under climate change has the potential to increase drought risk in Bangladesh. But the type and extent of drought varies according to the quantity and type of precipitation.

In Bangladesh context, Brammer (1987) defines drought as a period when supply of moisture in the soil is less than that which is required for satisfactory crop growth during a season when crops are normally grown. Moreover, Banglapedia mentioned that drought is the period when the moisture content of soil is less than the required amount for satisfactory crop growth during the normal crop growing season. Furthermore, an important sign indicates for drought condition may help individual or all to distinguish it at the field levels that include: (1) development of continually broken cracks on the dried up topsoil; (2) burnt-out yellowish foliage in the vegetation cover or "top yellow" syndrome, particularly, in betel nut trees and bamboo groves; (3) loosening of soil structure, ending up in the topsoil developing in to dust layer.

Drought is a recurrent phenomenon in some parts of the country, but the north-west region of Bangladesh is mostly drought-prone area because of high rainfall variability (Shahid and Behrawan 2008). This region is more prone to droughts as the area is relatively dry, receiving much lower rainfall compared to the rest of the country (Paul 1998). Therefore, drought occurs in this region in a regular basis. It is gradually being more reported in Rajshahi, Chapai Nawabganj, Naogaon, Rangpur, Bogura, Pabna, Dinajpur and Kustia regions because of its moisture retention capacity, infiltration rate characteristics.

The drought has a substantial impact on food production, affecting the food security of subsistence farmers. About 83 % of 12.49 million hectares of T-aman cultivable land and 9.32 million hectares of rabi crop land are affected by drought during pre-kharif season and rabi season, respectively (Climate Change Cell of MoEF in Bangladesh, 2009). Aside from this, Tanner et al. (2007) mentioned that about 2.7 million hectares of land in Bangladesh are vulnerable to annual drought. Furthermore, different degrees of drought cause different intensities of damage to the crops. For instance, the drought of 1994–1995 led to a decrease in rice and wheat production by 3.5 million tons (Rahman and Biswas 1995). Recent data show that drought has caused 25–30 % of crop reduction in the north-west part of Bangladesh (Rahman et al. 2008).

Based on the above, the main purpose of this book chapter is to provide the overall scenarios of drought in context of Bangladesh, its causes, impacts and various risk management practices carried out by different levels. Besides, an innovative approach is addressed here which help to minimize the risk for future drought mitigation.

7.2 Definition and Types of Drought

In general, drought can be defined as a temporary meteorological event, which stems from a deficiency of precipitation over an extended period of time compared to some long-term average conditions. In contrast, a broad definition of drought is a deficiency of precipitation over an extended period of time, usually a season or more, which results in a water shortage for some activity, group, or environmental sectors (UNISDR 2009). However, in Bangladesh context, Brammer (1987) defines drought as a period when supply of moisture in the soil is less than that which requires for satisfactory crop growth during a season when crops are normally grown.

In terms of typology, drought is often categorized into four types: meteorological, agricultural, hydrological, and socio-economic (Wilhite and Glantz 1985) (Fig. 7.1). Meteorological and hydrological droughts are physical events but agricultural drought refers to the impact of the first two on agricultural production. Agricultural drought happens after meteorological drought but before hydrological drought. So, it is important to understand these types and how it is interlinked with each other.

7.2.1 Meteorological Drought

Meteorological drought is usually an expression of precipitation's departure from normal over some period of time. The thresholds chosen, such as 50 % of normal precipitation over a 6-month time period may vary by location according to user

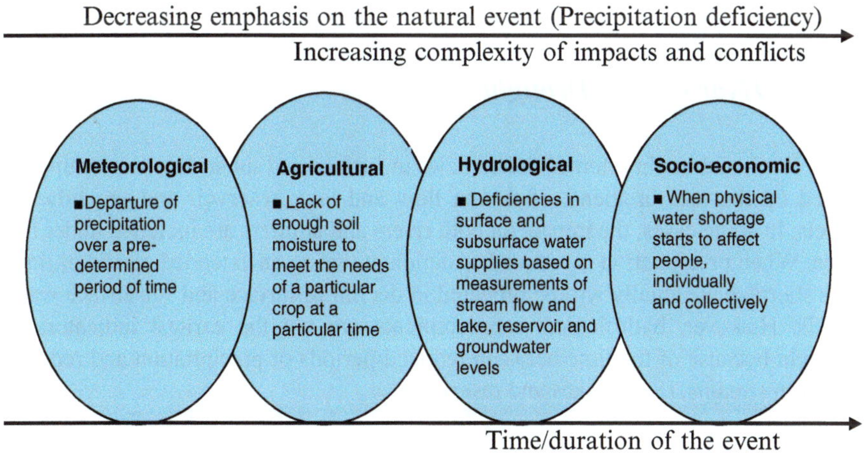

Fig. 7.1 Natural and social dimensions of drought (*Source*: Habiba 2012)

needs or applications. Meteorological droughts are usually region specific, and presumably based on a thorough understanding of regional climates. The variety of meteorological droughts definitions from different countries at different times show why it is a poor idea to apply a definition of drought developed in one part of the world to another:

- United States (1942): less than one tenth inch of rainfall in 48 h
- Great Britain (1936): 15 consecutive days with daily precipitation totals of less than one hundredth of an inch
- Libya (1964): when annual rainfall is less than 7 inch
- India (1960): actual seasonal rainfall is deficient by more than twice the mean deviation
- Bali (1964): a period of 6 days without rain
- Vietnam (1992): a period of 14 days without rain

7.2.2 Agricultural Drought

Agricultural drought occurs when there isn't enough soil moisture to meet the needs of a particular crop at a particular time. Crop water requirements depend on local weather conditions, soil and plants' characteristics and plant stage of growth. The extremity of an agricultural drought should therefore ideally be defined in terms of its impact on specific plant on a specific soil in a specific area, which makes it a difficult task to accomplish. In more general terms, agricultural drought exists when root zone soil moisture is insufficient to sustain crops between rainfall events. In this context, the status of soil water deficit in the top meter of a soil profile may be used as a drought measure (Tallaksen and Van Lanen 2004). In practical terms, some of the indices are used to monitor the developing deficits of water availability to crops (e.g. Crop Moisture Index-CMI and other indices of soil moisture condition).

7.2.3 Hydrological Drought

Hydrological drought refers to deficiencies in surface and subsurface water supplies based on the measurements of stream flow and lake, reservoir and groundwater levels. In some cases, the impact on both rivers and aquifers are included under this type. When precipitation is reduced or deficient during an extended period of time, this shortage eventually will be reflected in declining surface and subsurface water levels. However, hydrological measurements are not the earliest indicators of drought because of the time between reduced periods of precipitation and reduced water in streams, rivers, lakes and reservoirs.

Table 7.1 Characters of four types of drought

Type of drought	Characters
Meteorological drought	Meteorological drought is the amount of dryness and the duration of the dry period. Atmospheric conditions that result in deficiencies of precipitation change from area to area.
Hydrological drought	Hydrological drought is associated with the effects of periods of precipitation shortages on water supply. Water in hydrologic storage systems such as reservoirs and rivers are often used for multiple purposes such as flood control, irrigation, recreation, navigation, hydropower, and wildlife habitat. Competition for water in these storage systems escalates during drought and conflicts between water users increase significantly.
Agricultural drought	Agricultural drought mainly effects food production and farming. Agricultural drought and precipitation shortages bring soil water deficits, reduced ground water or reservoir levels, and so on. Deficient topsoil moisture at planting may stop germination, leading to low plant populations.
Socio-economic drought	Socio-economic drought occurs when the demand for an economic good exceeds supply as a result of a weather-related shortfall in water supply. The supply of many economic goods, such as water, forage, food grains, fish, and hydroelectric power, depends on weather. Due to variability of climate, water supply is sufficient in some years but not satisfactory to meet human and environmental needs in other years. The demand for economic goods is increasing as a result of increasing population. Supply may also increase because of improved production efficiency and technology.

7.2.4 Socio-Economic Drought

Socio-economic drought differs markedly from other types of drought because it reflects the relationship between supply and demand for some commodity or economic goods (such as water, livestock forage, or hydroelectric power) that is dependent on precipitation. Supply varies annually as a function of precipitation or water availability. Demand also fluctuates and is often associated with a positive trend as a result of increasing population, development and other factors (Wilhite and Glantz 1985).

However, Shaw et al. (2011) summarized the characters of each type of drought that is presented in Table 7.1.

7.3 Drought Events in Bangladesh

Bangladesh experiences of recurring drought than the past. From 1961 to 1991 Bangladesh faced 19 droughts (Climate Change Cell 2009). Major drought occurred in Bangladesh are 1973, 1978, 1979, 1981, 1982, 1992, 1994, 1995, 2000, 2006 and 2009. Table 7.2 represents the drought scenarios in Bangladesh.

Table 7.2 Years of severe droughts in Bangladesh

Year	Impact of drought
1791	• Droughts in Jessore; prices hiked two to three times
1865	• Droughts cause famine Dhaka
1872	• Sundarban suffers from droughts; scarcity of rain hampers rice cultivation
1951	• Droughts in the northwestern region damage rice
1973	• Droughts of 1973 lead to famine in 1974
1975	• Drought affected 47 % of the country and half of its population
1978–1979	• One of the worst droughts in recent times: 2 million tons less crops are harvested, 42 % of the cultivated land and 44 % of the population affected
1981	• Drought significantly lower crop yields
1982	• Drought in Jessore cause a hike in prices
1989	• Most of the rivers in the northwestern region dry up due to droughts, sandstorms in Naogaon, Chapai Nawabganj, Nilpahamari and Thakurgaon
1994–1995 and 1995–1996	• Another of the worst droughts of recent times. Damages to rice, jute and bamboo (another cash crop) in the northwestern region

Source: Climate Change Cell (2009)

7.4 Drought Periods in Bangladesh

Two critical dry periods occurred in Bangladesh (Karim et al. 1990), considering (1) the cumulative effect of dry days, (2) higher temperatures during pre-kharif (> 40 °C in March to May), and (3) low soil moisture availability that are as follows:

(1) ***Kharif***—droughts in the period June/July to October result from dry conditions in the highland areas especially in the Barind. Shortage of rainfall affects the critical reproductive stages of T. Aman rice, reducing its yield, particularly in those areas with low soil moisture-holding capacity. This drought also affects fisheries and other household-level activities.
(2) ***Rabi and pre-kharif***—droughts in the period January to May are due to: (1) the cumulative effect of dry days, (2) higher temperatures during pre-kharif (>40 °C in March/May), and (3) low soil moisture. This drought affects all the Rabi crops, such as boro, wheat, pulses and potatoes, and pre-kharif crops such as t. Aus, especially where irrigation possibilities are limited.

Karim et al. (1990) also prepared agricultural drought map by considering cumulative effect of dry days, higher temperatures during pre-monsoon period and soil moisture availability. Furthermore, Bangladesh Agricultural Research Council (BARC) has identified and mapped drought prone areas of Bangladesh for rabi and pre-kharif seasons based on the Agro Ecological Zones (AEZ) database and land resources inventory map at 1:1,000,000 scales (WARPO-EGIC 1996). Recently in 2001 BRAC has reviewed this concept and produced three different maps for pre-kharif, kharif and rabi seasons (Fig. 7.2). In these maps, the drought severity classes were defined as slight, moderate, severe and very severe related to the yield losses of 15–20 %, 20–35 %, 35–45 %, and 45–70 %, accordingly, for different crops (Karim and Iqbal 2001).

In addition, Iqbal and Ali in 2001 mentioned the drought severity areas (in million ha) of Bangladesh for different crop seasons which are given in Table 7.3.

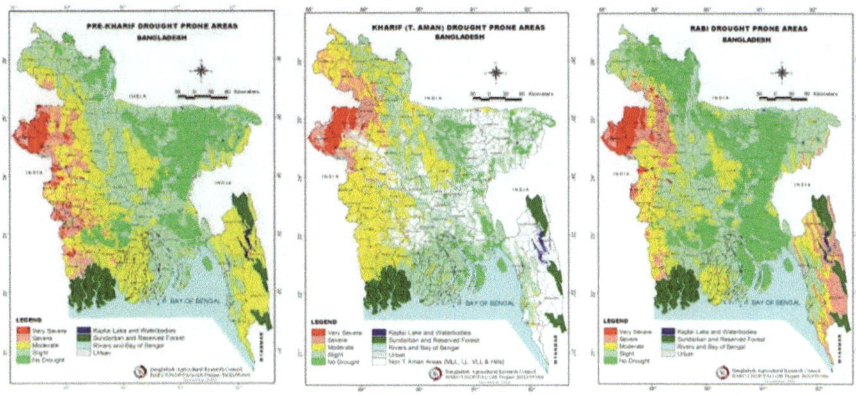

Fig. 7.2 Pre-kharif, kharif and rabi drought prone areas of Bangladesh (*Source*: Bangladesh Agricultural Research Council 2010)

Table 7.3 Summery of severity areas in Bangladesh by crop season (in million ha)

Drought class	Rabi season	Pre-Kharif season	Kharif season
Very severe	0.446	0.403	0.344
Severe	1.71	1.15	0.74
Moderate	2.95	4.76	3.17
Slight	4.21	4.09	2.9
No drought	3.17	2.09	0.68
No T-aman	–	–	4.71

Here, it has been reported that every year 0.45 million ha of land is affected by very severe drought during the rabi season, while 0.40 million ha and 0.34 million ha of land are affected during the pre-kharif and kharif seasons, respectively.

7.5 Causes of Drought in Bangladesh

Causes of drought in Bangladesh are related to climate variability and non-availability of surface water resources (Habiba et al. 2011a, b). Beside these, there are several issues that enhance drought severity in Bangladesh. These issues are briefly described under the following sub-headings:

7.5.1 Geophysical Feature

Bangladesh is a part of the Bengal Basin which has been filled by sediments washed down from the highlands on three sides of it, especially from the Himalayas. A network of rivers originated from this Himalayas and flows about 1,073 million acre feet/year water through Bangladesh of which 870 million acre feet/year comes from the neighboring country India. At the same time, it brings sediments about 1.5

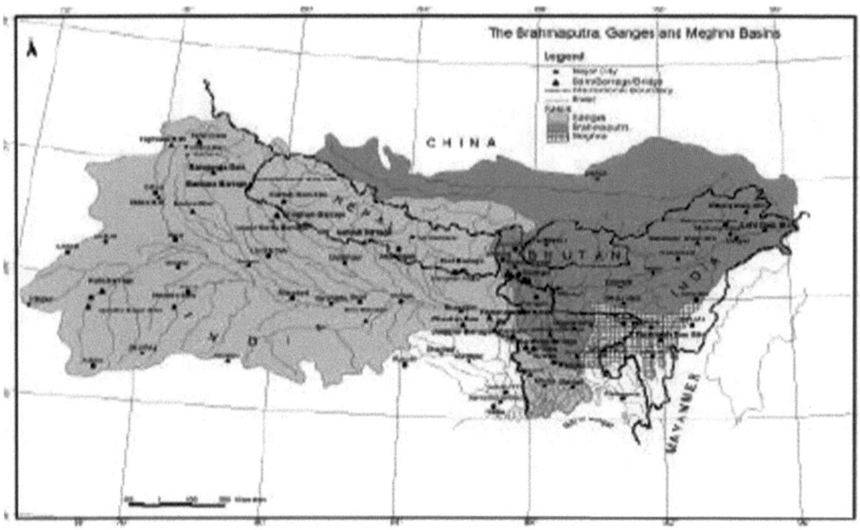

Fig. 7.3 Geographical location of Bangladesh (*Source*: Ministry of Environment and Forest, Bangladesh)

billion metric tons into the country every year that helps to build the blocks of the landmass of the delta (Yusuf 1991).

Bangladesh is the largest delta in the world and it is mostly used as "Ganges-Brahmaputra-Meghna Delta" (Fig. 7.3). According to Smith and Abdel-Kader (1988), a delta can grow only the seaward and upward against a rising sea level when the river-borne sediment influx is adequate. Therefore, this delta is important for the existence of Bangladesh because of carrying water and sediments from Himalayas. But the sediment supply is reduced in the current decades due to construct a barrage named Farakka on the Ganges in 1974 which curtailed delta growth and led to increase coastal erosion. Moreover, the reduced summer flows allow sediments to be deposited on the riverbeds downstream of the dam resulting in decreased water carrying capacity during the rainy seasons (Alexander 1989). This also leads not only to flow less water in the river during the dry season but also provide insufficient water supply that cause drought severity in Bangladesh especially in the north-west region.

7.5.2 *Rainfall Variability*

Bangladesh has a subtropical monsoon climate characterized by wide seasonal variations in rainfall, moderately warm temperatures, and high humidity. The mean annual rainfall of Bangladesh is about 2,300 mm. But, the average annual rainfall in north-west region is 1,329 mm, whereas the north-east part of the country is 4,338 mm (Shahid et al. 2005). Moreover, in terms of space and time, there has been observed wide fluctuations, with about 80 % of the annual rainfall occurs during

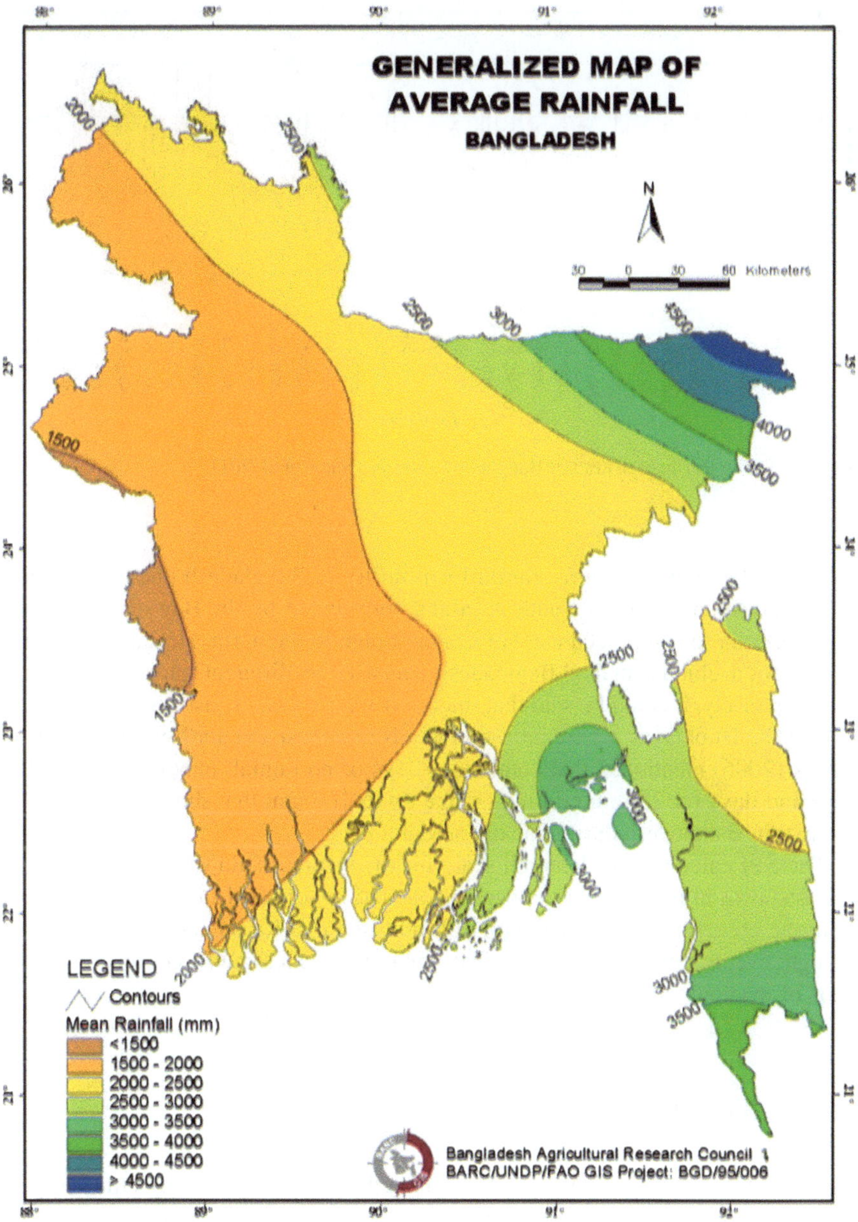

Fig. 7.4 Spatial distribution of rainfall in Bangladesh (*Source*: Bangladesh Agricultural Research Council 2010)

July to October and almost no rainfall during winter season (October to March) at national level (Hossain et al. 1987).

Figure 7.4 shows the spatial distribution of the average rainfall of Bangladesh, where it shows the rainfall variation in drought prone areas compared to the whole

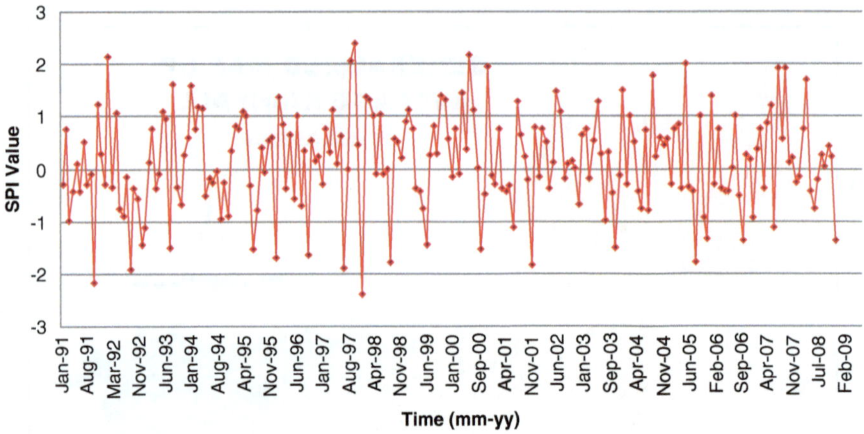

Fig. 7.5 SPI in Rajshahi district of Bangladesh (*Source*: Shaw et al. 2011)

country. The main reason for rainfall variability in Bangladesh due to it situates in the monsoon region of South Asia and surrounded by the Bay of Bengal and Indian Ocean to south, iced cap Himalayan range to the north. Hence, any change of monsoon climate makes Bangladesh sensitive to different natural calamities. More precisely, it could be said that the north-west region is just south of the foothills of the Himalayas, where monsoon winds turn west and north-west. Moreover, NAPA (2006) mentioned that continuous days of no rainfall may range up to 113 days and days for 25 mm rain may range up to 166 mm that show the severity of drought in Bangladesh during the dry season.

The SPI calculated in 1 month scale for the station of Rajshahi in Bangladesh that is shown in Fig. 7.5. The upper, middle and lower panels are for SPI obtained from the observations. The 1 month SPI indicated that Rajshahi experienced of 23 drought events from 1991 to 2009. This area receives less rainwater than the other parts of the country.

7.5.3 *Temperature*

Bangladesh has warm temperatures throughout the year, with relatively little variation from month to month. The average temperature ranges from 9 °C to 26.1 °C in this country. But in drought prone areas, the average temperature ranges from 25 °C to 35 °C in the hottest season and 9 °C to 15 °C in the coolest season. Even in the summer season, the temperature goes up to 45 °C or even more in this region. But in winter, temperature falls at 5 °C in some places. Based on GCM using MAGIC/SCENGEN, Agarwala et al. (2003) were carried out an analysis on changes in

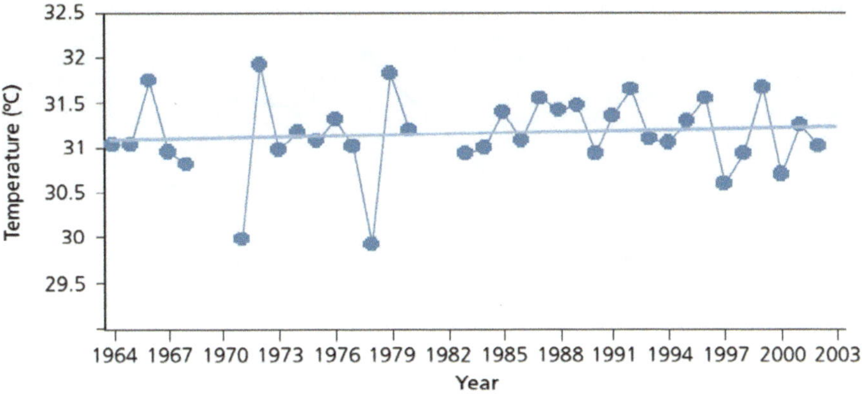

Fig. 7.6 Annual average maximum temperature variations (1964–2003) in north-west (Barind tract) Bangladesh (*Source*: Habiba 2012)

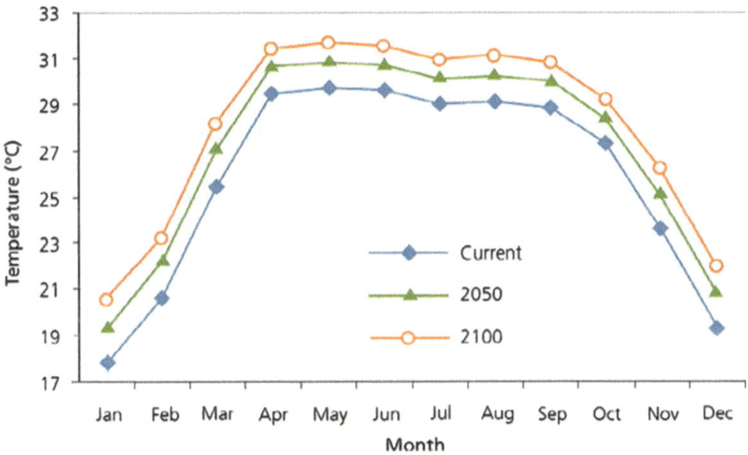

Fig. 7.7 Monthly mean temperature for current period (1964–2003) and projected for 2050 and 2100 in drought-prone areas of Bangladesh (*Source*: Habiba 2012)

average temperatures and precipitation over Bangladesh. According to GCM analysis, the average temperature of Bangladesh will increase by 1.4 °C (±0.16) by 2050. Figure 7.6 represents the year-to-year variation of average temperature from 1964 to 2003 in north-west Bangladesh. It shows a slight increase in temperature.

Moreover, the GCM data projects warmer for winter than summer months that are shown in Fig. 7.7. Based on the above projections, Bangladesh is likely to face more hot days and heat waves, longer dry spells and higher drought risk, especially in drought-prone areas are warmer and drier than the condition of 50 years ago.

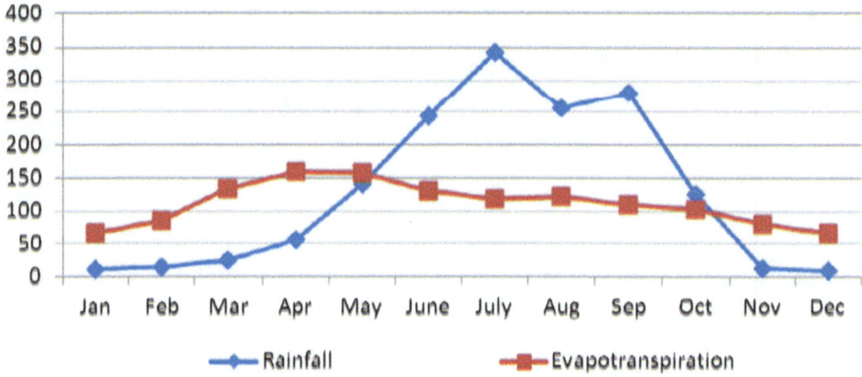

Fig. 7.8 Rainfall and evapotranspiration in northwestern area (*Source*: Habiba 2012)

7.5.4 *Evapotranspiration*

The total annual evapotranspiration is lower or equal to annual rainfall in some places, especially in drought-prone areas. In this area, crop water demand is not fulfilled because of rainfall variability along with high temperature. To meet up the water demand of the crops, the water deficit exists in the dry months of the year, as the demand is higher than the total rainfall. During monsoon season, the available rainfall may fulfill the crop water demand but during the dry season deficit is too large to meet the crop demand. That's why, agricultural drought occurs in this region. Figure 7.8 highlights that the evapotranspiration exceeded more than 0.5 times more during the dry season (of 7 months) than wet season in Rajshahi district of northwestern Bangladesh.

7.5.5 *Water Resources*

The Ganges, the Brahmaputra and the Meghna (GBM) basin discharges about 0.142 million m^3/s into the Bay of Bengal at peak period (Huq et al. 1999; Rahman et al. 2000). However in an average, Ahmad (2000) mentioned that availability of surface water flows is 1,350 billion cubic meters (BCM), of which only 1,160 BCM is available for in Bangladesh or flows to the Bay of Bengal. It is also noted that, availability of about 80 % of surface water is concentrated during the monsoon month (June to October). On the other hand, during the dry season, especially in the month of February, only 1 % of available water comes from standing water bodies (ponds, baors, haors, etc.) and seasonal wetlands (floodplains). The volume of surface water from such water bodies is about 3–3.5 BCM (Rahman et al. 2000). Other than in low-flow season, available water resources are found to be low to meet up the current demand. But due to the increase of population over the next 25 years, the per capita

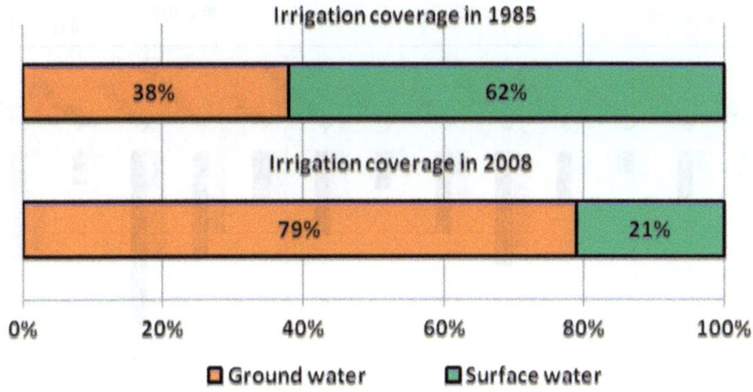

Fig. 7.9 Contribution of ground water compared to surface water in 1985 and 2008 (*Source*: Habiba 2012)

water availability in Bangladesh will progressively be reduced (Ahmad et al. 2001). However, such reduction of water availability will not only affect the huge population of this country but also causes water scarcity that will lead to drought severity during the dry season.

7.5.6 Misuse of Groundwater Resources

To combat with drought, it is essential to utilize its water resources properly. In Bangladesh, it has been seen that both surface and groundwater is used for irrigation and domestic purposes. Excessive groundwater exploitation is another issue that hastens drought severity in Bangladesh. According to governmental statistics, about 3 million ha (out of 8.02 million) of net cropped area is taken under irrigation, mostly in the dry season for cultivating irrigation dependent boro rice crop. The irrigation is accelerated through major canals (large scale irrigation), deep tube well, shallow tube well, lower lift pump etc. But, the contribution of groundwater is significantly increased than last two decades compared to surface water. Figure 7.9 reveals that in 1985, the contribution of groundwater was 21 % whereas in 2008, it reaches at 79 % which is more than triple than the previous year. Moreover, the rate of irrigated area in north-west region is higher than the rest of the country.

The groundwater is mainly recharged through rainfall during monsoon season. In a normal year, sufficient water replenishment takes place through rainwater during the monsoon season. But during the monsoon season low rainfall cannot recharge the groundwater level. Moreover, irrigation through deeper pump operated tube wells may help to supply water for a while in the dry season, but it causes the groundwater levels depletion deeper and deeper and irrigated land will turn into harder and harder. Excessive use of groundwater for irrigation as well as domestic uses caused depletion of groundwater level during the dry season in north-west Bangladesh (Habiba et al. 2011a, b). This causes a great threat to the irrigated

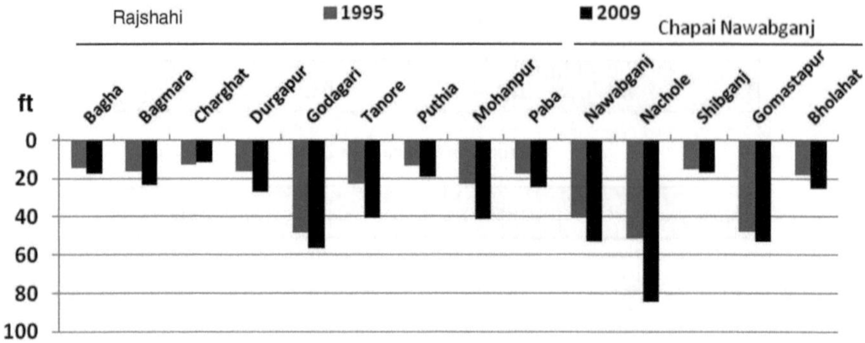

Fig. 7.10 Groundwater level in the study area in 1995 and in 2009 (*Source*: Barind Multipurpose Development Authority 2010)

agricultural system because of overdrawn aquifers, lowering of water tables and reduced stream flow. The Ministry of Environment and Forest (MoEF) of Bangladesh (2002) states that the ground aquifer level has gone lower by 8.95 to 18.56 m in some regions of north-west Bangladesh. This statement is in line with Fig. 7.10. From this figure, it is seen that almost all Upazilas of Rajshahi and Chapai Nawabganj districts experience depleted groundwater level when comparing the 1995 and 2009 Levels.

7.6 Drought Impacts in Bangladesh

7.6.1 Economic Impact

The extreme water events-drought causes suffering for the life and livelihoods of people and affects their socio-economic conditions. Its impact has been observed in different sectors. But especially agriculture, livestock and fisheries sectors are suffered most due to scarcity of water during the dry seasons.

7.6.1.1 Impact on Agriculture

In Bangladesh, drought affects rice crops in three different cropping seasons namely pre-kharif, kharif and rabi seasons. It is reported that every year 0.45 million ha of land is affected by very severe drought during the rabi season while 0.40 million ha and 0.34 million ha are affected during the pre-kharif and kharif seasons, respectively. It is noticed that T-aman alone hampered by various degrees of droughts during the season and affects 7.15 million hectares of land which is 91 % of the total T-aman land. For instance, in 1996–1997, with a production target of 8.7 million metric ton of aman rice, a total land of 4.9 million ha was cultivated. But due to drought condition, it caused damaged about 2.6 million ha of paddy land. Of which 2.2 million ha were

Table 7.4 Damage due to drought condition

Damaged areas/districts	Affected land (million ha)	Affected farmers/population (million)	Crop/resources damaged (US$ rate '000)
Rangpur, Bogra, Dinajpur, Thakurgaon, Panchhagarh, Kurigram, Nilphamari, gaibandha, Lalmonirhat, Serajgonj, Naogaon, Natore, Rajshahi etc. (total 16 district)	2.20	6.49	67.65
Barisal, Bhola, Barguna, Jhalakathi, Patuakhali	0.06	1.75	18.38
Noakhali, Lakshmipur, Feni	0.17	5.07	53.38
Sunamganj	0.04	1.17	12.35
Chakaria and Lama	0.02	0.78	5
Barind area	0.04	1.12	12.35

Source: BDPF (1998)

Table 7.5 Contribution of agriculture to GDP (%) at constant prices (Base: 1995–1996 = 100)

	Year								
Sector/sub-sector	1999–2000	2000–2001	2001–2002	2002–2003	2003–2004	2004–2005	2005–2006	2006–2007	2007–2008
Agriculture	25.58	25.03	23.99	23.47	23.08	22.28	21.85	21.37	20.88
A. Crops	14.59	14.7	13.75	13.43	13.23	12.51	12.28	12	11.7
B. Livestock	3.02	2.95	2.96	2.93	2.91	2.95	2.92	2.88	2.79
C. Forest & related services	1.88	1.87	1.88	1.86	1.83	1.82	1.79	1.76	1.75
D. Fishing	6.09	5.51	5.4	5.25	5.11	5	4.86	4.73	4.64

Source: Statistical yearbook of Bangladesh (2007)

in the 16 districts of north-west Bangladesh. Bangladesh Disaster Preparedness Forum (BDPF) made a compilation of damages to land, people and crops for 1996–1997 drought conditions in the country that is given in Table 7.4.

Nasreen and Hossain (2002) highlighted that drought affects the standing crops, water supplies and plant growth that leads to loss of productions, food shortages and famine. Drought not only affects rice crop but also has impacts on other crops like jute, wheat, corn, potatoes, sugarcane, different types of pulses and oilseeds and vegetables (Ahmed 2006). Recent 2006 data indicated that drought caused 25–30 % crop reduction in northwestern part of Bangladesh (Rahman et al. 2008).

7.6.1.2 Impact on GDP

Agriculture is the key economic driver in Bangladesh and especially the people in the rural community reliant as a critical source of livelihoods and employment. But contribution of GDP on agriculture is nearly 20.88 % in 2007–2008 which is continuously declined compared to its contribution in 2000 (Table 7.5). The sub sector's contribution is also declining such as crops, livestock and fisheries contribute 11.7 %, 2.79 % and 4.64 %, accordingly.

7.6.1.3 Impact on Food Security

According to Bangladesh Bureau of Statistics (2005), total food grains production was 19.32 million metric tons in 1991/1992 which has gradually increased to 29.77 million tons in 2007/2008, 6.13 % higher than previous year's production. Furthermore, National Food Balance sheet prepared by the Government of Food Planning and Monitoring Unit (FPMU) reported that total requirement of food grains (considering per capita per day requirement of 453.6 g) has increased from 18.79 million tons in 1991/1992 to 24.17 million tons in 2007/2008. In terms of production, boro exhibits higher than T-aman and aus in Bangladesh. For instance, the yield of boro was estimated as 17.76 million tons in 2007/2008 whereas aman and aus production were 9.66 and 1.51 million tones, respectively. Though boro production is higher than others but this crop depends on irrigation for consuming water at its different growth levels. So, if droughts sustain for long time, the production of boro declines and thereby increases household food insecurity. At the same time, food consumption as well household's food purchasing capacity is reduced due to short supply of food during drought period.

7.6.1.4 Impact on the Price of Rice

The crop loss impact was felt in the price of rice in the market. Generally every year, in the months of December, January and February, the price of rice normally remains low. However, due to the deficit production of T-aman rice, the price of rice showed an increasing trend from January 1998. As a result, the government and other policy makers became very cautious. The government quickly took the initiative to import food in an attempt to cover the deficit. Even then, in the following months food prices were higher than the other years (Rahman 1998).

7.6.1.5 Impact on Livestock

The livestock of Bangladesh includes 22.3 million cattle/buffaloes, 14.6 million sheep/goats, and 126.7 million poultry. Lack of grazing facilities and shortage of food constrains the mass rearing of cattle and goats during drought period. Livestock are affected by air temperature, humidity, wind speed and thermal radiation which influence their growth, milk production, reproduction, health and well-being. Even sometimes, cattle and poultry are suffered by heat stroke and caused diseases like black quarter, anthrax etc.

7.6.1.6 Impact on Fisheries

Most of the rivers and ponds dried up during the drought time causing unfavorable environment for the fishes. Fresh water fish hatching cannot survive under high salinity levels and fish growth also disrupted as the solubility of oxygen in water decreases with high temperature. Drought also hindered fishing activities. Haque (2007) mentioned that seasonal variations have diverse implications on fishing, hatchery operations, table fish production and livelihoods of a wide range of people directly and indirectly involved with fisheries and aquaculture.

7.6.2 Social Impact

7.6.2.1 Impact on Health

Drought cause major health problem for the household living in drought-prone area. In absence of drinking water, people bound to use pond water which accelerates various types of water-borne diseases. Drought along with heat stress causes different diseases like dysentery, diarrhoea, dengue and hypertension. Even, under high temperature and humidity, there will be problems of dehydration, especially affecting the elderly and children. Drought also leads to an increase in chronic energy deficiency among members of the agricultural labor force due to less food consumption.

7.6.2.2 Impact on Social Life

Drought not only has effect on agriculture, fisheries, livestock, but also has tremendous impact on social life's. People are facing safe drinking water problems during drought period. Even, sometimes confliction among the neighborhoods arises due to the scarcity of water during that period. Children are unwilling to go to school, so that during this period drop out is increased rather than other periods. Besides, agricultural labor has no work to do agricultural activities in their land. Therefore, they act as daily labor in that time. In extreme condition, farmers change their occupation and go to nearby village or cities to look for some jobs to meet household's basic demands.

7.6.3 Environmental Impact

7.6.3.1 Impact on Ground Aquifer

Surface and groundwater is the main source of fresh water in Bangladesh. But in drought prone areas, over exploitation of groundwater for irrigation purposes cause decline of the groundwater. For instance, the level of groundwater table of Nachole in dry season was 44 ft in 1995, where as in 2009 it reaches at 82 ft (Fig. 7.11). Ministry of Environment and Forest (MoEF) is agreed with this situation and they

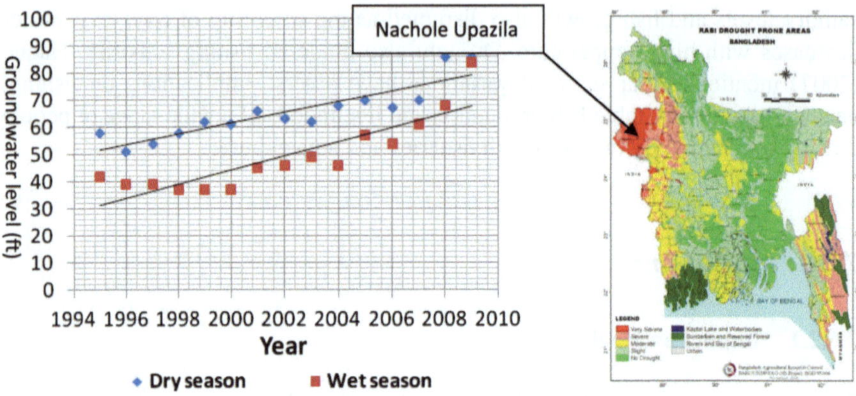

Fig. 7.11 Groundwater depletion in Nachole upazila in Chapai Nawabganj district (*Source*: Habiba et al. 2010)

reported in 2002 that the ground aquifer level goes below 8.95 to 18.56 m in some regions of north-west Bangladesh. Furthermore, it has been predicted that by 2018, the demand for irrigation may reach 58.6 % of the total supply (Selvaraju and Stephan 2007). Therefore, this tendency not only makes people bound to depend on groundwater extraction but also causes draw- down of groundwater level and lead to environmental problems such as heavy metal contamination and salinity.

7.6.3.2 Impact on Land Degradation

Drought acts as a catalyst of land degradation through reducing soil moisture and water retention capacity. As a result, land degradation is enhanced during drought periods consequent upon the drying out of topsoil and effective loss of soil structure and aggregation. In this condition, topsoil is easily blown away as a result of the erodibility of the soil and the erosive nature of wind and rainfall. Moreover, drought declines soil organic contents, reduce microbial activity and overexploitation of sparse vegetation.

7.6.3.3 Impact on Biodiversity

Droughts also cause damaged to plant and animal species, increase incidence of the diseases by bringing pathogen and parasites. It accelerates to dry out the natural water bodies and thus causes loss of wild and cultural stocks. It reduces water levels in reservoirs, lakes and ponds, increased salt concentration, water temperature, air and water quality, degrades the quality of landscape. Moreover, it causes loss of biodiversity by hampering the microbial activities, extinction of some fish species and diminished the reproductive capacity of cattle and poultry.

7.7 Drought Risk Management Practices in Bangladesh

7.7.1 At Institutional Level

At national level, the Ministry of Food and Disaster Management (MoFDM) has the mandate to co-ordinate all disaster management activities within the country. This ministry is responsible for formulation and implementation of national DRM (Disaster Risk Management) policy along with co-ordination of all disaster management activities within the country. Besides, Disaster Management Bureau (DMB) and Directorate of Relief and Rehabilitation (DRR) are also belonging to this ministry. DMB coordinates drought relief work with local governments. Moreover, DMB receives early warning messages from BMD (Bangladesh Meteorological Department) by fax and email, and disseminates information also by fax and email through District Relief and Rehabilitation Officers (DRROs) under Deputy Commissioners (DC) in 64 districts of Bangladesh. The information also starts following down through other various levels up to field levels (that is District Disaster Management Committee, Upazila Disaster Management Committee and Union Disaster Management Committee) under the framework of disaster management guidebook that is, Standing Orders on disaster.

Furthermore, in order to increase resilience at all levels, from the national to community level and to reduce damage and losses from natural disasters and the impacts of climate change, the Government of Bangladesh approved Comprehensive Disaster Management Programme (CDMP) to assist MoFDM in achieving government vision, mission and objectives. The program encompasses all aspects of risk management with an objective to strengthen the capacity of Bangladesh disaster management system. Adding together, Climate Change Cell under Ministry of Environment and Forest (MoEF) was designed to address climate change adaptation issues in disaster management.

7.7.1.1 Early Warning System

Bangladesh yet could not develop powerful institutions like those in Western Europe and North America to forecast a complex climate related disaster like El-Nino, but the institutions established at the national level such as the Space Research and Remote Sensing Organization (SPARRSO), the Bangladesh Meteorological Department (BMD) and The Bangladesh Water Development Board (BWDB) by the government of Bangladesh are responsible for monitoring and forecasting disasters. At national level, BMD generates weather and climate related information relevant to drought risk management. It also creates public awareness on the basis of severity and risks associated with projection that will better prepare at-risk communities and public officials to mitigate the consequences. Mass media plays an important role in disseminating weather and climatic information to the farmers, farmers group and community associations at local level.

Fig. 7.12 Vegetative Growth of Mung Bean for Green Manuring (*Source*: LACC Project, DAE 2010)

7.7.1.2 Other Activities Taken by Government and NGOs

In terms of drought risk reduction, government also undertakes relief measures by providing drinking water, food grains and food subsidies to special groups and through food-for-work programs. The rural works program of the GoB (Government of Bangladesh) offers employment to the population affected by drought and helps mitigate the drought severity. To meet the increased demand for food; the government, research institute and NGOs have initiated modern agricultural technology and practices such as High Yielding Variety (HYV) seeds, increased irrigation coverage, and introducing drought tolerant varieties. Among them, DAE (Department of Agricultural Extension) provides some training programs to the farmers on how to utilize supplementary irrigation in T-aman crops, alternate wetting and drying (AWD) method for rice crops; quality seeds production, collection and distribution at farmer's level.

Besides, LACC (Livelihood Adaptation to Climate Change) project completed by Food and Agriculture of United Nations and the DAE under CDMP of Ministry of Food and Disaster Management where they executed different viable adaptation options for drought areas and found various adaptation practices beneficial for the livelihood such as mini ponds for rain water harvesting, homestead gardening, mango and jujube cultivation etc. (Fig. 7.12).

Bangladesh Agricultural Research Institution (BARI), Bangladesh Rice Research Institution (BRRI) has developed some drought tolerant rice varieties such as BRRI DHAN-56, BRRI DHAN-57. But these are still waiting to release as a drought resistant variety. Mainly the farmers of drought-prone area practice short duration crop varieties like pulse crop, oil seed crop that distributed by research institutes. Even in 2009, government offered 100 h free irrigation for millions of farmers by pumping up underground water after a severe lack of monsoon rains.

Despite governmental activities, various NGOs like BRAC, PROSHIKA, CARITAS, TMSS, ASA and TRINOMUL have extended their activities in drought

Fig. 7.13 Irrigation water supplied by deep tube well through BMDA

affected areas. Although different NGOs are extended their work in drought-prone areas but this is limited and only deal with micro-credit program. In addition, very little amount of drought adaptation practices are done through CARITAS like rain water harvest, demonstration of drought tolerant crops, re-excavation of pond to facilitate domestic use of water, irrigation in kitchen garden, use drip irrigation etc. at community and house hold level.

7.7.1.3 At Infrastructure (Physical) Level

In the event of drought, the government of Bangladesh has undertaken irrigation project through installing deep tube well to increase agricultural productivity particularly in the northern region. Irrigation is necessary for many types of crops, especially for boro rice cultivation. Records show that the irrigation coverage was less than 2 million ha in the early 1980s and took almost two decades to reach about 5 million ha. It also reveals that the growth rate was a bit higher in the last decade (starting from 1994) than the earlier decade (starting from 1982).

Several governmental agencies like BMDA (Barind Multipurpose Development Authority), BADC (Bangladesh Agricultural Development Corporation) and RDA (Rural Development Academy) has undertaken irrigation project in drought affected areas. Among them, BMDA plays the key role and has successfully expanded their executing area through construction of cross dams and water control structures; re-excavation of canals and ponds; installation of DTWs; afforestation; improve surface water augmentation; construction of irrigation canals and roads; electric connection for the DTWs; drinking water supply through the over head tank and production of fine and aromatic rice etc. (Fig. 7.13).

Fig. 7.14 Intercropping of rice with mango

7.7.2 At Community and Household (Social) Level

Community people are the first victims who suffer most with the consequential impact of drought. They undertake diverse coping methods by which they can deal with this adverse situation arise through drought. The people of the drought affected region have changed their cropping pattern and practiced diversified crops to endure with this situation. For instance, potato is a winter crop which takes the lead role as a replacement for rice cultivation in some Upazilas (sub-district) of Rajshahi district. Furthermore, vegetables, tomato, sugarcane, pulse crops, spicy crops are successively grown in this drought-prone areas. Even, in some areas, agricultural fields are now transferred into different fruit tree cultivation fields like mango, jujube, guava etc. (Habiba et al. 2011a, b) (Fig. 7.14).

To irrigate the field, farmers supply water through nearby sources, such as ponds, khari/canals, bils, haors, rivers and install hand pump tube wells as well as shallow tube well (Fig. 7.15). Moreover, the villagers who are financially able to spend money for irrigating water in their field; they personally sink deep/shallow tube well in their cultivated field. Otherwise, they provided money to get irrigation facility from the institution named BMDA. But there is no certainty that these well will have water because decreasing trend in groundwater table. Therefore, farmers of this region feel that lack of water is the principal reason for non-adoption of agricultural adjustment. On the other hand, majority of rural women not only engage in agricultural farming but also do post harvest activities. Few of them are keeping busy themselves into sewing stuff locally called *nakshi katha* during their leisure time which helps them to earn additional money by selling these goods during the

Fig. 7.15 Withdraw of groundwater through shallow tube well

Fig. 7.16 Rearing of goose (*left*) and making cow dung sticks (locally called *ghutey*) (*right*) (*Source*: Habiba et al. 2011a, b)

drought time. Adding together, household involved themselves into other income generating activities like business, wage labor, services and construction works, and other non-agricultural farming like rearing livestock, poultry, dairy farm, cattle fattening, fish culture, making cow dung as fuel etc. (Fig. 7.16). Among different coping methods related to livelihood activities, people primarily try to use whatever savings they have. They use reserved food grain and cut down on their daily consumption during drought period. Those who have no savings, or no livestock or poultry, they often borrow money from relatives, friends, neighbors, and even in extreme cases, they mortgage their own land or loan money from NGO through fulfilling some terms and conditions such as membership, high interest rate, mode of installment, etc. Nevertheless, the migration is limited in this region whereas seasonal migration takes place for want of employment and to meet household expenditure.

7.8 Innovative Approach for Drought Risk Reduction

Drought is a development issue in many countries, and especially in the Asian monsoon region, it is a water management issue (Nguyen et al. 2009). When drought occurs in rural areas, it not only has impact on agriculture sectors, but also has its social impacts like health issues, student drop outs, social conflicts on water usage, and livelihoods. Therefore, drought mitigation needs physical measures like irrigation, infrastructure and proper land use mapping. Besides, the water management practices have strong relations to the institutional dimensions like drought policy, management skills of local governments and coordination among different stakeholders including government and non-governments (Nguyen and Shaw 2009). Likewise, the timely early warning issue is very much related to physical as well as institutional issues (Prabhakar and Shaw 2007).

Considering all the issues related to drought, Habiba et al. (2011a, b) developed a holistic approach named SIP approach that helps to measure the existing level of different socio-economic, institutional and physical conditions of the drought-affected area. This approach has three specific dimensions (socio-economic, institutional and physical) and each dimension is then divided into different sets of primary and secondary indicators. The details are given in Table 7.6. All together, there are 11

Table 7.6 Dimensions and indicators used in SIP approach

Dimensions	Primary indicators	Secondary indicators
Socio-economic	Education and awareness	• Literacy rate •Knowledge about drought •Having predictability •Taking preventive measure •Awareness about drought
	Health	• Access to safe water •Time period having access of safe water •Extent of diseases •Primary health care facility •Recovery
	Usage	• Food consumption •Reserved food grain •Migration •Changing occupation •Non agricultural farming
	Social capital	• Social cohesion •Participation •Water related conflict •Build consensus •Acceptance of leader
	Economic aspect	• Income source •Other income generating activities •Use of savings •Sell off assets, land or live stocks • Credit, subsidy
Institutional	Policy	• Incorporation into plan •Effectiveness of the plan •Support by GO and NGO •Water management activities •Public awareness program
	Management	• Collaboration •Co-ordination •School/college •Highlighting drama •Community leader/imam
	Co-ordination	• Training •Demonstration •Credit/loan •Aids •Subsidy
Physical	Infrastructure development	• Electricity supply •Fuel supply •Dam •Water reservoir •Drought warning system
	Irrigation	• Irrigation system •Irrigation facilities (DTW, STW, over head tank etc.) •Supplemental irrigation •Dependency on rainfall •Rain water harvest
	Land use	• Built up area •Vegetative area •Water bodies •Drought tolerant crop •Fruit tree plantation

Source: Habiba et al. (2011a, b)

primary indicators (5 for socio-economic, 3 for institutional, and 3 for physical), and 55 secondary indicators (each of the primary indicators are divided into 5 secondary indicators). For example, in institutional dimension, policy is used as one of the primary indicator, in which incorporation into plan, effectiveness of the plan, support by GO and NGO, water management activities, and public awareness program are used as secondary indicators in SIP approach. Data on each of these primary and secondary indicators are collected through a comprehensive set of questionnaires. Each question should be ranked between one (poor, not sufficient/existent) and five (good) in a five-point rating scale.

7.8.1 Application of SIP Approach: Insight from a Case Study

The north-west part is designated as drought-prone areas of Bangladesh (Banglapedia 2006). Drought takes place more frequently in this area than the other part of the country because of high rainfall variability. The average annual rainfall of this area is 1,329 mm whereas the country's average annual rainfall is 2,300 mm. This rainfall shortage accompanied with high temperature hastens drought severity of north-west region. Furthermore, Shahid et al. (2005) mentions that the dryness index of this part of Bangladesh is close to that of a dry region.

For this reason, the SIP approach was carried out at 14 Upazilas of two severe drought-prone districts in northwest part of Bangladesh (9 Upazilas from Rajshahi district and 5 Upazilas from Chapai Nawabganj district). It was done with the help of the Deputy Commissioner (bureaucratic head of the district) of Rajshahi and Chapai Nawabganj district, and the questionnaire was filled by the Upazila Nirbahi Officer (who is responsible for supervising all Upazila level administrative/development approach works and preparing as well as coordinating Upazila development plans of the respective Upazila) with the consent of other officials such as Upazila Agriculture Officer, Upazila Live Stock Officer, Upazila Fisheries Officer and so on.

The scores of indicator and dimension for SIP approach has obtained using simple arithmetic functions such as weighted mean index (WMI) and aggregate weighted mean index (AWMI), respectively. Depending on the score, the drought resilience level are divided into five categories from very high (4.0>3.4), high (3.4>2.8), medium (2.8>2.2), low (2.2>1.6) to very low (1.6>1.0). Figure 7.17 shows drought resilience from overall (a), socio-economic (b), institutional (C) and physical (D) perspectives of the targeted sub-districts (upazilas). In terms of overall resilience, eight upazilas are categorized as high, five upazilas as medium and one upazila as very high resilience that are highlightied in Fig. 7.17a. The highest overall drought resilience was found in Mohanpur upazila because of its high socio-economic, institutional and physical resilience values. On the contrary, the lowest overall drought resilience was found in Tanore upazila because of its low socio-economic, institutional and physical resilience values. Moreover, comparing the socio-economic, institutional and physical resilience levels of SIP approach, most of the upazilas have high socio-economic resilience than institutional and physical resilience.

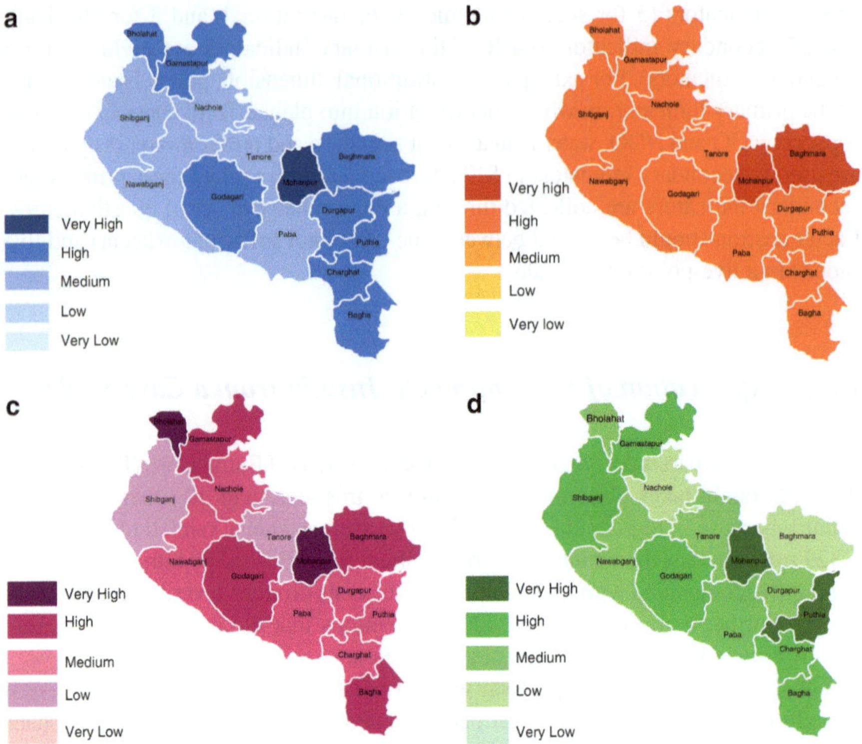

Fig. 7.17 Drought resilience map of overall (**a**), socio-economic (**b**), institutional (**c**) and physical (**d**) perspectives in Rajshahi and Chapai Nawabganj districts (*Source*: Habiba et al. 2011a, b)

7.8.2 Importance of SIP Approach Implication at Micro Level

Successfulness of macro level planning depends on a proper combination of grass root level planning. This is because any development program to be implemented over a state or region needs micro level information and levels of planning such as identifying the right areas, allocation of funds, monitoring the activity, and assessing the results among them. Macro level analysis cannot express the overall demands, needs, necessities and limitations of an area. It has limitations in finding the needs and gaps in a particular situation and it cannot highlight the root causes. In that case, micro level analysis is important that can not only provide the existing conditions of a locality but also add the requirement of grass root level. This understanding can be achieved through research at micro-level because overall analysis cannot clarify the root causes in a wider scale. Therefore, it needs a narrow-scale assessment at the grassroots level.

In this context, SIP is a suitable pragmatic approach that provides overall socio-economic, institutional and physical situation of a drought affected area which then facilitates into action planning. It is expected that this approach will help to have a better understanding of a locality and also help to know about which locality is more resilient and how people build resilience to drought. Furthermore, the approach can provide the development method/policy in an affected area by which households can cope with drought. It can assist the government to make a quick decision about any locality which is severely drought prone and also help to allocate budget during a crisis period. It is hoped that this initiative can make a contribution for addressing grassroots level's problem and potential assessment and thus help to protect livelihoods and assets during drought period. Furthermore, it helps upazilas and households to become more resilient towards drought.

7.8.3 *Enhancing Drought Risk Management Policy and Action Through a Framework*

After discussing the approach, this section attempts to talk about a framework for drought risk management policy and action that mainly derived from SIP approach. Among 55 secondary indicators of SIP approach, only 15 secondary indicators have the capacity where farmers' can point out their voice regarding drought risk reduction. Because, enhancing farmer's resilience towards drought, it is imperative to identify the suitable drought adaptive practices (DAP) that help farmers to minimize the drought risk. Therefore 15 secondary indicators were used as 15 drought adaptive options of farmers where they can give their opinion based on their needs and local circumstance as well.

Followed by the resilience assessments, thus, another intensive field research was carried out at the same district considering irrigated and non-irrigated village of each upazila by using a semi-structured questionnaire. This questionnaire mainly focuses on various adaptation practices of farmers toward drought and serves as viable action- and implementation-oriented tool for making an upazila resilient to drought disaster. The results came from this study highlight a total of 60 drought adaptive practices (30 drought adaptive practices for individual and family level, and 30 drought adaptive practices for community level) from the study area. Among 30 drought adaptive practices, 15 originated from irrigated village and 15 from non-irrigated village of that area at each level.

Farmers adaptive capacity does not automatically translate into successful adaptation to climate change (O'Brien et al. 2006). It requires greater institutional capacity at all levels of government and more efficient coordination between different levels of government. Regarding this, the study further considers community perspective. Because, community is believed as the centre of disaster risk reduction, and community actions make the disaster reduction most effectively. Therefore, community participation and control is essential for any successful implementation, orientation and maintenance of any disaster risk reduction project (Habiba and Shaw 2012). Table 7.7 represents farmer's community level prioritized drought adaptive practices (DAP) in irrigated and non- irrigated areas, respectively. From

Table 7.7 Incorporation of SIP approach and farmers' recommended top-most priority Drought Adaptive Practices (DAP)

SIP approach			Top-most prioritized drought adaptive practices at community level of farmer	
Dimension	Primary indicator	Secondary indicator/Name of the drought adaptation option	Irrigated area	Non-irrigated area
Socio-economic	Education and awareness	Having predictability	Received information from TV, Radio	Received information from TV, Radio
		Taking preventive measure	Dairy farm	Community savings and credit system
	Health	Primary health care facility	Community health care service	Community health care service
	Economic	Other income generating activities	Establish dairy farm	Establish dairy farm
		Use of savings	Save money	Save money
		Selling of assets, land or livestock	Sell agricultural goods	Sell agricultural goods
Institutional	Management	Public awareness program	Public awareness program	Public awareness program
		Community leader/Imam	Extension worker	Extension worker
		Aids	Cash	Electricity
Physical	Infrastructure development	Electricity supply (alternative energy sources for irrigation)	Use of diesel for irrigation purposes	Use of diesel for irrigation purposes
		Water reservoir	Drag the river and use of river water	Drag the river and use of river water
		Drought warning system	Establish drought information center	Establish drought information center
	Irrigation	Supplemental irrigation	Use of plastic pipe for irrigation	Use of plastic pipe for irrigation
	Land use	Drought tolerant crop	Vegetable gardening	Vegetable gardening
		Fruit tree plantation	Establish mango orchard	Establish mango orchard

Legend:
Means different adaptive practices at community level in both irrigated and non-irrigated area

Table 7.7, it has been seen that among 15 drought adaptive practices, both irrigated and non-irrigated farmers were agreed with 13 adaptive practices. They differed only two drought adaptive practices: one is taking preventative measure before drought period and other one is aids from institution. In terms of taking preventative measure, irrigated farmer's prioritized dairy farm which would be a suitable measure for them to conquer with drought. It helps them to earn extra money by selling of its various products such as milk, meat and dairy products during the drought period. On the other hand, non-irrigated farmers recommended community savings and credit system which will help them to start business or other income generating activities by saving money in a communal process.

After assembling farmer's recommended and prioritized drought adaptive practices (DAP), this study further accumulated farmer's barriers which hinder them to adopt these DAPs toward drought. Therefore, total 81 barriers evolved from the farmer's level through focus group discussion. Four focus group discussions were undertaken in both irrigated and non-irrigated villages of the study area. Irrigated farmers mentioned total 40 barriers for 15 drought adaptive practices while non-irrigated farmers have 41 barriers to implement these drought adaptive practices into action. For example, in case of vegetable gardening, farmers in irrigated areas noted that lack of quality seed, unavailability/crisis of seed at the planting time, low organic matter content in soil, lack of irrigation facility, dynamism of retailer might occur. On the other hand, lack of quality seed, poor knowledge of dealer about how to use insecticide or pesticide, crisis of seed at the planting time, low organic matter content soil and lack of irrigation facility are mentioned by non-irrigated farmers. In most cases, the farmers faced the same barriers in both irrigated and non-irrigated areas which impede them to implement these drought adaptive practices (DAP).

According to OECD (Organization for Economic Co-operation and Development) in 2008, the adaptation efforts can be made more effective with enhanced coordination and institutional support. Moreover, it is increasingly recognized that effective climate change adaptation needs to be incorporated into other development initiatives such as livelihood enhancement, poverty alleviation, environmental management and sustainable development. For the successful implementation of drought adaptive practices, this study assembled these DAP into short-term, mid-term and long-term basis. It was done through workshop where the key aim was to harmonize the drought adaptive practices and to state the role of stakeholders along with specific timeframe which can yield in enhancing drought resilience of farmers in both irrigated and non-irrigated villages of each upazila in Rajshahi and Chapai Nawabganj districts. Therefore, it has been observed that at the community level, 18 adaptive practices are identified as short-term, 3 adaptive practices as mid-term and 9 adaptive practices as long-term among 30 adaptive practices. Furthermore, it assessed stakeholders' role where it has been found that national government and community as the primary stakeholder to implement these drought adaptive practices (DAP) for future drought risk reduction.

Based on the findings from the SIP approach, farmers prioritized Drought Adaptive Practices (DAP), focus group discussions and various workshops; Fig. 7.18 aims to deliver a framework which could serve for the development of

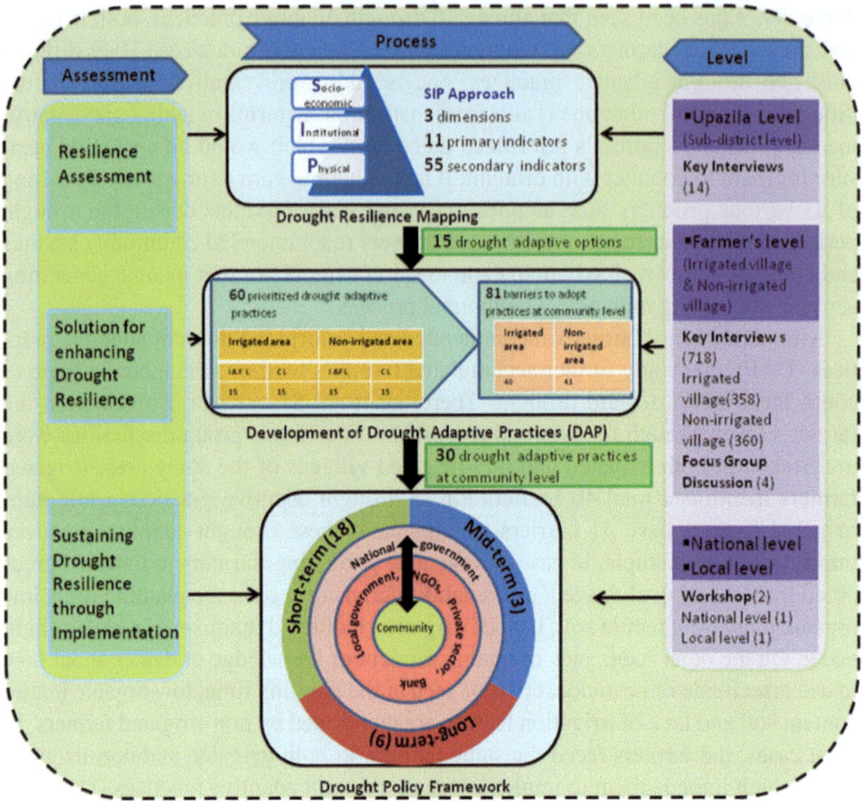

Fig. 7.18 Framework of enhancing drought risk management policy and actions through farmer's adaptive practices

drought risk management policy and actions as well as boost northwestern farmer's adaptive capacity towards drought in future. The framework developed in this study has a potential to enhance drought risk management policy and actions for northwestern Bangladesh, because it provides the improved understanding of the scopes and barriers of drought risk management. Thus, finally it helps to facilitate policy makers to gain the outcomes of the study and what is needed to formulate and enact appropriate policies and actions.

7.9 Conclusion

In context of global warming, most of the climatic models project a decrease in precipitation in the dry season and an increase during the monsoon season in south Asia (Christensen et al. 2007). This will cause a mean combination of more extreme floods and droughts in the region. According to the IPCC (2007), Bangladesh will experience increased droughts outside the monsoon season and face severe floods due to glacier melting and more intense monsoons. Moreover, too much water during the monsoon season and too little water during dry season make Bangladesh more vulnerable to flood and drought disaster, respectively. Although, Bangladesh is familiar with flood and cyclone affected country all over the world and made a good progress in case of flood and cyclone disasters. Contracting with drought, there is no drought policy compare to flood and cyclone in Bangladesh. Even, governmental and other organizational activity towards drought is very limited. It needs more governmental initiatives to deal with drought successfully.

Regarding drought risk, this chapter mentions a number of causes of drought; various economical, social and environmental impacts, and how different levels perform a wide range of management practices towards drought in context of Bangladesh. Furthermore, to minimize the drought risk, an innovative approach named SIP is applied at upazila (sub-district) level which not only measures the existing level of socio-economic, institutional and physical condition of a drought affected area but also helps to compare which upazila is more or less drought resilience than other upazilas. Besides, when SIP approach is placed on the spatial scale, it can be used as rapid planning tool, which can identify the overall needs of the upazila. Therefore, building resilience against drought needs to be started at local or micro level and become part of other long-term considerations and an integral part of policies related to agriculture, water, food security, and hazard planning. In this regard, the SIP approach will help to build resilience at different levels like local, regional, or national and can play a significant role in decision making for future drought risk reduction plan. On the other hand, to adapt with drought, this study further suggested a drought risk management policy and action framework, which provides how to gather and utilize farmer's drought adaptive practices, and how these adaptive practices would be translated into action through the involvement of stakeholder within a specified timeframe.

As population of Bangladesh is still growing by two million every year and may increase by another 24 million over the next 12 years. Therefore, Bangladesh will require about 27.26 million tons of rice for the year 2020 to feed the increased population. Since, drought destroys food chain, food stock and agro-based production system of the country, so, it is utmost important to take policy and actions for drought mitigation at local level to national level. Therefore, it is expected that this baseline study will not only beneficial to a number of stakeholders in the country, particularly in the drought management sector, but also will help agricultural organizations, development authorities, researchers and risk insurers to improve their understanding of drought in the northwestern region of Bangladesh. Likewise, this study assists policy makers to utilize the outcomes and what are gaps and challenges to formulate and enact appropriate policies and actions in terms of drought risk reduction in context of Bangladesh.

Acknowledgements The financial support provided by the GCOE-ARS program of Kyoto University is highly acknowledged. The first author specially thanks the Graduate School of Global Environmental Studies for supporting her research at Kyoto University, Japan.

References

ADB (Asian Development Bank) (2005) Country Strategy and Program 2006–2010. Bangladesh Resident Mission, Dhaka
Agarwala S, Ota T, Ahmed AU, Smith J, Van Aalst M (2003) Development and climate change in Bangladesh: focus on coastal flooding and the Sundarbans. OECOD, Paris
Ahmed AU (2006) Bangladesh climate change impacts and vulnerability: a synthesis. Climate Change Cell, Department of Environment (DOE), Department of Environment and Forests, Comprehensive Disaster Management Program (CDMP), Dhaka
Ahmad QK (ed) (2000) Bangladesh water vision 2025: towards a sustainable water world. Bangladesh Water Partnership (BWP), Dhaka, 73 p
Ahmad QK, Ahmed AU, Khan HR, Rasheed KBS (2001) GBM regional water vision: Bangladesh perspectives'. In: Ahmad QK, Biswas AK, Rangachari R, Sainju MM (eds) Ganges–Brahmaputra–Meghna region: a framework for sustainable development. University Press Limited, Dhaka, pp 31–80
Alexander D (1989) The case of Bangladesh. Amherst, MA: University of Massachusetts, Amherst, p 15
Bangladesh Agricultural Research Council (BARC) (2010) New Airport Road, Farmgate, Dhaka 1215, Bangladesh
Bangladesh Bureau of Statistics (BBS) (2005) Compendium of environment statistics of Bangladesh. Dhaka: Planning Division, Ministry of Planning, Government of Bangladesh
Banglapedia (2006) Drought in Bangladesh. Retrieved from http://banglapedia.search.com.bd/HT/D_0284.htm . Accessed on July 12, 2012
Barind Multipurpose Development Authority (2010) Ministry of Agriculture. Dhaka, Bangladesh
BDPF (1998) Disaster Report, 1997, Dhaka: Bangladesh Disaster Preparedness Forum
Brammer H (1987) Drought in Bangladesh: lessons for planners and administrators. Disasters 11(1)
Christensen JH, Hewitson B, Busuioc A, Chen A, Gao X, Held I, Jones R, Kolli RK, Kwon WT, Laprise R, Magaña Rueda V, Mearns L, Menéndez CG, Räisänen J, Rinke A, Sarr A, Whetton P (2007) Regional climate projections. In: Solomon SD, Qin M, Manning Z, Chen M, Marquis KB, Averyt MT, Miller HL (eds) Climate change 2007: the physical science basis. Contribution of working group I to the fourth assessment report of the intergovernmental panel on climate change. Cambridge University Press, Cambridge
Climate Change Cell (2009) Department of Environment, Ministry of Environment and Forest, People's Republic of Bangladesh, Bangladesh
Habiba H (2012) Enhancement of drought risk management policy and actions incorporating Farmer's adaptive practices in Northwestern Bangladesh. Ph. D. Thesis, Kyoto University, Kyoto
Habiba U, Shaw R (2012) In: Shaw R. (ed), Bangladesh experiences of community based disaster risk reduction: Community, Environment and Disaster Risk Management., Emerald Group Publishing Limited, 10:91–111
Habiba U, Shaw R, Takeuchi Y (2011a) Socio-economic impact of droughts in Bangladesh. In: Shaw R, Huy N (eds) Droughts in Asian Monsoon Region: community, environment and disaster risk management. Emerald Group Publishing Limited, UK,8:25–48
Habiba U, Shaw R, Takeuchi Y (2011b) Drought risk reduction through a socio-economic, institutional and physical approach in the northwestern region of Bangladesh. Environ Hazards 10(2):121–138
Habiba U, Takeuchi Y, Shaw R (2010) Overview of drought risk reduction approaches in Bangladesh. In: Shaw R, Pereira J (eds) Climate change adaptation and disaster risk reduction: issues and challenges. Emerald Group Publishing Limited, UK, 5:37–58
Haque MM (2007) Decentralized Fish Seed Network in Northwest Bangladesh: impacts on rural livelihoods. PhD Thesis, Institute of Aquaculture, University of Sterling, Sterling
Hossain M, Islam ATMA, Saha SK (1987) Floods in Bangladesh. Recurrent disaster and people's survival. University Research Centre, Dhaka, pp 64–66

Huq S, Karim Z, Asaduzzaman M, Mahtab F (1999) Vulnerability and adaptation to climate change for Bangladesh. Kluwer Academic Publishers, Dordrecht

Intergovernmental Panel on Climate Change (2007) Climate change 2007: impacts, adaptation and variability. In: Parry ML, Canziani OF, Palutikof JP, van der Linden PJ, Hanson CE (eds) Contribution of working group II to the Fourth Assessment Report of the Intergovernmental Panel on Climate Change. Cambridge University Press, Cambridge

Iqbal A, Ali MH (2001) Agricultural drought: severity and estimation. A Report prepared under the project BGD/95/006 under the Ministry of Agriculture, Government of the People's Republic of Bangladesh, Dhaka, April 2001

Karim Z, Iqbal MA (eds) (2001) Impact of land degradation in Bangladesh: changing scenario in agricultural land use. BARC, Dhaka

Karim Z, Ibrahim A, Iqbal A, Ahmed M (1990) Drought in Bangladesh agriculture and irrigation schedule for major crops, Bangladesh Agricultural Research Center (BARC). Publication No. 34, Dhaka, Bangladesh

LACC Project, DAE (2010) Ministry of Agriculture. Dhaka, Bangladesh

MOEF (Ministry of Environment and Forest) (2002) Government of the People's Republic of Bangladesh, Bangladesh

Nasreen M, Hossain KM (2002) Sustainable development: Bangladesh perspective. In: Paper presented at the NGO session of World Summit on Sustainable Development (WSSD), held in Johannesburg, August 26–September 4, 2002

National Drought Mitigation Center (NDMC) (2006) What is drought? Understanding and defining drought. Retrieved from http://www.drought.unl.edu/whatis/concept.htm. Accessed on July 23, 2012

Nguyen H, Prabhakar SVRK, Shaw R (2009) Adaptive drought risk reduction in Cambodia: reality, perceptions and strategies. Environ Hazards 8:245–262 (Springer Publication)

Nguyen H, Shaw R (2009) Water resource management, climate change and indigenous knowledge. In: Shaw R, Sharma A, Takeuchi Y (eds) Indigenous knowledge and disaster risk reduction: from practice to policy. Nova Publisher 59–72

O'Brien G, O'Keefe P, Rose J, Wisner B (2006) Climate change and disaster management, Disasters, 30 (1):64–80

Paul BK (1998) Coping mechanisms practiced by drought victims (1994/95) in North Bengal, Bangladesh. Appl Geogr 18:355–373

Prabhakar SVRK, Shaw R (2007) Climate change adaptation implication and drought risk mitigation: a perspective from India. Clim Change 88:113–130

Rahman A, Biswas PR (1995) Devours resources. Dhaka Courier 11(42):7–8

Rahman AA, Huq S, Conway GR (eds) (2000) Environmental aspects of surface water systems of Bangladesh. The University Press Limited, Dhaka

Rahman A, Alam M, Alam SS, Uzzaman MR, Rashid M, Rabbani G (2008) Risks, vulnerability and adaptation in Bangladesh, Human Development Report 2007/08, Human Development Report Office Occasional Paper, 2007/13

Rahman RI (1998) Food security, increase food production and ownership: recent experience of Bangladesh. "Bangladesh Dev. Studies, No. 16, Annual No. 1405, BIDS

Selvaraju R, Stephan B (2007) Climate variability and change: adaptation to drought in Bangladesh. A resource book and training guide. Asian Disaster Preparedness Center, Food and Agriculture Organization of the United Nations, Rome

Shahid S, Behrawan H (2008) Drought risk assessment in the western part of Bangladesh. Nat Hazards 46:391–413

Shahid S, Chen X, Hazarika MK (2005) Assessment aridity of Bangladesh using geographic information system. GIS Development 9:40–43

Shaw R, Nguyen H, Habiba U, Takeuchi Y (2011) Overview and characteristics of Asian Monsoon Drought. In: Shaw R, Huy N (eds) Droughts in Asian Monsoon Region, vol 8. Emerald Group Publishing Limited, Bingley, pp 1–24

Smith SE, Abdel Kader A (1988) Coastal erosion along the Egyptian delta. J Coastal Res 4(2):245–255

Statistical yearbook of Bangladesh (2007) Policy and Planning Support Unit, Ministry of Agriculture. Dhaka, Bangladesh. Adopted from http://www.moa.gov.bd/sattisctics/statistics.htm. Accessed on July 12, 2012

Tallaksen LM, Van Lanen HAJ (eds) (2004) Hydrological drought – processes and estimation methods for streamflow and groundwater, vol 48, Developments in Water Sciences. Elsevier B.V, Amsterdam

Tanner TM, Hassan A, Islam KMN, Conway D, Mechler R, Ahmed AU, Alam M (2007) ORCHID: Piloting climate risk screening in DFID Bangladesh. Detailed Research Report. Institute of Development Studies, University of Sussex, Sussex

UNISDR (United Nations International Strategy for Disaster Reduction) (2009) Drought risk reduction framework and practices: contributing to the implementation of the Hyogo framework for action. UNISDR, Geneva, pp 213

WARPO-EGIC (1996) An Atlas of Selected maps and Spatial Data in Bangladesh. Jointly published by Water Resources Planning Organization, and Environmental and Geographic Center, Dhaka, Bangladesh

Wilhite DA, Glantz MH (1985) Understanding the drought phenomenon: the role of definitions. Water Int 10:111–120

Yusuf SA (1991) Disaster preparedness and management in Bangladesh. A research mimeo, Dhaka

Chapter 8
Landslide and Flashflood in Bangladesh

Abdul Awal Sarker and A.K.M. Mamunur Rashid

Abstract Flashflood and associated landslide are become major disasters in the hilly regions of Bangladesh, occurring almost every year. Flashflood and landslide disasters are caused by a set of preliminary and triggering factors which determine their location, frequency and magnitude. Excessive rainfall in the piedmont area with a high intensity is the main source of flashflood in the hilly area and resultant landslide specifically in the areas composed of unconsolidated rocks. The annual rainfall ranges from 2,200 mm along the western boundary to 5,800 mm in the north-east corner and even higher in different catchments due to monsoon depression. Being the higher order basin of the three largest basins Brahmaputra, Ganges and Meghna all the water discharges through the three major rivers of the country. Due to the higher intensity of rainfall with a shorter period of time these channels could not manage to discharges this deluge amount of water to the downstream to Bay of Bengal resultant sudden flood and associated landslide. Apart from physiology, hydrology and climatology, the changes in the geomorphology in relation to land use changes as well as deforestation, hill cutting and unplanned infrastructural development influences flashflood and landslide disaster impact further. Among the hilly region part of the north-eastern, north–south and northern regions are susceptible to and facing these disaster almost each year. Lack of proper landuse planning, weak enforcement by the local authorities increases risks and progress vulnerability significantly. Although the authority has taken initiatives of integrated planning including structural and non-structural measure to mitigate the damages by the disasters, lack of capacity of forecasting appropriately does not provide with sufficient lead time to reduce damages of personal property and economic assets. Further initiatives

A.A. Sarker
Poverty Environment Climate Mainstreaming Project, National Planning Commission, Dhaka, Bangladesh

A.K.M.M. Rashid (✉)
Poverty Environment and Climate Mainstreaming Project, UNDP, Dhaka, Bangladesh
e-mail: nemoibsru@gmail.com

R. Shaw et al. (eds.), *Disaster Risk Reduction Approaches in Bangladesh*,
Disaster Risk Reduction, DOI 10.1007/978-4-431-54252-0_8, © Springer Japan 2013

requires for basin wide integrated water resources management, increasing water retention capacity, early warning, landuse zoning and public awareness for effective management of flashflood and landslide disasters. The main objective of this chapter is to identify mountain risks and vulnerability of Bangladesh due to flashflood and landslide hazard and their underlying causes and effects. This discussion will also provide with some recommendations and policy implication for effective adaptation and mitigation measures.

Keywords Climatology • Geomorphology • Hydrology • Land use zoning • Orographic enhancement • Physiographic

8.1 Mountain Risks of Bangladesh

Mountains of Bangladesh are located in the southern, eastern and northern part of the country including the Chittagong Hill Tracts and Chittagong Region; Mymensingh and Jamalpur; and Sylhet Region. Northern and eastern hills occupy almost 12 % area of Bangladesh (about 44,90,150 acre) and underlain by unconsolidated or little-consolidated beds of sandstones,[1] siltstones[2] and shale[3] and the higher hill ranges in the Chittagong Hill Tracts, Chittagong and Sylhet regions are underlain by rocks of the Surma and Tipam formation (Brammer 1996; BBS 2009a, b). These hills have been uplifted and folded into a series of anticline[4] and syncline[5] during the collision between the Indian and the Eurasian plates in Miocene period (Khan 1991; Mannan 2002). The composition of soil of these hilly regions is complex and the young rocks have higher contents of easily-weatherable feldspars.[6] Therefore, the soil of this hilly region is very much susceptible to the landslide risks during heavy rain carried by monsoon wind system.

The annual rainfall ranges from 2,200 mm along the western boundary to 5,800 mm in the north-east corner and even higher in different catchments (BBS 2009a, b). During monsoon season the strong moist air blows at lower level from the Bay of Bengal and creates feeder clouds over the hilly area due to orographic enhancement. This process washed out large amount of water from the lower level cloud caused increasing rainfall over the hilly area at north-east and southern

[1] Sandstones are sedimentary rock, mostly composed of sand-sized grains (0.063–2 mm in diameter) like, quartz or feldspar is weathered easily.

[2] Siltstone is hardened sedimentary rock that is composed primarily of angular silt-sized particles (0.0039–0.063 mm in diameter) and is not laminated or easily split into thin layers.

[3] Shale is any of a group of fine-grained, laminated sedimentary rocks consisting of silt- and clay-sized particles.

[4] Anticline is a fold that is convex up and has its oldest beds at its core.

[5] Syncline is a fold with younger layers closer to the center of the structure.

[6] Feldspars are group of rock forming minerals make up as much as 60 % of the earth crust.

part of the country. In addition to the rainwater from inside the country, the flood plain of the north-eastern part receives large volume of water from Brahmaputra–Ganges–Meghna Basin with high intensity caused sudden inundation of vast area. Recently, the human made deforestation in the upstream areas, hill cutting and unfriendly agricultural system further increase the risk of flash flood and associated landslide disasters.

8.2 Extreme Events and Natural Hazards

Mountains are typically exposed to multiple hazards (Kohler and Maselli 2009). Mountain regions of Bangladesh included many of world's most sensitive resources are particularly vulnerable due to high relief, steep slope, shallow soils, adverse climatic condition and geological variability. Among the common natural and human-made hazards the mountain ecosystem of the country are susceptible to earthquake, water logging, flash flood and landslide etc. Climate change and induced variability further increases the frequency and intensity of the hydro-meteorological hazards and also introduces number of new hazards including the convective storm, drought, scarcity of the fresh water and destroying biodiversity.

In addition to the factors associated with the geology and climatology, human activities also influences the impact of hazards in the hilly regions. Destruction of mountain forests or inappropriate farming practices accelerates erosion and exposure of land to the risk of landslides, environmental degradation and sudden floods. Unfriendly development interventions including construction of roads and infrastructures without considering local geographical and soil characteristics influences secondary disasters and increasing the risks of damages and losses. Increasing population due to migration from mainland to the hilly areas is a significant factor that has led to greater concentrations of population in valleys, increasing environmental degradation and vulnerability to large-scale disasters. Often, migration is linked to poverty, which by itself exposes people to higher risk, especially where access to safe housing and safe land is left unconsidered (UN/ISDR 2002).

8.3 Risk and Vulnerability of Mountain Region

Vulnerability of mountain people in Bangladesh is commonly attributed with poverty, landlessness, inaccessibility to the natural resources base, illiteracy, isolation, backwardness and fragility stemming from rugged terrain and harsh climates; in turn, risks are linked with greater proneness to geophysical extremes, unfriendly agricultural practice, increasing population and scarce resources. Hilly area of Chittagong, Mymensingh and Sylhet region consisted of 1,57,818 households of 7,83,044 population (BBS 2009a, b). Most of these tribal people's livelihood directly or indirectly connected to the natural resources base.

Economic backwardness is one of the major features of vulnerability index of the people of this region. The poverty map by World Bank-World Food Programme-BBS joint initiatives (2005) shows the highest incidence of extreme poverty in the south-east and north-east hilly region. In the Chittagong Hill Tracts (CHT) regions about three-fourth of the households (74 %) are living below the lower poverty line (<Tk.866/person/month) and 86 % below the upper poverty line (<Tk.1,025/person/month); and the households living below lower and upper poverty lines are 78 % and 89 % respectively among indigenous people and 69 % and 83 % respectively among Bengali (UNDP 2009). Among these region the Chittagong Hill Tracts shows About 62 % households irrespective of ethnicities, according to direct calorie intake (DCI) method, are living below the absolute poverty line (below 2,122 kcal), while 36 % are hardcore poor (below 1,805 kcal) (UNDP 2009).

High rate of inequality is very common scenario in the hilly region and prevalence of poor and vulnerable people increases with elevation (Hassan et al. 2005; Huddleston and Ataman 2003). Mountain people of Bangladesh are generally suffering disproportionately from poverty and low level of development than the plain land area. The area specific constraints intensifying the prevalence of poverty including fragility of ecosystems; remoteness; inadequate accessibility and marginalization of mountain communities from the mainstream; lack of equity in terms of access to basic facilities such as water, sanitation, health care, education, and physical infrastructure, as well as to markets, political power, and representation; lack of employment opportunities; and proneness to natural disasters.

Dependency of hilly people on the natural resources for their livelihood and other daily necessities are one of the major factor make them prone to climate change. Majority of the people of the hilly area engages in some form of agricultural activity and is thus highly dependent on natural resources. Among the total population of the Chittagong Hill Tracts (CHT) 18 % are involved in agriculture as their primary source of income and additional 12 % are involved as secondary occupation (UNDP 2009). The 65 % of the population of this area owned their agricultural land, however the total landless people shows approximately 34.20 % (Gunter et al. 2008). The impact climate change limiting the alternative livelihood options of these poor and marginalized indigenous people and may further limit the capacity of the local community to cope with and adapt to relevantly new and extreme environmental events resulted force more people below the poverty level. Climate change and associated hazards may significantly influence in the slow economic development, high rate of poverty, marginalization, lack of health & nutrition, and changed livelihood options of the people (UNEP-WCMC 2002).

Furthermore, the ecosystem systems and people are subject to variety of drivers of change including development policies, political rivalry and increasing pressure on land and resources due to economic growth, increasing population and livelihood patterns. Despite high degree of uncertainty, the biophysical fragility of the ecosystem of the hilly region has direct consequences for the socio-economic vulnerability of the people. Although around 8 % of people of the mountain region living in urban area (BBS 2009a, b), the majority of the people still live in remote rural settings. Indigenous people of the country living in this region consisted of diversity of traditional knowledge, languages and cultures. Lack of modern technology and support from government along with cultural barriers contributes in the progression of their vulnerability to different disasters significantly.

8.3.1 Landslide Risk

Landslide is a natural phenomenon and accelerated by human interventions. This is a major natural hazard in recent years and received great concern in south-eastern part of Bangladesh as Chittagong and Chittagong Hill Tracts region. During the last five decades, Chittagong suffered about 12 times landslides (BWDB 2005). By the devastation of this disaster 17 people died in 1999, 13 in 2000, 91 in 2007, 54 in 2010 and 17 causalities occurred in 2011 in Bangladesh (BWDB 2005). During the last three decades the death toll is approximately 200 and occurred massive economic and property loss. The factors or causes of landslide is topography, weakening of slopes through saturation by water, steeping of slope by erosion, properties of soil (sandy soil), torrential rain, and high velocity surface runoff. It is remarkable that the natural factors of landslide are more accelerated by human activity as, indiscriminately hill cutting for housing and brick field, deforestation, lack of vegetation cover in hills, lack of appropriate laws and legislation and its implementation by respective authorities.

8.3.2 Landslides Incidence of Bangladesh

The physiographic characteristics (see Fig. 8.1) of Bangladesh make the country more susceptible to landslide. Physiographically, most of the area of Bangladesh is floodplain and only 18 % is hilly and tract area (Islam and Uddin 2002). According to geological time scale, hilly area of Bangladesh developed in tertiary age and mainly composed of unconsolidated sedimentary rocks such as sandstone, siltstone, shale and conglomerate (Rashid 1991). The hill areas are underlain by tertiary and quaternary sediments that have been folded, faulted and uplifted, then deeply dissected by rivers and stream (Brammer 1996). These areas consist of sandstones and shale and also with siltstone which have less stability. Having the acid properties of the hilly soil is easily saturated with water. The bedrock and soil structure of these hills are not stable, for which reason these areas are highly prone to landslide. Unsustainable landsuse and alteration in the hills including indiscriminate deforestation and hill cutting are two major factors in Bangladesh that aggravated the landslide vulnerability in the hilly areas (Table 8.1).

The susceptibility of the soil erosion in the Chittagong Hill Tracts is very high among the vast hilly region of Bangladesh. However most of the causalities due to landslide disasters occurred near the Chittagong, Bandarban and Cox's Bazar area due to large exposed people of this area including Lebubagan, Kasiaghuna and

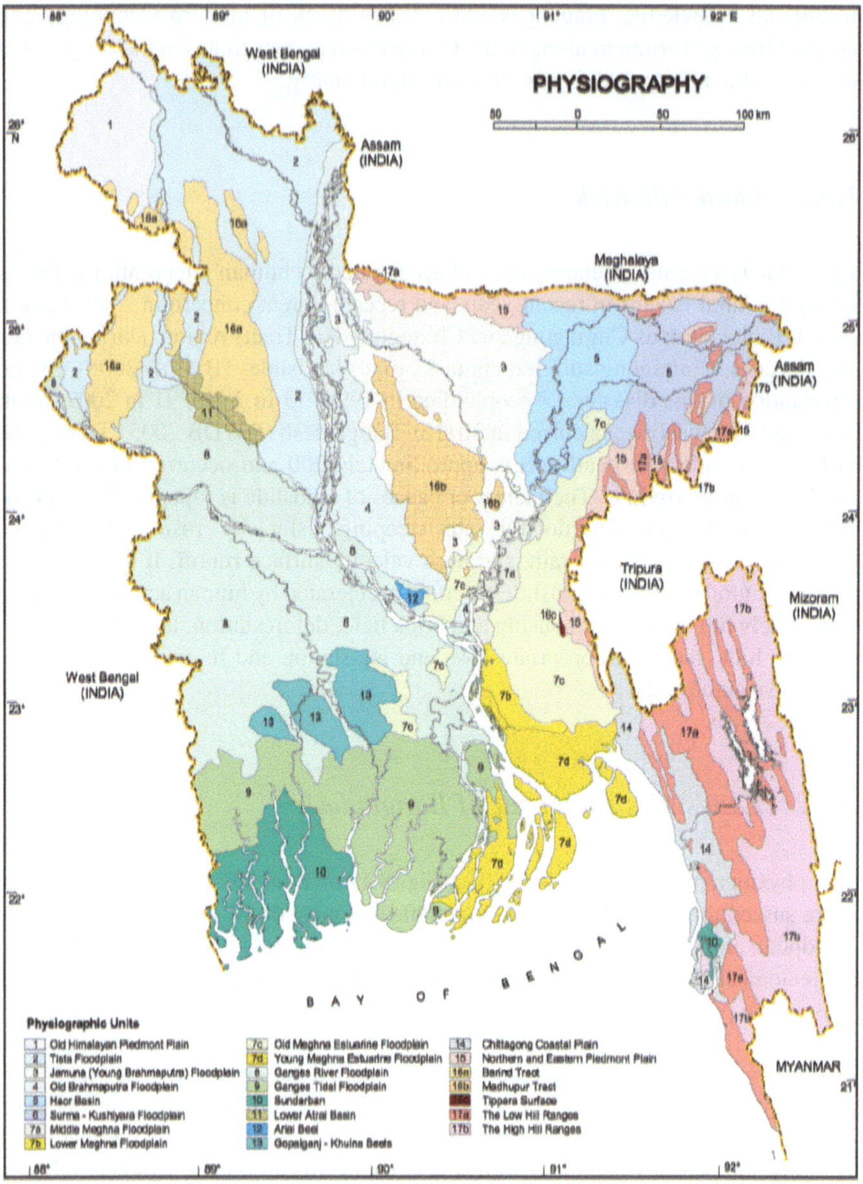

Fig. 8.1 Physiographic characteristics of Bangladesh (*Source*: Modified from SRDI 1997; Rashid 1991; Riemann 1993)

Workshop in Cantonment area; Kusum Bagh area at Dampara; Baijud Bostami area; Sekandar Para; Chittagong University area; Batali Hill and Moti Jharna, Lama in Bandarban; and district town and Himchhari of Cox's Bazar. During the devastated landslide of 1998, 1999, 2000, 2007, 2011 and 2012 all these area were affected.

Table 8.1 Land susceptibility to different soil erosion in the hill areas of Bangladesh (Khan 2011)

Area	Moderate susceptibility	High susceptibility	Very high susceptibility	Total
Chittagong Hill Tracts	350	1,814	10,765	12,929
Chittagong and Cox's Bazar	414	949	954	2,317
Greater Sylhet District	161	462	964	1,587
Others (Comilla, Netrokona, Jamalpur)	–	35	102	137
Total	925 (5 %)	3,260 (20 %)	12,785 (75 %)	16,970 (100 %)

The historical landslide occurred in Bangladesh are given below (Mahmood and Khan 2010; Reuters 2007; BBC 2010; CNN 2011; Hossein 2012; USGS 1976) (Table 8.2):

8.3.3 *Causes of Landslide Disasters*

The mechanism of landslide is occurred by both internal forces and external factors. When the gravitational force acting on a slope exceeds its resisting force, slope failure and mass wasting occurs. The slope material's strength and cohesion and the amount of internal friction between materials help maintain the slope's stability and are known collectively as the slope's shear strength. The steepest angle that cohesion-less slope can maintain without losing its stability is known as its angle of repose. When a slope possesses this angle, its shear strength perfectly counterbalances the force of gravity acting upon it. Landslides occur when the stability of a slope changes from a stable to an unstable condition. A change in the stability of a slope can be caused by a number of factors, acting together or alone.

There are two primary categories of causes associated with the major landslides in Bangladesh, including natural causes and human induced causes (Fig. 8.2). Natural causes are the major triggering mechanism for landslides either singly or in combination with water pressure and seismic activity. Effects of these causes varied widely and depend on steepness of slope, morphology or shape of the terrain, soil type, and underlying geology and whether there are people or structures on the affected area. Human induced causes increase the intensity of the impact of landslide mainly identified as unplanned settlement in the risk zone, modern agricultural practice without considering soil characteristics, cutting hill for roads and other construction with considering proper slope and deforestation.

Natural causes: Slope saturation by water is the primary cause of the landslide. Saturation can be occurred due to intense rainfall, changes in ground-water levels, and surface-water level changes along coastlines, earth dams, and in the banks of lakes, reservoirs, canals, and rivers (Highland and Bodrowsky 2008). Precipitation and run-off are closely related to the sudden inundation of hill slope and landslide.

Table 8.2 Historical landslide in Bangladesh and associated issues

Year	Location	Soil characteristics	Mechanism	Impact
1968	Kaptai-Chandraghona Road	Deep, well drained loams derived mainly from tertiary sandstones and shale which are permeable and exposed to massive erosion	Removal of protective vegetation make the bondage of soil loose which then exposed to the monsoon rain and eroded rapidly	Reservoirs silted up but no causalities occurred
1970	Ghagra-Rangamati road	Well exposed sandstone on the anticline; yellow and gray medium grained occasionally pebbly sandstone and clayey sandstone with inter-beds of mottled clay	Deforestation make the bondage of soil loose which then exposed to the monsoon rain and eroded rapidly and washed away houses and properties	Roads and properties have been damaged, houses collapsed
1990	Jhagar beel, Rangamati	Well exposed sandstone on the anticline; shale, sandstones and siltstones in the underlying and overlying strata	Due to the deforestation and heavy & prolonged rain during the monsoon period, this disaster occurred	Roads and properties have been damaged
1997	Charaipada, Bandarban	Well exposed sandstone; lower portion is variegated coarse grained cross-bedded sandstones	Heavy rainfall creates excessive pressure of water which caused the soil washed away	90,000 m^2 area was affected but not causalities occurred
1999	Lama and Aziz Nagar Union, Bandarban; and Chittagong	Well exposed sandstone; sandstone and minor amounts of clay-stone and silt in the Chittagong area	Heavy and incessant rainfall at that time was one of the causes of sliding	17 people killed, houses damaged and 893 ha of cultivated land and household garden and 50 km road were damaged
2000	Chittagong University and Southern Chittagong	Well exposed sandstone; some small area are also sandy or rocky with steep hill slope are highly susceptible to erosion	Deluge of mud and water that swamped various part of the port city amid torrential rain	13 people were killed and 20 injured; and property damaged

2007	Cantonment and Lebubagan, Chittagong	Sandstone and shale with very steep slope	Hill cutting made the soil bondage loose which is washed away due to heavy and prolonged monsoon rain	128 persons died of which 59 were children; 100 of peoples injured and lots of property damaged
2010	Cox's Bazar and Bandarban	Deep, well drained loams derived mainly from tertiary sandstones and shale; some of the area also consisted of well exposed sandstone and Boral formation	Settlement was developed at the foot hill area by cutting the hill side and the prolonged torrential rain caused severe flood and mudslides	53 persons died and a number of houses and properties damaged; roads have been blocked
2011	Batali Hill, Chittagong	Sandstone and shale with very steep slope	Settlement was developed at the foot hill without any precautionary measures and the prolonged torrential rain caused severe flood and mudslides	15 persons died including women and children; property damaged and houses are washed away
2012	Lama and Naikhanchari, Bandarban; Chittagong Port Area; and Moheshkhali, Cox's Bazar	Deep, well drained loams derived mainly from tertiary sandstones and shale; some of the area also consisted of well exposed sandstone and Boral formation	Prolonged torrential rain due to the influence of monsoon depression over the Bay of Bengal washed away houses and settlements located on the foot hill area	108 people died; houses, roads and properties damaged and the communication with the impacted area collapsed

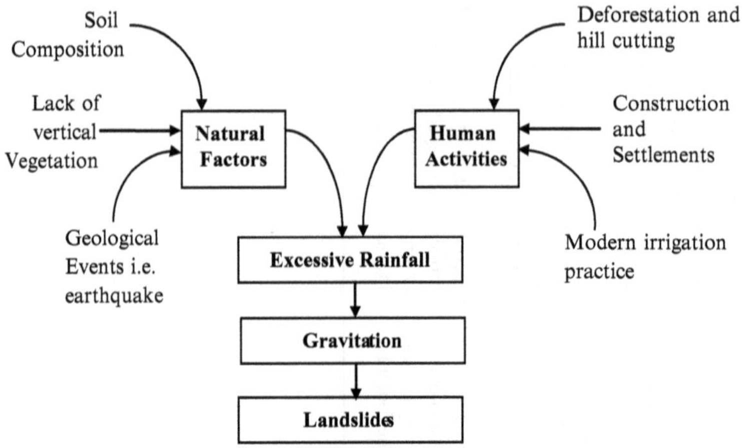

Fig. 8.2 Landslides and associated natural factors and human activities

Previous geological events or modern irrigation practice may cause losing soil particle ease the saturation process; where the deforestation helps in removal of topsoil from the surface of the hill. Few major natural causes of landslide are given below:

- Formation of hill soil with loose and weathrable particles i.e. feldspar and undercut by the agent like water or air.
- Previous geological events i.e. earthquake that reduced the shear strength of soil particles. Earthquakes in steep landslide-prone areas greatly increase the likelihood that landslides will occur, due to ground shaking alone, liquefaction of susceptible sediments, or shaking-caused dilation of soil materials, which allows rapid infiltration of water. Shear strength is controlled by a number of variables such as friction, which itself is proportional to the normal force, as well as many other variables such as the roughness, cohesion, and dryness of the material.
- Absence of vertical vegetation due to natural events (i.e. wild fire) decreases the bondage of soil particles and stability of soil.
- Excessive rainfall that results increase of ground water pressure to destabilize the slope of the hill.
- Mass wasting is caused by gravity. On a mass of material gravity exerts a force downward proportional to the amount of mass. Saturation of the pore spaces increased the weight of the mass increasing the sheer force as well as intensity of the hazard.

Human Activities: Expanding population in the landslide prone area due to migration are the primary means by which humans are contributing to the landslide disasters. The new settlements create disturbing to the natural drainage system and destabilizing slopes. Unsustainable and unplanned use of land for industrial purpose and irrigation practice by the migrated people is also reducing stability of the soil for containing moisture and losing bondage of top soil. Landslide related human induced factors are given below:

- Construction of new settlements in the foothill of land slide prone area that increases exposure of people to the risk.
- Construction of industries or roads or settlements without considering the proper slope of the hill. Some of the slopes are very steep as more than 90° compare to the ground surface. Earthworks alters the shape of a slope, or which imposes new loads on an existing slope.
- Deforestation of the hilly area especially deep rooted vegetation reduces the strength of the soil particle to make bondage among them. In addition some of the afforestation practices are also responsible for the destabilization of soil particles if the local soil characteristics are not considered.
- Modern irrigation practice without considering the traditional knowledge loosens the bondage of soil particles.
- Lack of awareness of the people about the landslide risks and avoiding the early warning are two most important factors that increase the loss and damages by the land slides.

8.4 Case Study: Chittagong Landslide

Chittagong is the second largest city of Bangladesh located in the south-eastern coast. With a population of five million and housing the main sea port of the country gives it an economic significance second only to the capital Dhaka. A north–south central hill range extends into the urban zone from the north and gradually loses height as it comes closer to the river. The city area comprises of small hills and narrow valleys, bounded by the Karnaphuli to the south, the coastal plain and the Bay of Bengal to the west and the floodplain of the Halda to the East. The highest ground level within the city area is about 60 m above Mean Sea Level.

The coastal areas of the city are highly prone to cyclone and tidal surge. Water-logging is a perennial problem but is more of a man-made hazard. The hilly areas are highly exposed to landslide caused by heavy rainfall and unsustainable hill cut. Growth of informal settlements on these unstable slopes has exacerbated the problem. Chittagong experienced a devastating cyclone in 1991 but since then, major urban disaster has been land-slide which occurred in 2007 and 2008. Both of these disaster events drew major attention from policy-makers, media, NGOs, humanitarian agencies and local society.

8.4.1 Chittagong Land-Slide, June 2007

The landslide on 11 June 2007 which occurred in several areas of Chittagong city was the severest of such occurrences in the country's history. Major city areas

Table 8.3 Number of death as a result of the landslide of 11 June 2007

		No. of death			
Area	Thana	Female	Male	Children	Total
Lebubagan (Chittagong Cantonment)	Hathazari	17	14	41	72
Baizid Bostami	Baizid	3	1	1	5
Kushumbag Lalkhanbazar	Khulshi	6	2	6	14
Sekandarpara	Hathazari	4	1	4	9
Chittagong University	Hathazari	2	1	2	5
Motizharna		1		1	2
Missing identity					20
Total number of deaths					127

Source: Report of 6 August 2007, District Administration, Chittagong

Table 8.4 Number of affected families by landslide of 11 June 2007

		No. of affected family		
Area	Thana	Severely affected	Partially affected	Total
Lebubagan (Chittagong Cantonment)	Hathazari	235	1,173	1,554
Baizid Bostami	Baizid	1	320	321
Kushumbag Lalkhanbazar	Khulshi	87	110	197
Total affected family				2,072

Source: Report of 6 August 2007, District Administration, Chittagong

affected were the foothill settlements consisting of the slums of Lebubagan (Chittagong Cantonment), Baizid Bostami, Kusumbag, Motizharna, and Lalkhan bazaar hilly area. Death toll and number of families affected from the event is described in Tables 8.3 and 8.4. This disaster primarily affected the urban poor who had built informal settlements on hazardous locations.

8.4.2 Political Economy of Hill-Cutting

National and Local media reports highlighted the political economy of hill cutting as a major cause of landslide in Chittagong city. National dailies such as Prothom Alo, Daily Star, Financial express, Independent and Dhaka Mirror as well as local dailies such as Azadi, Dainik Purbakon, and Dainik Karnophuli, have published reports on the issue. Reports highlighted both the indiscriminate nature of hill-cutting for residential and commercial purposes as well as the involvement of influential persons and groups in the process. Even brick-fields are also driving the propensity towards hill-cutting to use the soil in the making of bricks. Most of the developmental and regulatory bodies turn a blind eye to the problem. Many of the reports allege their active connivance in the process. Some of the reports highlighted the allocation of hill land which is in government ownership to influential individuals at negligible prices. These allocations are often obtained on the promise of tree-planting and forest preservation but once in hand, are cut to build houses or commercial properties.

A look at judicial records indicates that legal redress rarely sees the day. A writ petition filed in 1997 was still found to be making the paper round. A Show Case notice was served on the government at one stage but no specific results have emerged. However, it is fair to say that a process of judicial activism remains even if not much of results can be shown. Between 2007 and 2010, 49 cases were registered in this regard against individuals and real-estate companies.

8.5 Flash Flood Risk

Flashflood are one of the most common forms of natural disaster in the hilly region of Bangladesh occurred almost every year. This sudden inundation of water also carried silt, rocks, sediments and other debris caused damage to agricultural products and causalities. The hills of Bangladesh are located near south-eastern part of the Chittagong Hill Tracts region and created long, linear ridges running approximately NNW-SSE, except the hills located in the north of Mymensingh and Sylhet region along the foot of the Shillong Plateau (Brammer 1996). The location, topography and climate of the region lead hydro meteorological hazards including frequent floods and flash floods. The hilly area specifically north-eastern Sylhet Basin suffered by the sudden inundation of water almost every year during the monsoon period when large amount of water received due to torrential rain. Additional water from outside the country intensifies the event and creates huge devastation. Prediction of flashflood is very difficult and provides very short duration of time for taking measures which increase the damage and loss.

8.5.1 Flashflood Incidence of Bangladesh

Mountain region of Bangladesh are suffer by the flashflood triggered by excessive torrential rainfall especially during the monsoon period. Average rainfall is very high around 2,300 mm per year, majority of which occurred during monsoon period (BDM). Three major rivers flowing through the country originated from outside the country and received nearly 1,200 bm^3 water every year (BWDB). During the monsoon period sometimes huge precipitation occurred within a very short duration caused sudden flood different region of the country including the north-east *Haor*[7] region consisted of Sylhet, Shunamgonj and Habigonj in the Meghna Basin. This area is low-lying flat with gentle slope towards south, so the rainwater from the hills passes over the agricultural lands and large area of the *Haor*. Along with the north-eastern part, eastern, south-western, south eastern, southern and north-western region are also suffered by the flashflood hazard (Fig. 8.3).

[7] A *haor* is a wetland ecosystem in the north eastern part of Bangladesh which physically is a bowl or saucer shaped shallow depression area.

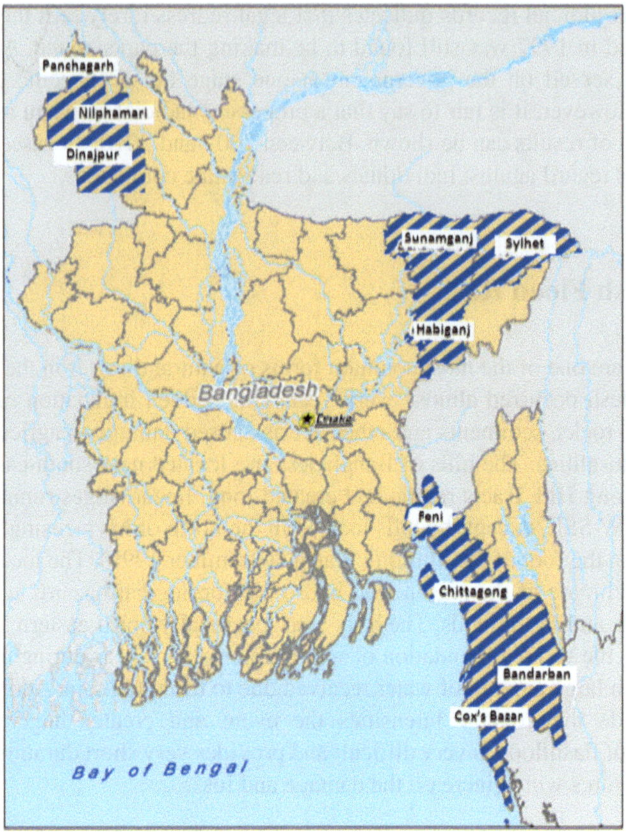

Fig. 8.3 Flash flood and landslide prone areas of Bangladesh (WFP 2012)

As the flashflood prediction earlier is difficult, allow little time to react and take proper preparedness for protecting life and properties. Flashfloods may last from several minutes to several days and may happened hilly as well as coastal area of Bangladesh, but are more commonly in north-east and south-east hilly region. Although flashfloods generally affect smaller areas and populations than riverine floods, their unexpected and intense nature pose significant risk to people and infrastructure, leading to death and destruction.

Historical Flashflood in different regions of Bangladesh with associated factors is given below (ADRC 2003; CNN 2001; IFRC 2003; Oxfam 2010; BWDB 2005; Reuters 2012) (Table 8.5):

Flashflood is regular phenomena in the north-eastern region of Bangladesh were recorded in 1954, 1961, 1966, 1970, 1987,1988, 1993, 1999, 2004, 2006, 2007 and 2010 (BWDB 2005). The intensity of flashflood is changing with increasing the devastation due to increasing population in the risk zone (piedmont of hilly area). The trends of occurrence of flashflood event are also changing due to climatic

Table 8.5 Major regional flashflood in Bangladesh and associated issues

Region	Location	Year (month)	Mechanism	Impact
North-Eastern	Sylhet, Sunamgonj, Moulvi Bazar, Habiganj and Netrakona	2003 (June)	Flashfloods triggered by heavy rain and water gushing across the border from neighboring India	Houses, roads and embankment submerged; 3,000 households homeless
	Sylhet, Sunamganj, Moulavibazar and Habiganj	2004 (April)	12.30 m of water flows 5–6 m height of the embankment (during peak) with 25,000 m^3/s volume due to prolonged rainfall (2–3 days) associated with cross-boundary run-off	Damage estimated USD 1,000 million; 8,000 km^2 haor area (Surma Basin) affected; roads, school and houses destroyed; many people displaced
	Sylhet, Sunamgonj, Hobigonj, Kishorgonj and Moulavibazar	2010 (May)	Heavy rain in upstream (Jan–April) 3,543 mm create huge amount of water flow and raise of riverbed increase the height of water level (April was 8.05 m). Embankment has breached in 128 points due to poor maintenance. Early warning system appears ineffective	Due to early flood crop destroyed; many people displaced and migrated
Eastern	Comilla	1999 (July)	14.7 m of water flows 95 cm above the danger level (during peak) with 603.08 m^3/s volume due to prolonged rainfall (3–4 days) associated with water from river Gomuti	Damage estimated USD 100 million; 890 km^2 area of Meghna basin affected; 75,000 people displaced
South-West	Meherpur, Chuadanga, Kushtia, Jhenaidah, Jessore and Satkhira	2000 (September)	12.43 m water flows (during peak) due to precipitation of 600 mm/day associated with flood water from West-Bengal breaching polder-embankment	10,000 km^2 area of Ganges Basin affected; fisheries and agricultural products damaged; diseases outspread; communication disrupted

(continued)

Table 8.5 (continued)

Region	Location	Year (month)	Mechanism	Impact
South-Eastern	Rangamati and Khagrachari	2003 (July)	Heavy rainfall during the past 24 h, with 24.2 cm falling in many areas	20 km^2 in the district of Rangamati and Khagrachari affected; 65 people died; houses and roads washed away
	Bandarban, Chittagong and Cox's Bazar	2012	Prolong rainfall for five days (463 mm in 24 h) caused sudden inundation and mudslide	108 people died and many injured; houses, rail, roads collapsed
Southern	Chittagong	2000 (June)	Prolonged torrential rain caused mudslide	80 km^2 in the district of Chittagong city affected; 11 people killed; many people injured
Northern	Nilphamari, Rangpur, Lalmonirhat and Kurigram	2001 (August)	Prolonged rain for four days associated with the water from neighboring country	Villages and embankments washed away; 50,000 people affected
North Western	Rajshahi and Naogaon	2004 (October)	12.30 m of water flows (during peak) with 1,100 m^3/s volume due to waters rolling down from across the border of India	75 km^2 area of Ganges Basin affected; fisheries and agricultural products damaged; 50,000 people displaced and homeless

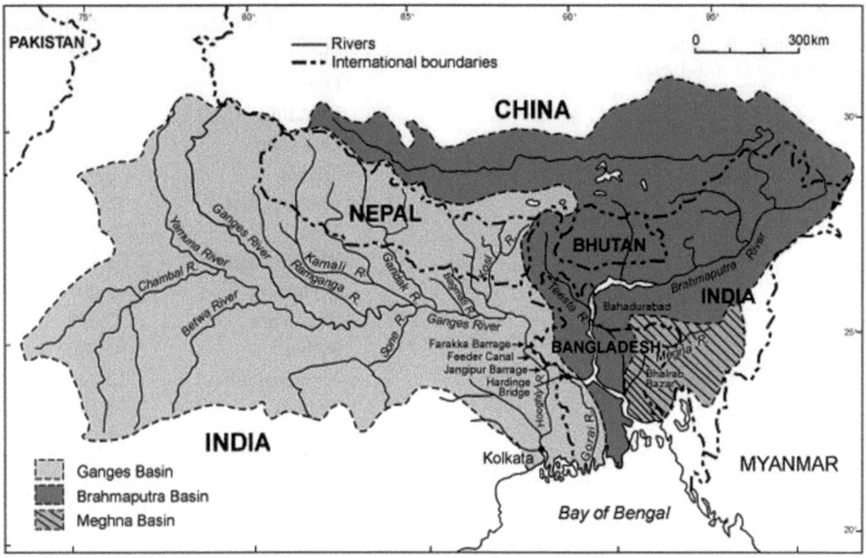

Fig. 8.4 Brahmaputra–Ganges–Meghna basin area (*Source*: Mirza et al. 2001)

reason and associated human induced development failure. On 12 July 1999 the severe flash flood caused by breaching the left bank of the river Gomuti due to a sudden rise in the water level in the river to about 92 cm above the danger level. The primary reason was excessive rain associated with the dam release water from the neighboring country. The flood inundated about 890 km of area including about 350 km^2 croplands. The damage has been estimated by BWDB approximately USD 100 million due to damaged embankment and other protective works of the BWDB. The frequent flashflood associated with landslide in the eastern hilly area also caused by deforestation.

Flashflood in the western part of the south-west region was occurred in 2000, after nearly 200 years due to unusual rainfall (BWDB 2005). The disaster affected large area of Meherpur, Chuadanga, Kushtia, Jhenaidah, Jessore and Satkhira districts and damaged vegetable gardens, croplands, roads and other infrastructures. As per BWDB, the major reason behind the flood was tremendous water inflow from West Bengal in India through the river Ganges into its tributes which spillover the Meherpur and intensified after breaching of polder-embankments in Satkhira district.

8.5.2 *Causes of Flashflood Disasters*

Flashflood in Bangladesh associated with variety of factors. The major cause in the hilly region are intense and prolonged rainfall events, landslide dam outburst, sudden intrusion of water from the upstream region and human induced development failure (Fig. 8.4).

Table 8.6 Mean annual rainfall in and around Bangladesh (1960–1991)

Name of the station	Rainfall (mm)
Shillong	2,271
Dhubri	2,436
Lumding	1,161
Guwahati	1,538
Silchar	3,018
Cherrapunjee	11,131
Tura	3,500
Williamnagar	3,207
Nongstoin	3,207
Baghmara	3,698
Sylhet	7,000
Mymensingh	2,432
Tangail	1,440
Dhaka	1,400

Source: CNRS (2009)

8.5.2.1 Intense and Prolonged Rainfall

The intense and prolonged rainfall caused in Bangladesh by the orographic enhancement of cloud in the hilly region, cloudburst, stationary monsoon troughs and the monsoon depression in the Bay of Bengal. In addition to the in-country rainfall huge amount of water flows from the upstream region during the monsoon season.

During monsoon season of each year a well marked low pressure zone formed over the Bay of Bengal in the month of April to June, moves towards the north-west of Bangladesh. The air consisted of high moisture content obstructed near the north-eastern hills and forced to orographic uplift, lead the air to be saturated and condensed and formed orographic cloud around the hill top. Sometimes during the monsoon season the lower cloud constantly replenished by strong surface flow of damp air, caused prolonged period of rainfall. The annual rainfall in the western boundary has been recorded around 2,200 mm while 5,800 mm to even higher in different catchments of the north-east corner of Bangladesh caused massive flash-flood (Table 8.6).

8.5.2.2 Topographic Characteristics

Due to the steep and unstable slopes of the hilly area of Bangladesh is prone to recurrent and devastating flashflood associated with landslide. The low capacity of the soil of hilly region for holding water forced the water to flow over the surface. In addition the aggradations of the riverbed due to siltation resultant overflow of the river bank and adjacent area with very short notice.

8.5.2.3 Dam Outburst

Weak geological formation, tectonic activity and rugged topography associated with the excessive rainfall caused large amount of debris flow sometimes blocked the river channel. This situation prevents the outflow of water and leading to a temporary reservoir or lake above the upstream. Eventually the unstable dam breaks either as a result of pressure from the volume of water, or when the water rise high enough to flow over the top of the dam and destabilize it. Occasionally secondary landslides falling into the reservoir leads to a combination of pressure and overtopping with a sudden catastrophic failure of the dam. The resultant outburst of water can have effects far downstream. Such outburst events are generally random and cannot be predicted with precision, although when a landslide blocks a river the likelihood of an outburst at some time is clear.

8.5.2.4 Human Induced Development Failure

A lack of maintenance of drainage sluices, including intake and outfall canals, in the coastal polders is another major reason for drainage congestion and the delay in drainage. Human interventions on the flood plains in the form of roads, homesteads, polder embankments etc., creates barrier to natural drainage system. Unplanned infrastructure development through deforestation and without considering proper drainage of natural water flow reduces the moisture content capacity of the soil and increase the overflow of water through the surface.

8.5.2.5 Climate Change Related Extreme Events

Changing climate have significant influence over the extreme event like unusual rainfall, excessive rainfall and increased water due to hot weather in the upstream region. Due to the changing environment the prediction of extreme climatic event become more difficult for the weather forecaster. In addition the ineffective and outdated early forecasting and warning technologies increase the damages and losses. In 2004 the north-eastern flashflood was the worst among all as the rainfall was 1.5 times higher than normal April month of the year. The flashflood caused damages of 1,000 million US. Dollar and inundated 8,000 km^2 *haor* area of Sylhet, Sunamganj, Moulavi Bazar and Habiganj district.

8.6 Landslide and Flashflood Risk Reduction Approaches

Most of the landslide and flashflood events take place in remote and isolated catchments which is difficult to reach. Therefore the external helps may take several days to search and rescue affected communities, during which time they have to cope

Table 8.7 Structural and non-structural measures for landslide and flash flood risk management

			Catchment-wide interventions (agriculture and forestry actions and water control work)
			River training
			Other flood control measures (passive control, water retention basins and river corridor enhancement, rehabilitation and restoration)
			Tolerance
			Emergency response system
Structural measures	Risk acceptance	Tolerance strategies	Insurance (personal properties i.e. crops and houses)
Non-structural measures	Risk reduction	Prevention strategies	Delimitation of flood areas and securing flood plains
			Implementation of flood areas regulations
			Application of financial measures
		Mitigation strategies	Technology innovation and extension especially for early cropping or flood responsive cropping
			Reduction of discharge through natural retention
			Emergency action based on monitoring, warning, and response systems (MWRS)
			Public information and education

Source: Modified from Colombo et al. (2002)

with the community resources and arrangements. So, it is essential to build the capacity of communities to manage the risks from disaster by themselves. Individual households usually have strategies in place, but the effectiveness of these individual efforts can be enhanced many fold if they are coordinated (Table 8.7).

Furthermore, risk reduction of landslide and flashflood disaster requires knowledge of specific areas which are subject to the hazard and if possible, the ability to predict the time of occurrence (UNDRO 1979). Creation of available information on this natural extreme events and dissemination for awareness building may reduce number of at risk people. The zoning of land through proper landuse planning and enforcement of rules can ensure the rational management. Early warning and adequate response mechanism with involving community people may be a sustainable salutation for creating a resilient community facing the disaster.

Community risk management committee is a very non-structural mechanism for making the resilient community to landslide and flashflood hazard. The use of local and indigenous knowledge can be an important part of community-based landslide and flash flood risk management; it is essential that gender aspects are taken into account and the effects on all members of a community considered during planning (ICIMOD 2010). Due to the difficulty in prediction of exact location, magnitude and extent of the landuse and flash flood disaster the structural measures are rarely taken. Along with the structural measures, non structural measures should be in place to reduce the risks of massive damage and loss. There are some specific measures that could be able to help reduction of risks are given in the following section.

8.6.1 Community Based Risk Reduction

The community based risk reduction strategies are very important for community resilience to landslide and flash flood disaster. Community based early warning system should be developed and functional within the community. The community should be equipped properly with available resources for risk reduction and emergency response and should be trained for effective response. Indigenous knowledge in the risk reduction approach requires especially considerations for well acceptance by the local people.

8.6.2 Land Use Vulnerability Assessment and Zoning

Landuse vulnerability assessment and zoning based on the physiographic & geomorphological characteristics, demographic and socio-economic features are important for the sustainable management of the disaster. The hilly area specially in the urban settings of the Chittagong composed of complex characteristics including foothill settlements (high class, middle class and slum), few are with vegetation, few are completely barren, few contains both hill top and foothill settlement. Socio-economic and environmental aspects should be considered in the rules and landuse plan and the compliance of legal and policy instruments should be in place for effective enforcement.

After the massive landslide during 2007, a national level Ministerial Order No. 07/01/03 came to the Divisional Commissioner (DC) office to look into the major causes of landslides and prepare mitigation measures (CDO 2008). A technical committee has been created consisted of government agencies, Chittagong City Corporation (CCC), CDA, researchers, engineers and NGO workers for identifying priorities for action. The committee also emphasize on the integration of hill management policy with Environmental Conservation Act 1995 and Building Construction Act 1952 and other urban planning and development policies.

8.6.3 Proper Structural Mitigation Measures

Structural measures considering the safety measures, building codes and better drainage facilities are inevitable for mitigation of landslide disaster risk. The mitigation policies for landslide disaster allowed the informal settlers for hill slope settlements upon fulfilling some prerequisites including satisfactory risk and vulnerability assessment by the city level vigilance team and taking required structural measures. A committee involving representatives from CCC, CDA and Institute of Engineers Bangladesh (IEB) also created for establishing cost-effective engineering measures like reinforcement walls along hill slopes. Specific measures should be taken for increasing the capacity of river channel and the bridges and culverts should have more clearance to allow for the drainage of floodwaters. Water reservoir for storing water could be better solution for solving flash flood problem and shortage of water during dry season. Although river training is expensive but can be taken for guiding water flow to control sudden inundation.

8.6.4 Relocation of the Foothill Settlements

The relocation of houses and structures are one of the major instrument for reducing the vulnerability and risk of landslide disaster. A relocation policy has been prepared for extremely vulnerable people living along the hill slope informal settlements (Ahammad 2009). The authority has been provided to the City Corporation and Divisional Commissioner for selection of relocation point. The relocation process is in process and two sites are primarily selected for relocation of 2,00,000 people of the Chittagong city. However, no such initiatives have been taken for the Chittagong Hill Tracts region.

8.6.5 Controlling Hill Cutting Through Proper Enforcement of Legal Provisions

Hill cutting which is one of the major causes of landslide in Bangladesh required to be prohibited. Sustainable structural and non-structural measure should be taken compliance with existing legal provisions. Construction of retaining wall in the hill cutting area could be possible structural measures. A national hill management policy preparation is in process to ban on setting up brick kilns and reducing cutting hill for settlements and other purposes (Newage 2012). The building construction rules by Chittagong Development Authority also mentioned the hill cutting issues which requires more clarification and proper enforcement.

8.6.6 Real Time Monitoring and Early Warning

Prediction of the hazardous event can be able to reduce impact of disaster through issuing early warning and taking proper preparedness measures. Real time monitoring and prediction of weather including rainfall and discharge forecasting is very important for early warning and effective response measures. A flood forecast and warning system was developed by the FFWC of the Bangladesh Meteorological Department (BMD) however the capacity of the BMD through technical instruments and expertise requires. The early warning system is activated in the Chittagong city and enforced, however during the recent year response from the people were absent caused massive destruction and loss of lives.

8.6.7 Technological Innovation and Extension

Technological innovation is very important for disaster resilient agricultural production and extension. Early cropping varieties considering the seasonal climatic nature is inevitable for protecting crops from impact of flash flood and associated landslide.

8.6.8 Enhancement of Public Awareness

Awareness of the people for reduction of landslide disaster risks is one of the major instruments for risk reduction of landslide in Bangladesh. Along with the awareness building different social motivation activities can be organized. The Community Based Risk Reduction Organization may be created involving local people including the most vulnerable group of the landslide prone area. The landuse and management; social afforestation, hill cutting issues should be incorporated within the comprehensive awareness programme.

8.6.9 Establishment of the Emergency Response and Recovery Team and Facilities

The lack of capacity of the local government in the early response and relief facilities is one of the major causes of landslide disaster large impact. Special team including the local people should be formed, trained and drilled for the capacity development. The technological capacity along with the administrative capacity also required to be developed. Local volunteer group has been formed under different NGOs and CBOs who are working in different incident requires more training and instrumental support for effective response. Insurance can be provided to the affected people for early recovery and government can facilitate motivating insurance companies in this purpose.

References

ADRC (2003) Bangladesh: flash flood: 2003/06/14. Asian Disaster Reduction Center (ADRC), Dhaka. URL: http://www.adrc.asia/view_disaster_en.php?NationCode=50&lang=en&KEY=603. Accessed 29 Jun 2012

Ahammad R (2009) Understanding institutional changes for reducing social vulnerability to landslides in Chittagong City. Bangladesh, Ecosystem, Governance and Globalization, Stockholm Resilience Centre (SRC), Stockholm University

BBC (2010) Landslides and floods kill 53 in Bangladesh. British Broadcasting Center (BBC), London. URL: http://www.bbc.co.uk/news/10315737. Accessed 28 Jun 2012

BBS (2009a) Compendium of environment statistics of Bangladesh. Bangladesh Bureau of Statistics (BBS), Statistics Division, Ministry of Planning, Government of Bangladesh (GoB), Dhaka

BBS (2009b) Statistical yearbook of Bangladesh. Bangladesh Bureau of Statistics (BBS), Statistics Division, Ministry of Planning, Government of Bangladesh (GoB), Dhaka

Brammer H (1996) The geography of the soils of Bangladesh. University Press Limited, Dhaka

BWDB (2005) Flash flood events and mitigation management in Bangladesh. In: Workshop on managing flash floods and sustainable development in the Himalayas, 23–28 October, 2005, Lhasa, Bangladesh Water Development Board (BWDB), Dhaka

CDO (2008) Identification of landslide causes and recommendation for risk reduction. Chittagong Divisional Office (CDO), Chittagong

CNN (2001) Flash floods hit 50000 people in Bangladesh. Cable News Network (CNN), Atlanta

CNN (2011) Mudslide in southeastern Bangladesh kills 15. Cable News Network (CNN), Atlanta

CNRS (2009) Adopting early warning system to address flash flood in the deeply flooded Haor (Wetland) basin in North-East Bangladesh. Center for Natural Resources Studies (CNRS), Dhaka

Colombo AG, Hevas J, Arllam ALV (2002) Guidelines on flash floods prevention and mitigation. NEIDES, Ispra

Gunter BG, Rahman A, Rahman AFMA (2008) How vulnerable are Bangladesh's indigenous people to climate change? Bangladesh Development Research Center: Dhaka, Bangladesh

Hassan R, Scholes R, Ash N (eds) (2005) Ecosystems and human well-being: Current state and trends. Volume 1: Findings of the condition and trends, Working Group of the Millennium Ecosystem Assessment. Washington: Island Press

Highland LM, Bodrowsky P (2008) The landslide handbook – a guide to understanding landslides. United States Geological Survey (USGS), Virginia

Hossein F (2012) Landslides, floods kill 76 in southern Bangladesh. The Daily Star, Chittagong. URL: http://www.dailystar.com.lb/News/International/2012/Jun-27/178383-thirty-dead-150000-stranded-in-bangladesh-floods.ashx#axzz1yz4BehDI. Accessed 28 Jun 2012

Huddleston B, Ataman E (2003) Towards a GIS-based analysis of mountain environments and populations, Environment and Natural Resources Working Paper No. 10. Rome: Food and Agriculture Organization of the United Nations

ICIMOD (2010) Managing flash flood risk in the Himalayas. Information Sheet # 1/10. International Centre for Integrated Mountain Development (ICIMOD), Kathmandu

IFRC (2003) Bangladesh flash floods and Landslides. International Federation of Red Cross and Red Crescent (IFRC), Dhaka

Islam MN, Uddin MN (2002) Country paper on hydrogeology section. In: International workshop on arsenic issue in Bangladesh, Dhaka

Khan FH (1991) Geology of Bangladesh. University Press Limited, Dhaka

Khan MS (2011) Strategies for arresting land degradation in Bangladesh. In: Sarker D, Azad AK, Singh SK, Akter N (eds) Strategies for arresting land degradation in South Asian Countries. SAARC Agriculture Centre, Dhaka

Kohler T, Maselli D (eds) (2009) Mountains and climate change – from understanding to action. Geographica Bernensia and Swiss Agency for Development and Cooperation, Berne

Mahmood AB, Khan MH (2010) Landslide vulnerability of Bangladesh hills and sustainable management options: a case study of 2007 landslide in Chittagong City. SAARC Workshop on Landslide Risk Management in South Asia. SAARC Disaster Management Centre, New Delhi

Mannan A (2002) Stratigraphic evolution and geochemistry of the Neogene Burma Group, Burma Basin, Sleet, Bangladesh. Faculty of Science, University of Oulu, Oulu

Mirza MMQ, Warrick RA, Erickson NJ, Kenny GJ (2001) Are floods getting worse in the Ganges, Brahmaputra and Meghna basins? Glob Environ Chan B Environ Hazards 3(2):37–48

Newage (2012) Relocation of families from Chittagong hill slopes decided. Newage, Dhaka. URL: http://www.newagebd.com/detail.php?date=2012-05-10&nid=9805. Accessed 29 Jun 2012

Oxfam (2010) In-depth assessment report early flash flood in Hoar area 2010. Oxfam, Dhaka

Rashid HE (1991) Geography of Bangladesh. University Press Limited, Dhaka

REUTERS (2007) Bangladesh landslide toll reaches 128. Reuters. URL: http://www.reuters.com/article/2007/06/15/environment-bangladesh-landslide. Accessed 28 Jun 2012
REUTERS (2012) 100 dead, 250,000 stranded in Bangladesh floods, London, Reuters. URL: http://in.reuters.com/article/2012/06/27/bangladesh-floods-idINDEE85Q03T20120627. Accessed 28 Jun 2012
Riemann KU (1993) Geology of Bangladesh. Girder Borntraeger, Berlin
SRDI (1997) Physiograpy of Bangladesh. Soil Resources Development Institute (SRDI), Dhaka
UNDP (2009) Socio-economic baseline Survey of Chittagong Hill Tracts. Chittagong Hill Tracts Development Facility (CHTDF), United Nations Development Programme (UNDP), Dhaka
UNDRO (1979) Natural disaster and vulnerability analysis, report of Expert Group Meeting, 9–12 July 1979, Office of the United Nations Disaster Relief Coordinator: Geneva
UNEP-WCMC (2002) Mountain watch: environmental change and sustainable development in mountains. United Nations Environmental Programme (UNEP), Nairobi
UN/ISDR (2002) Newsletter Issue 5, ISDR Informs: Latin America and the Caribbean disaster reduction for sustainable mountain development. United Nations-International Strategy for Disaster Reduction
USGS (1976) Lightweight aggregate production from clay-stone and shale in Bangladesh. United States Geological Survey, Rolla
WFP (2012) History of shocks. Food Security Atlas, World Food Programme (WFP), Dhaka

Chapter 9
Arsenic Contamination in Bangladesh: Contemporary Alarm and Future Strategy

Md. Anwarul Abedin and Rajib Shaw

Abstract For thousands of years, groundwater has served as a unique and reliable source of potable water in developed as well as developing countries. But, at the end of the twentieth century, groundwater contamination of arsenic in Bangladesh is a serious national problem. Sixty-one districts out of sixty-four districts currently have been affected by arsenic contamination and up to 77 million people in Bangladesh have been exposed to toxic levels of arsenic from drinking water. To save the lives of millions in Bangladesh: a great challenge for the Government of Bangladesh is to provide safe drinking water for the urban and rural population. Hence, this chapter provides an insight into the historical background to the problem of safe drinking water in Bangladesh from surface water, groundwater, and rainwater sources. It includes special reference to the occurrences, nature and extent and causes of arsenic contamination in groundwater aquifers of Bangladesh, which has emerged as a major crisis of the present decade. In addition, this chapter put its special attention on social problems due to the presence of arsenic in the drinking water, its impact on human health, agriculture, food chain and environment, Government and other organizations initiatives against arsenic contamination and the present status of technological achievement of arsenic removal at individual and community level in Bangladesh and finally this chapter serves as a ground for the common readers including researchers, policy makers, practitioners and academia.

Keywords Arsenic contamination • Community perception • Ground water usage • Strategy and policy • Surface water

Md.A. Abedin (✉) • R. Shaw
Graduate School of Global Environmental Studies, Kyoto University, Kyoto, Japan
e-mail: masumagriculture@yahoo.com

R. Shaw et al. (eds.), *Disaster Risk Reduction Approaches in Bangladesh*, Disaster Risk Reduction, DOI 10.1007/978-4-431-54252-0_9, © Springer Japan 2013

9.1 Introduction

Water quality will become the principal limiting factor for sustainable development in many countries early in this century (Ongley 1999). "Everything living is created from water" is an ancient quotation, which closely describes the importance of water (Anonymous 1997). Alarming information has emerged in recent decades about the widespread presence of arsenic (As) in groundwater used to supply drinking water in many countries on all continents. Of these regions, Bangladesh and West Bengal are most seriously affected in terms of the size of the population at risk and the magnitude of the health problems (Khan et al. 2000; ANL 2003).

A member of the nitrogen family, arsenic is an odorless and tasteless semi-metal that occurs naturally in rock and soil. It can combine with other elements to form organic and inorganic arsenicals, the latter being generally more toxic and more prevalent in water. WHO says consumption over long periods of time of drinking water containing arsenic in excess of 10 μg/L can lead to arsenicosis, a chronic illness that produces skin disorders, gangrene and cancer of the kidneys and bladder.

The contamination of groundwater in Bangladesh is often called the biggest mass poisoning in history. It began in the 1970s, when the United Nations Children's Emergency Fund (UNICEF) initiated the construction of millions of tube-wells with the aim of providing Bangladeshis with clean and safe drinking water—an alternative to the surface water contaminated with diarrhea-causing bacteria that killed a quarter of a million Bangladeshi children each year. But as the tube-wells eliminated one problem, they, in turn, brought about a new tragedy of unimagined proportions. Arsenic-contaminated water from the wells started poisoning millions, bewildering the scientists trying to understand how the poisoning works as well as how to control it. Meanwhile, the 8–12 million contaminated wells across the country continue to be the main source of water for most Bangladeshis, presenting a growing danger to their lives. The scope of the calamity has drawn worldwide concern and triggered a broad effort to remedy the problem.

Arsenic is a heavy metalloid that is toxic to plant and animal and carcinogenic to Human. Therefore, the existence of arsenic in groundwater is of global concern because it is a serious threat to human health and many of its compounds are especially potential poisons. In Bangladesh, the arsenic contamination of groundwater was confirmed in 1993, with groundwater arsenic concentrations of some areas reaching ≥2 mg As/L (Tondel et al. 1999; BGS 1999). Karim (2000) reported that groundwater of the majority wells in 60 of 64 districts, covering approximately 118,000 km^2 (nearly 80 % of the country), had arsenic concentrations exceeding the WHO's limit of 0.01 mg/L, and 70 % groundwater contained arsenic above 0.05 mg/L (Ahmed 2001). The severely affected areas are in the southeast and southwest zones, with a report of isolated "hot spots" in the other parts of the country. As per BGS report (1999), 35 % of the samples were above 0.05 mg/L (the Bangladesh Drinking Water Standard) and 51 % samples above 0.01 mg/L (the WHO Guide Value). Rashid et al. (2004) observed that arsenic concentration started to increase with the beginning of dry season and continued up to May/June and decrease with the beginning of the monsoon and reached at the minimum level after the monsoon. They found arsenic concentration to be the highest with hand tube-wells (HTWs) (27 %), followed by shallow tube-wells (STWs) (21 %) and the least with deep tube-wells (DTWs) (7 %). The middle layers (i.e. those between 40 and160 ft) reflected the highest levels of arsenic contamination in groundwater, the shallower layers up to 35 ft and the deeper layers below 160 ft below the surface showed uniformly low (safe) levels of arsenic. In case of lateral zoning with a 4 km assigned distance, most of the unsafe wells were within the first zone and decreased gradually with the increase of distance from the river Padma (Rashid et al. 2004).

At the end of the twentieth century, the arsenic contamination in groundwater in Bangladesh has been documented as a serious environmental health disaster with severe socioeconomic consequences; a great challenge for the Government of Bangladesh is to provide safe drinking water for the urban and rural population. Hence, this chapter provides an insight into the historical background to the problem of safe drinking water in Bangladesh from surface water, groundwater, and rainwater sources. Then the focus shifts to the occurrences and causes of arsenic contamination in groundwater aquifers of Bangladesh, which has emerged as a major crisis of the present decade. In addition, social problems due to the presence of arsenic in the drinking water, its impact on human health, agriculture, food chain and environment, Government and other organizations initiatives against arsenic contamination and the present status of technologies available for the removal of arsenic in Bangladesh and finally conclusion and recommendations are briefly discussed.

9.2 Potential Scale of the Problem

This section focuses on origin of the problem, extents and how groundwater used for drinking, cooking and irrigation becomes contaminated with arsenic. Worldwide this is the most prevalent route of exposure and the primary cause of arsenicosis.

9.2.1 Origin and Natural State

Arsenic is a ubiquitous allotropic trace element and the 52nd most common element in the earth's crust (Ahmed 2003). It is tasteless and odorless. It occurs naturally in many mineral ores and is the main constituent of more than 200 mineral species. Arsenopyrite (FeAsS), realgar (As_4S_4) and orpiment (As_2S_3) are minerals commonly found in ore zones (Selinus et al. 2005).

Arsenic occurs in the environment in several oxidation states but in water is mostly found as the inorganic forms, arsenite (+3) and arsenate (+5). Organic forms of As rarely significant in groundwaters but may become more important in water affected by industrial pollution. The toxicity of different arsenic species varies in the order arsenite > arsenate > monomethylarsonate > dimethylarsinate. Trivalent arsenic is about 60 times more toxic than arsenic in the oxidized pentavalent state, and inorganic arsenic compounds are about 100 times more toxic than organic arsenic compounds (Jain and Ali 2000). Unlike other heavy metalloids and oxyanion-forming elements, arsenic can be mobilized under a wide range of oxidizing and reducing conditions at the pH values typically found in groundwater (pH 6.5–8.5).

Arsenic is one of the oldest poisons known to men and its applications throughout history are wide and varied (Vahidnia et al. 2007). Because arsenic in the bedrock is easily dissolved into surrounding water, inorganic arsenic is frequently present at elevated concentrations in groundwater (IARC 2004; WHO 2001). The presence of large-scale arsenic contamination of groundwater in Bangladesh was first confirmed by the Department of Public Health Engineering in 1993. This raised much concern. However, given the nature and magnitude of the problem, so far there has been a dearth of reliable estimates of the extent of arsenic exposure and its related health problems.

The high arsenic groundwaters of the region are mainly from aquifers of Holocene age which comprise unconsolidated grey micaceous sands, silts and clays deposited as alluvial and deltaic sediments associated with the Ganges, Brahmaputra, and Meghna river. The sediments are derived from the upland Himalayan catchments and from basement complexes of the northern and western part of West Bengal, India. Many studies have observed that the highest concentrations of arsenic in the shallow Holocene aquifer of Bangladesh occur at depths typically around 15–50 m (Harvey et al. 2002; Klump et al. 2006). Therefore, the Holocene sediment is the main host of As-contaminated groundwater in Bangladesh (e.g., Smedley and Kinniburgh 2002; Ravenscroft et al. 2005; BADC 1992), i.e., As- contaminated groundwater occurs in Holocene sediments along the channels of the main rivers, the Ganges, Brahmaputra, and Meghna.

Arsenic contamination of groundwater occurs predominantly in shallow aquifers. The middle layers (i.e. those between 12 and 49 m) show the highest levels of arsenic contamination in groundwater, the shallower layers up to 11 m and the deeper layers below 49 m below the surface shows uniformly low (safe) levels of arsenic (BGS 1999). Arsenic concentration starts to increase with the beginning of dry season and continues up to May/June and decreases with the beginning of the monsoon and reaches at the minimum level after the monsoon (Rashid et al. 2004).

9.2.2 Causes (Mechanism of Arsenic Release in Groundwater)

The arsenic in groundwater is of natural origin, and the distribution of arsenic contaminated groundwater is related to the geology, with most of the arsenic contaminated tube-wells drawing water from the Middle and Upper Holocene sediments. From current literature, it appears that the actual causes of high arsenic concentration in the groundwater of Bangladesh have not yet been clearly pinpointed. Out of the few hypotheses initially proposed to explain the possible mechanism of arsenic release, the scientific community appears to have converged on two major variants:

1. Arsenopyrite and Arsenic-rich Pyrite (Oxidation Process):
   Oxidation of arsenic bearing sulfide minerals [such as arsenopyrite (FeAsS) and pyrite (FeS_2)] in aquifer can release arsenic into underground water. The rate of oxidation of sulfide minerals is limited by the presence of an oxidizing agent, most commonly atmospheric oxygen (as O_2). Relatively deeper ground water is isolated from the atmosphere and the availability of oxygen in deep aquifers is limited by the amount of oxygen present in recharge water. Human activity that can significantly influence sulfide mineral oxidation and arsenic release into the aquifer is increased pumping of under ground water. Increased pumping and reducing recharge can greatly accelerate oxidation rates of arsenic-bearing minerals by lowering water table and exposing minerals to atmospheric oxygen. The whole process is also known as "Oxidation Process".

   In the absence of oxygen, nitrate (NO_3) can also act as an oxidizing agent and can promote oxidation of arsenic-bearing sulfide minerals. High nitrate concentrations from agricultural activities can therefore enhance arsenic release into under ground water.
2. Arsenic rich Iron Oxi-hydroxides (Reduction Process):
   Arsenic derived from weathering of arsenic-rich base metal sulfides is often found to be associated with iron oxyhydroxides in downstream sediments. Arsenic (both arsenite and particularly arsenate) has high affinity for iron oxyhydroxides and becomes associated with them as a result of adsorption. Sediments in the Ganges delta region are known to have iron oxyhydroxides grains on the mineral coatings on the mineral grains and at many places these coatings have been found to be rich in arsenic.

   Arsenic can be released from these arsenic rich iron oxi-hydroxides as a result of dissolution and desorption. A reducing redox environment (oxygen-deficient conditions) in the subsurface can promote dissolution of iron oxi-hydroxides and release of associated arsenic into under ground water. The normal burial of the alluvial sediments during the development of the delta leads to strongly reducing conditions due to the microbial consumption of oxygen during the process of organic matter oxidation. Introduction of organic waste into an aquifer can also promote a reducing environment. In addition, lowering of pH can also promote dissolution of iron oxi-hydroxides and subsequent release of associated arsenic. This process in general is known as the "Reduction Process".

The above two hypotheses may be operative in different parts of a country at a time. However the BGS report "Arsenic contamination of groundwater in Bangladesh" strongly supports the second hypothesis—the reduction process associated with iron oxides, although it does state that it is still a hypothesis and requires further study. The BGS study found that high arsenic concentrations were associated with strongly reducing conditions rather than oxidizing conditions. The oxidation hypothesis is not getting support in the absence of widespread arsenopyrite in arsenic-prone areas of Bangladesh, and arsenic concentrations showed a broad negative correlation with sulphate concentrations.

9.2.3 Scale of the Problem

Usually, Bangladesh is very much dependent on groundwater resources both for drinking and irrigation purposes. Until the detection and identification of Arsenic, groundwater was considered safe for drinking. Tube-wells have, in the majority, replaced the traditional surface water sources and diarrhoeal disease has reduced significantly. An estimated 97 % of drinking water of the rural population in Bangladesh is now supplied by groundwater. About 80 % of the population is covered by manually operated shallow tube-wells and 6 % by manually operated deep tube-wells. It has been estimated that about 8.0 million hand pump tube-wells have been installed under private initiatives and government has sunk about 1.2 million tube-wells. In 1993 DPHE first identified high concentration of arsenic in shallow tube-well in Chapai Nawabganj adjacent to an area of West-Bengal which had been found to be extensively contaminated in 1988. Extensive contamination was confirmed in 1995 when additional surveys showed contamination of shallow tube-wells across much of southern and central Bangladesh. WHO declared arsenic contamination as a "Major Public Health Issue" in 1996 and informed Bangladesh Government to deal with emergency basis.

Department of Public Health Engineering (DPHE) along with British Geological Survey (BGS) and Mott MacDonald Ltd. survey (approximately 3,500 samples) throughout Bangladesh, but excluding the Chittagong Hill Tracts, revealed that 27 % of the shallow tube-wells are contaminated with arsenic above the level of 0.05 mg/L (50 ppb) and 46 % of the shallow tube-wells tested are contaminated with arsenic above the WHO guideline 0.01 mg/L (10 ppb). Eight of the sixty-one sampled districts had no samples exceeding the Bangladesh standard for arsenic (0.05 mg/L) and all districts except Thakurgaon had at least one well exceeding WHO guideline value.

According to the study finding the worst affected districts (percent of sampled wells with greater than 0.05 mg/L arsenic) were Chandpur (90 %), Munshiganj (83 %), Gopalganj (79 %), Madaripur (69 %), Noakhali (69 %), Satkhira (67 %), Comilla (65 %), Faridpur (65 %), Shariatpur (65 %), Meherpur (60 %), Bagerhat (60 %) and Lakshmipur (56 %). The least affected districts were Thakurgoan, Barguna, Jaipurhat, Lamonirhat, Natore, Nilphamari, Panchagar, Patuakhali (and 0 %), Rangpur (1 %), Dinajpur (2 %), Noagoan (2 %) (NAISU 2002). Figure 9.1 illustrates the severity of arsenic contamination in Bangladesh.

DPHE/BGS/MML estimated that the population exposed to arsenic contamination more than 0.05 mg/L (>50 ppb) would lie in the range 18.5–22.7 million. However the BGS-DPHE studies finally gave an estimation of the number of population exposed to arsenic concentration above 0.05 mg/L (50 ppb) and 0.01 mg/L (10 ppb) to be 35 million and 57 million respectively [based on upazilla-averaged statistics the exposure levels to arsenic exceeding 0.05 mg/L (50 ppb) and 0.01 mg/L (10 ppb) were computed as 28 million and 46 million respectively, but the BGS report states that they consider the larger figures to be more reliable].

Dhaka Community Hospital and School of Environmental Studies (SOES), Jadavpur University, Calcutta Trust tested water from 64 districts of Bangladesh.

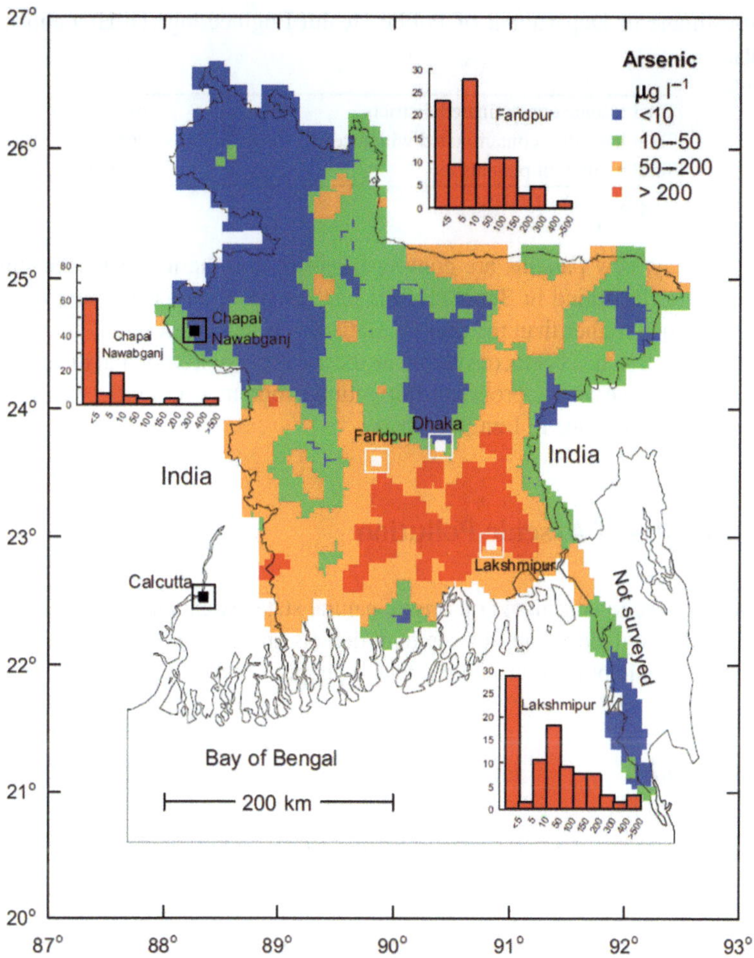

Fig. 9.1 Arsenic contamination in Bangladesh (adopted from Abedin 2011)

Their finding up to February 2000 shows that in 47 districts arsenic in ground water is above 0.05 mg/L and in 54 districts above 0.01 mg/L.

Normally, continual contact to high doses of arsenic cause dermatologic, neurologic, vascular and carcinogenic effects. The most common symptoms of arsenicosis is dermatological which include dark/white pigmentation (melanosis) and gradual hardening of palms and soles along with appearance of hard nodular lesions over these areas (keratosos). Exposure to arsenic from drinking water increases the risk of skin, lung and bladder cancer and possibly that of other sites also. In a report WHO has predicted that in most of the southern part of Bangladesh almost one in every ten adult deaths will be a result of cancer triggered by Arsenic poisoning in the next decade (Smith et al. 2000). From the experience of Taiwan it has been forecasted that almost two million of people are at risk of developing cancer in the next decades.

An estimate of Department of Public Health Engineering (DPHE) and DGHS reveals:

Total arsenic contaminated districts	61
Total Upazillas contaminated with arsenic	268
Total number of patients	13,333

August 2002—DGHS

Day by day, new patients are coming in notice and the number of arsenicosis patients so far identified in Bangladesh is just the tip of an iceberg and the actual number is much higher than the present statistics. Hence, the current arsenic situation of Bangladesh has been considered as the greatest environmental disaster of the world. So effective and appropriate mitigation program is urgently warranted for tackling this catastrophe in Bangladesh.

9.3 Impacts of Arsenic Pollution

The previous section discussed on origin, natural state, extent and severity of arsenic pollution of Bangladesh and this segment depicts about the devastating effects and impacts of arsenic poisoning of the affected people.

9.3.1 *Health*

The data collected by the governmental bodies, NGOs and private organizations reveal that a large number of populations in Bangladesh are suffering from melanosis, leucomelanosis, keratosis, hyperkeratosis, dorsum, non-petting oedema, gangrene and skin cancer18. Melanosis (93.5 %) and keratosis (68.3 %) are the most common presentations among the affected people. Patients of Leucomelanosis (39.1 %) and hyperkeratosis (37.6 %) have been found in many cases. Few cases of skin cancer (0.8 %) have also been identified among the patients seriously affected by the arsenicals (arsenite and arsenate). Figure 9.2 show skin lesions on back, palm and soles, respectively.

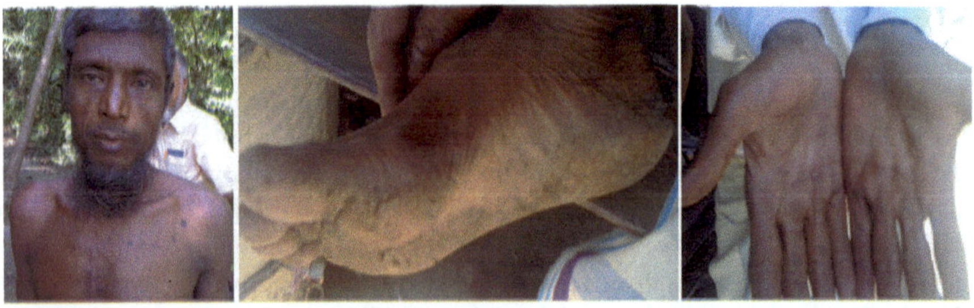

Fig. 9.2 Arsenic-affected diseases in Bangladesh

There are several factors may have been responsible for triggering off the arsenic related diseases in Bangladesh. The primary reason appears to be the malnutrition, a state that describes 80 % of the population of Bangladesh. Having less immunity, a huge number of people are suffering from the chronic arsenic poisoning. Many People have died, many are dying and many will die of arsenic diseases. In brief, the majority of the people in Bangladesh are grappling with the massive health crisis caused by the arsenic diseases.

9.3.2 Agriculture and Food Safety

Over the last 20 years there has been a large increase in the extraction of groundwater for irrigation in Bangladesh. Of the four million ha of land under irrigation, about 2.4 million ha are now being irrigated from shallow tube-wells. In fact, approximately 95 % o f the groundwater extracted is for irrigation, mainly for *Boro* [dry season] rice production. The remaining 5 % is used for domestic purposes. If we consider that a quarter of tube-wells used for drinking water in Bangladesh are thought to be contaminated by arsenic, then a very high percentage of contaminated irrigation tube-wells can be expected. One study we reviewed estimated that each year, the amount of arsenic added to arable soil, mainly paddy fields, through irrigation amounts to around 1,000 tones. In the west-southwest of Bangladesh, where the highest concentrations of arsenic are found in soil, irrigated land had higher levels compared to adjacent non-irrigated fields. From the limited data available, there are indications that soil concentrations of arsenic are increasing over time because of irrigation. But is unclear under what conditions and in what time frame this takes place, which makes it very difficult to quantify the risks. Figure 9.3 represents the arsenic transport in soil-water-plant systems.

Use of high arsenic content groundwater for irrigation has resulted in the accumulation of arsenic in crops and plants. Contamination was even greater in leafy vegetables—in amaranthus and spinach, arsenic content can be double or triple the levels found in rice. Based on the studies carried out so far, it is significant to note that high concentrations of arsenic in vegetables and rice in affected areas, such as Chandipur village in Laximpur (Table 9.1), indicate that dietary habits are also responsible for arsenic problems for the people of Bangladesh (Huq et al. 2001).

Fish constitutes an important part of the diet in Bangladesh. Concentration of arsenic in water is not always responsible for the arsenic level in fish species. Most important is how a metal is transferred tropically through the food webs and, for example, whether it is biomagnifiable or biodiminishable. Although there are no values for the arsenic concentrations in fish in Bangladesh, it is important to note that much of the arsenic present in fish and shellfish is in either the fat-soluble or water-soluble organoarsenic form that is essentially non-toxic (Lunde 1973).

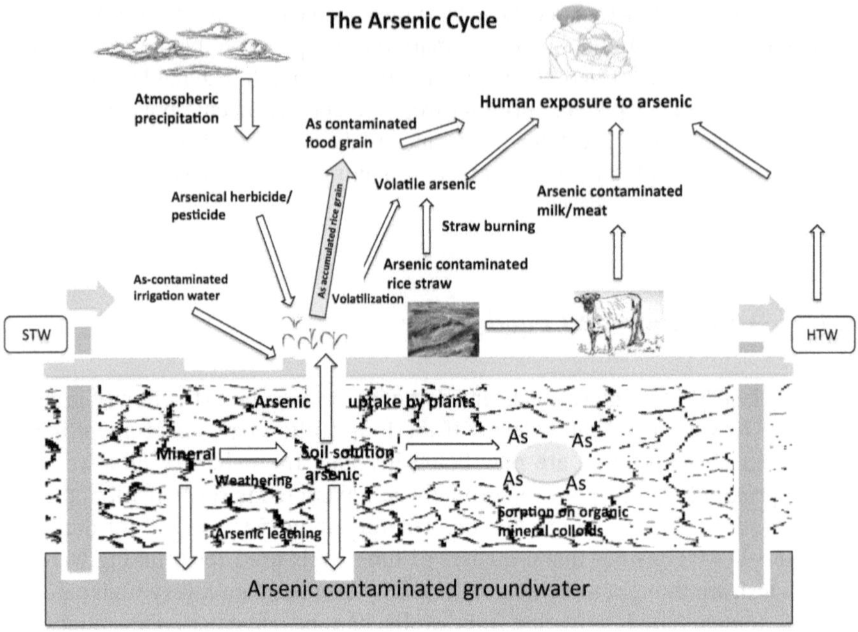

Fig. 9.3 The arsenic cycle in soil-water-plant interfaces (redraw from DPHE 2000)

Table 9.1 Arsenic concentration (μg/g) in vegetables in an affected area in Bangladesh (Huq et al. 2001). Figures rounded

	Arsenic concentration (μg/g)		
Samples	Minimum	Maximum	Average
Rice (*Oryza sativa*)	0.44	0.50	0.47
Kochu, leaf (*Colocasia antiquorum*)	0.11	0.11	0.11
Kochu, stem (*C. antiquorum*)	0.74	0.92	0.83
Kochu, creeper (*C. antiquorum*)	0.21	0.23	0.22
Gourd, leaf (*Lagenoria siceraria*)	1.88	2.01	1.94
Brinjal (*Solanum melongena*)	0.18	0.20	0.19
Sweet potato, leaf (*Ipomea batatus*)	0.38	0.41	0.40
Pui shak (*Basilia alba*)	0.30	0.33	0.31
Kalmi shak (*Ipomea aquatic*)	0.29	0.35	0.32
Papaya (*Carica papaya*)	0.19	0.26	0.22
Red pumpkin (*Cucurbita maxima*)	0.20	0.27	0.24

9.3.3 Water

Generally, drinking water and the food chain are the main pathways by which one can be exposed to the poisonous effects of arsenic. It has already been discussed how the arsenic level in groundwater in Bangladesh varies from one region to

Table 9.2 Estimated daily intake of arsenic in the affected areas of Bangladesh

Population at risk of the affected districts[a]	75 million
Drinking water habits including rice water (L/d)[b]	5
As concentration (μg/L)	>50–1,000
As intake per person per day (μg/d/person)	>250–5,000
Intake for affected population (ca. 75 million) (t/a)	5.48–11

[a]Dainichi Consultant, Inc. (1999)
[b]BGS/MML (1999)

another because of geological and geomorphological nature of the areas as well as the fact that the aquifers are shallow. The information indicates how groundwater arsenic concentration varies with shallow aquifers (<150 m). Even old and new types of wells generate different results, which cannot be explained. However, different studies indicate that in the affected areas, groundwater arsenic concentration varies between >50 and 1,000 μg/L (maximum permissible limit in Bangladesh: 50 μg/L). Occasionally, very high concentrations of arsenic in shallow tube-wells in Pabna (14 mgAs/L) and in Kushtia (9 mgAs/L) were reported (The New Nation 1996; DCH 1998). It is possible to estimate the daily intake of arsenic from drinking water in the affected areas and this is shown in Table 9.2. Daily inorganic arsenic intake for the adult population in the affected areas varies between >250 and 5,000 μg/day as compared to 5 μg/day in the United States (US EPA 1982). This high intake originates not only from drinking water, but also from drinking rice-water and water used for cooking.

9.3.4 Environment

Arsenic has not only vast impact on health, livelihood and food chain, water of Bangladesh but also has large amounts of arsenic end up in the environment. Water and land-living plants and animals show a wide range of sensitivities to different chemical forms of arsenic. Their sensitivity is modified both by biological factors and also by their surrounding physical and chemical environment. Furthermore, arsenic compounds cause short-term and long-term effects in individual plants and animals and in populations and communities of organisms. These effects are apparent, for example, in aquatic species at concentrations ranging from a few micrograms to milligrams per litre. The nature of the effects depends on the species and time of exposure. The effects include death, inhibition of growth, photosynthesis and reproduction, and behavioral effects. Environments contaminated with arsenic contain only a few species and fewer numbers within species. If levels of arsenate are high enough, only resistant organisms, such as certain microbes, may be presented.

9.3.5 Social

Arsenic related disease creates massive social problems like social hazards, poverty, social instability, superstition, ostracism, and marriage related problems.

This problem is also linked with poor mental health and developmental disabilities. These issues based on various studies are discussed below.

Social Hazards and Poverty: Arsenic is not only a physical but also a social phenomenon that directly and indirectly hits the social life (Nasreen 2003). In addition, arsenic toxicity and arsenicosis diseases, arsenic poisoning creates extensive social implications for its victims and their families in affected areas. A number of socio-economic problems like social uncertainty, social injustice, social isolation and problematic family issues are reported due to arsenicosis (Bhuiyan and Islam 2002; Hassan et al. 2005). Arsenicosis is found to be more prevalent among the poor (Khan et al. 2006) who suffer from dietary deficiency, who have no alternative sources of safe drinking water and who are unable to get proper care and treatment because of financial constraints. The effects of long-term arsenicosis are also severe in the poor and could cause social problems, interrupt the societal ties (Chowdhury et al. 2006) and trigger the social problems. Unfortunately, these social problems in affected areas are still not fully recognized and understood (Chowdhury et al. 2006).

There is a strong link between poverty and arsenicosis diseases. Arsenicosis enhances the economic burden of the poor. The majority of victims are considered as a burden to their family and society. Most of the poor arsenicosis patients remain untreated due to financial restraints. Again, if the poor arsenicosis patients go for treatment, they need to spend a big proportion of their money on this, which finally diminishes the household income and increases the economic burden on the poor victims and their families. Moreover, the cost of obtaining arsenic free water also diminishes household income (Chowdhury et al. 2006).

Social Instability: In Bangladesh, arsenicosis causes extreme instability of social life which is producing social stigmatization and discrimination (Chowdhury et al. 2006; Nasreen 2003). Unaffected people are generally scared of arsenicosis, therefore they tend to avoid and isolate arsenic victims (Chowdhury et al. 2006; Nasreen 2003). Social conflicts over contaminated water destroy the social harmony and network relationships (Chowdhury et al. 2006; Nasreen 2003). Arsenicosis disease hampers socialization by social stigmatization and discrimination (Nasreen 2003). For instance, arsenic patients often remain ostracized in all age-groups and barred from social activities. Children of arsenicosis patients are not allowed to attend social and religious functions as well as denied to take water from a neighbour's tube well and students debarred from school (Nasreen 2003). Affected families are also not allowed to take baths in any of the village ponds (Nasreen 2003). Some unaffected people behave in a hostile manner and think that patients should either stay in their homes or leave the village (Hassan et al. 2005).

Ostracism and Marriage Related Problems: Most of the cases, arsenic victims are abandoned, not only by society but also by their family members. There are some instances that arsenicosis leads to a break-down of the marital relationships. For instance, wives were divorced or separated or sent back to their parents' house because of the arsenicosis disease (Chowdhury et al. 2006). There are also some evidences that wives left arsenic affected husbands because they were afraid of arsenicosis (Nasreen 2003). Problems before marriage are also notable. For example, it is difficult to find a spouse for an arsenic victim. Generally people are reluctant to establish marital relationships with those families suffering from arsenicosis. Young women and men in the affected families are advised to remain unmarried (Chowdhury et al. 2006). Affected women also experience socially undesirable events like dowry, physical torture, and polygamy (Chowdhury et al. 2006). Due to the patriarchal system and lower socio-cultural position of women in the society, unmarried women and women abandoned by husband and families live inhumanly (Chowdhury et al. 2006).

9.4 Current and Proposed Activities

After describing the impact of arsenic on all most all aspect of Bangladesh, this section emphasis on the initiatives and activities of different sector viz. international organizations, government, NGOs, and private sectors i.e. "who is doing what" on arsenic mitigation in context of Bangladesh.

Department of Public Health Engineering (DPHE) and Local Government Division (LGD) are both key departments under the Ministry of Local Government Rural Development and Co-operation (LGRD&C). DPHE has a number of different arsenic activities at various levels of implementation and is working with a wide variety of development organizations. Furthermore, other governmental organizations viz. Water Development Board, Directorate General Health Services and international organizations viz. World Health Organization, World Bank, Japan International Cooperation Agency etc. Table 9.3 summarizes major activities of Bangladesh Government along with different organizations regarding arsenic. Beside these organizations, Asia Arsenic Network, Arsenic Crisis Information Centres, BRAC (Bangladesh Rural Advance Committee), CARE Bangladesh, Dhaka Community Hospital, Dhaka Ahsania Mission, Grameen Bank, Water Aid Bangladesh, Proshika, NGO Forum for Drinking Water Supply and Sanitation, World Vision Bangladesh, International Centre for Diahorreal Disease Research, Bangladesh, Bangladesh University for Engineering and Technology (BUET), Dhaka University still doing research on arsenic and try to solve the problem for the people of Bangladesh.

9.5 Methods of Arsenic Detection in Both Field and Laboratory

Detection and determination of arsenic is a key component of any arsenic mitigation program. Detection is carried out to find arsenic-affected areas on a large scale (screening testing) and also to determine which sources are contaminated within individual communities (blanket testing). Testing and/or detection is not a one- off

Table 9.3 Summary of major activities of Bangladesh Government and national and International organizations

Organizations	Major activities
Bangladesh Government	The National Arsenic Mitigation Programme (NAMP) is a new initiative of the GoB to develop a more coordinated response to the Arsenic crisis than before, through a programme through partnership approach
Department of Public Health Engineering (DPHE)/World Bank/Swiss Agency for Development and Cooperation (SDC)	a) Screening, Community Development and Mitigation Program in 188 Upazilas b) Screening and Mitigation in 100 Pourashava c) Training of 2,000 Doctors and 11,000 Health Workers d) Strengthening of existing DPHE Zonal Laboratories and establishment of three new zonal labs
DPHE/UNICEF	a) Testing of 51,000 tubewells in 1998 using field test kits b) Comprehensive communication campaign including radio and television
DPHE/DANIDA	a) Screening for tube-well b) Installation of about 950 deep tube-wells up to June 2002 c) Evaluation of Bucket Treatment Unit (BTU) as arsenic removal technology d) Pond Sand Filter e) Rain Water Harvesting System
Directorate General of Health Services/UNICEF	a) Active patient identification in UNICEF working areas by house to house screening b) Compilation of a patient profile c) Distribute medicine
DPHE/WHO	This includes technological aspects of arsenic removal and an epidemiological review of the health effects and immediate actions required for mitigation
Australian Agency for International Development (AusAID)	AusAID has initiated an arsenic mitigation program named "Australian Arsenic Mitigation Program (AAMP)" with funding of \$A3 million over the period 2001–2004
Department for International Development (DFID)	a) Groundwater Studies for Arsenic Contamination b) Rapid Assessment of Household Level Arsenic Removal Technologies
Japan International Cooperation Agency (JICA)	a) Training on Arsenic, arsenicosis and how to use test-kit b) Test all tubewells c) Operation of a Mobile Arsenic centre to conduct awareness education with visual Aid d) Identify arsenicosis patients e) To install safe water options f) Distribute medicine

activity, periodic testing is an important part of understanding the nature of the arsenic contamination problem and assessing mitigating efforts. See Module 10 for more information on the role of testing within overall mitigation programs.

It is more difficult to analysis for arsenic than for other contaminants in water, for which a simple strip test or a meter give an instant result. However, the various methods available for testing for arsenic have improved in recent years. Both field and laboratory testing methodologies can be used.

9.5.1 Field Methodologies

Field kits are very important given the scale of testing required but are not accurate for determining the exact arsenic concentration in water at the low concentrations important for human health. A semi quantitative result of arsenic concentration can be obtained using the field kits, most of which depend on comparing the color obtained during the test to a color chart with the kit. Field kits users are usually from a non-technical background, and may feel pressurized to read the resulting color to tend towards expected results. However, proper training to the field workers does improve results. Some crosschecking of field results is important.

Mercury Bromide stain method: Recently, most of the field-test kits (e.g. Merck, Asian Arsenic Network -AAN, General Pharmaceuticals Limited -GPL, NIPSOM, HACH) are based on the "Gutzeit" method. This process involves the reduction of arsenite and arsenate by zinc to give arsine gas which is then used to produce a stain on mercuric bromide paper.

There have been many studies on the sensitivity and reliability of these kits. The most extensively field tested of these kits are the Merck, AAN and GPL kits. The evaluations have generally shown these kits to be reasonable at detecting high concentrations (greater than 100 ppb) but less reliable at lower concentrations.

The newly developed HACH kit is currently undergoing extensive field-testing and to date has produced encouraging results on both reliability and accuracy when cross-checked with laboratory testing.

Calorimetric methods: Another field test kits use the SDDC (Silver Diphenyl Dithio Carbamate) method which relies on arsine generation and the colour reaction with SDDC. Arsenic hydride is absorbed into a solution of silver diphenyl dithio carbamate; the orange to red-violet soluble compound that is produced is analyzed by absorption spectrophotometry. The absorption line is measured to find the arsenic concentration. If no substances that obstruct the process are present then detection of arsenic concentrations to below 50 ppb is feasible.

9.5.2 Laboratory Methodologies

The following are the different methodologies for determining arsenic concentration in the laboratories that are currently used in Bangladesh:

1. Atomic Absorption Spectrophotometry
   (a) Flow Injection Hydride Generation AAS
   (b) Graphite Furnace AAS
   (c) Flame AAS
   (d) Electrothermal AAS
2. Neutron Activation Analysis
3. Inductively Coupled Argon Plasma Emission Spectrometry (ICP)

4. Anodic Stripping Voltametry
5. Spectrophotometry
6. X-ray spectroscopy
    (a) Particle-induced X-ray emission spectrometry (PIXES)
    (b) X-ray fluorescence (XRF) spectroscopy
    (c) X-ray absorption fine structure (XAFS) spectroscopy

The minimum detection level of Flow Injection Hydride Generation Atomic Absorption Spectrophotometer is 0.003 mg/L and that of Spectrophotometer is 0.03 mg/L.

9.6 Arsenic Mitigation Options

Information about location and extent of arsenic contamination of groundwater is the primary requisite for undertaking any mitigation initiatives. On the other hand, safe water for drinking as well as domestic and irrigation purposes is the prerequisite for improved public health situation and sustainable development of a country. In addition to the arsenic problem, numerous water borne diseases toll lot of lives every year in Bangladesh. Both the GoB and NGOs have been working with great emphasis in this field since last few decades for ensuring safe drinking water to the people. This section depicts the arsenic mitigation options currently available in Bangladesh by the following sub-headings:

9.6.1 General Mitigation Approach

Safe drinking water can be obtained from surface water, groundwater or rainwater sources. Each source has different characteristics relating to quality, quantity, reliability, user acceptability and costs that will determine use. When considering sources and water supply technologies for arsenic mitigation, selection should be on the basis of avoidance or of a substantial and consistent reduction of the ingestion of arsenic. In assessing best alternative water options and/or arsenic removal technologies the following basic criteria should be evaluated:

- Water Quality (i.e. does the system consistently provides bacteriologically and chemically safe water?)
- Water Quantity (e.g. flow rate, access to water at peak times)
- Affordability (capital, operation & maintenance)
- Reliability
- Life expectancy (e.g. how does one know when to change filter media)
- Convenience (e.g. time & effort involved)
- Time considerations
- Gender issues (e.g. ergonomically appropriate, division of labour)

- Environmental risks (e.g. sludge disposal, excess water/drainage issues)
- Operational safety (e.g. user accidental misuse, physical and chemical safety, robustness)
- Risk substitution (e.g. introduction of bacteriological contamination)
- Logistical sustainability of system (e.g. are reagents available locally, life time of system, market base, involvement of private sector)
- User acceptability
- Necessary operation and maintenance training
- Information, Education & Communication

Following two are the main approaches for mitigation of arsenic contamination of drinking water:

1. Filtration process (e.g. Sonofilter, three pitcher filter, shapla filter etc.)
2. Alternative water supply options (e.g. Deep Tube-well (DTW) water, Rain water and Surface water etc).

Filtration process: The household level arsenic removal technologies that are commonly used in the affected areas of Bangladesh those have affordability to buy this type of technology.

- Passive Sedimentation
- DPHE/DANIDA Bucket Treatment Unit
- Stevens' Institute Technology
- Adarsha Filter
- GARNET home-made filter
- SONO- 3 kolshi method
- BUET Activated Aluminium Filter
- Alkan Activated Aluminium Filter
- Tetra Hedron
- Ion exchange resins
- Rajshahi University/New Zealand iron hydroxide slurry
- SORAS (Solar Oxidation and Removal of Arsenic)

Alternative Safe Water Options: Alternative safe water options can be provided at either household or community level.

The household level options include:

- Accessing water by sharing safe (green) tube-wells,
- Using protected dug wells,
- Rainwater harvesting,
- Treating surface water (e.g. by use of household filter or solar disinfection).

Community level alternative options include:

- Deep tube well with hand pump,
- Deep tube well with motorized pump, overhead tank and series of stand posts (below tank or distributed in the area),
- Rainwater harvesting,
- Surface water treatment through pond sand filters,

Pond water Rainwater harvest Pond sand filter

Fig. 9.4 Individual and community level alternative safe water options

- Other surface water filters or treatment technologies,
- Disinfection systems.

Figure 9.4 shows some individual and community level alternative safe water options available in the arsenic affected rural areas.

9.6.2 SIPE (Socio-Economic, Institutional, Physico-Chemical and Environmental) Approach

Water is the key to life and civilization. But even today, about 1.1 billion people are living without access to safe drinking water (Shaw and Thaitakoo 2010). Bangladesh also faces worst condition due to arsenic contamination alone with salinity and drought make water scarcity particularly in southwest part of Bangladesh. It has significant impacts on health, food chain but also has social impacts indirectly viz. students drop outs due water collection, social conflicts on water usage and livelihoods of the people. Therefore, arsenic mitigation needs physical, social, environmental and institutional measures like water quality, infrastructure, water supply and access, water policy, budget and so on. Besides, the water management practices and mitigation process have strong relations to the institutional dimensions like arsenic policy, management skills of local governments and coordination among different stakeholders including government and non-governments. In addition, environment is the key component to govern the entire cycle of human and water management.

Abedin and Shaw (2012) developed a holistic approach named SIPE (Socio-economical, Institutional, Physicochemical and Environmental) approach that helps to measure the existing level of different physico-chemical, socio-economic, institutional and environmental conditions of the arsenic-affected area. This approach has four dimensions (physico-chemical, socio-economic, institutional and environmental) and there are 20 primary indicators (4×5) (five each for physico-chemical, socio-economic, institutional and environmental, respectively) and 100 secondary indicator ($4 \times 5 \times 5$) (each of the primary indicators are divided into five secondary indicators). The details are given in Table 9.4. Data on each of these primary and

Table 9.4 Measures of safe water adaptability index (SIPE)

Dimension	Primary indicator	Secondary indicator
Physico-chemical	Water quality	Arsenic level
		Salinity level in surface water
		Salinity level in groundwater
		pH level
		Level of iron
	Water quantity	Surface water reserve (m^3)
		Ground water reserve (m^3)
		Recharge rate
		Annual rainfall (mm)
		% of safe water availability
	Infrastructure	Existence of water distribution network
		Water treatment centre
		Water conservation
		Water reservoir (tank, dam)
		Rain water harvest
	Land use	% of built up area
		% of water bodies
		% of vegetation
		% of area under shrimp cultivation
		% of cultivable land
	Water supply and access	% of population having access to safe drinking water
		% of area having access to irrigation for agriculture
		% of arsenic free tube well
		Piped water supply
		Distance of nearest safe fresh/drinking water source
Socioeconomic	Education-awareness	% of population having education on water security
		% of drop out due to water insecurity
		Taking preventive measure
		Information dissemination
		Safe water awareness-raising activities
	Social capital	Leadership
		Social norms
		Social network
		Social abnormalities (divorce)
		Migration (regional shifting)
	Social facilities	Domestic use
		Participation on community water system
		Water reuse
		Existence of Pond sand filter
		Household expenditure on water
	Health	Arsenic affected people
		Diseases related to water scarcity
		Health care facilities & Health awareness-raising activities
		Health care worker visit
		Level of preparedness on health issues
	Water dependent livelihood	Agriculture
		Shrimp culture and fisheries
		Employment for water construction & maintenance
		Drinking water supplier
		Salt production activities

(continued)

Table 9.4 (continued)

Dimension	Primary indicator	Secondary indicator
Environmental	Biodiversity	Extinction of fresh water loving fishes
		Biodiversity (diversity within species)
		Species (change in number of species and population size)
		Ecosystem (change in natural habitats)
		Natural resource management program
	Soil degradation	Soil pH
		Soil structure
		Drainage
		Fertility (nutrient balance)
		Accumulation of salts
	Salinisation	People affected
		Area affected
		Impact of shrimp cultivation on environment
		Crop loss (yield)
		Introduction/cultivation of saline tolerant varieties
	Extreme event	Cyclone
		Flood
		Drought
		Heat wave
		Cold wave
	Contamination/ pollution	Presence of Cd (Cadmium)
		Presence of Na (Sodium)
		Presence of microorganism (*Coliform* Bacteria)
		Level of BOD (Biological Oxygen Demand)
		Level of COD (Chemical Oxygen Demand)
Institutional	Policy	Arsenic mitigation policy
		Salinity mitigation policy
		Surface water quality standard
		Ground water quality standard
		National water(drinking) policy
	Management	Training program
		Public awareness program
		Warning system
		Monitoring and evaluation
		Support/subsidies
	Co-ordination	Interrelation between GO& NGO
		Interrelation between GO & community
		Interrelation between NGO & community
		Interrelation between GO and Int. donor agencies
		Public private partnership
	Budget	Allocation for water sector
		Proper utilization of budget
		External fund sourcing
		Cost-effectiveness of expenditure
		Budget sharing
	Institute	Knowledge sharing
		Municipal wastewater treatment
		Government agencies capacity
		Water activities by village organization/community based organization
		Devolvement of function

Source: Abedin and Shaw (2012)

secondary indicators are collected through a comprehensive set of questionnaires. Each question should be ranked between one (poor, not sufficient/existent) and five (good) in a five-point rating scale.

Therefore, the SIPE approach was carried out at 16 upazilas of two arsenic, salinity and drought affected districts in southwestern part of Bangladesh (9 upazilas from Khulna and 7 upazilas from Satkhira district). With regard to the overall safe water adaptability results, five upazilas are categorized as having medium, ten upazilas having low and one upazila having very low adaptability. There is no very high and high level of adaptability found in this study. This result reflects the upazila's low to medium preparedness and adaptive capacity against safe water supply and access due to insufficient safe surface and groundwater bodies, inadequate infrastructure, lack of surface and groundwater standards, poor economic condition and lack of education-awareness, poor health-care facilities, high frequency of extreme events and natural contamination, lack of budget, poor salinity, arsenic and drought policy at upazila level, among other. Moreover, it is noted that SIPE approach not only covers water scarcity caused by arsenic but also by salinity and drought too.

9.7 Lesson Learned

Arsenic contamination in drinking and irrigation water is predominantly a public health problem. Therefore, understanding the public health dimension of the problem is an essential prerequisite to successful arsenic mitigation programme. Failure to understand the complicated nature of the arsenic problem at the policy level complicates the mitigation efforts in a country, such as Bangladesh, with very limited resources. This is further complicated by lack of coordination between health and drinking water supply authorities. As this is a public health problem, public health experts should play the leading role in arsenic mitigation programmes in Bangladesh.

Provision of arsenic free, safe drinking water remains at the core of any arsenic mitigation efforts. In addition, arsenic contaminated irrigation water that used to irrigate crop field especially *Boro* paddy cultivation also important which has passive ingestion through food chain. Although this issue is still not paid that much attention in compare to drinking water. In the absence of any single suitable alternative to tube wells that could supply more than 80 % of the rural Bangladeshi population, we need to consider region- specific water supply options. Given the variations in geohydrological situations, soil types and water chemistry, development and provision of region-specific water supply options are likely to be more effective. Information on water chemistry is particularly important to develop arsenic removal technologies because the performance of most of these removal technologies substantially depends on water qualities, especially pH (Milton et al. 2006). There are a number of filters that were used as arsenic mitigation option, with the SONO filter being one of those. Technical and social evaluation of it revealed that the users are reluctant to repair the broken filter by their own. Maintenance problem, lack of proper sludge disposal guidance, slow flow rate and lack of ownership were other problems of the filter (Shafiquzzaman et al. 2009). Other filters had similar drawbacks to the SONO filters. So, adequate research works need to be completed as a priority to identify the region-specific water supply options to plan an effective arsenic mitigation options. Although this was supposed to be completed at the initial phase, unfortunately appropriate scientific approach has been lacking in the whole arsenic mitigation efforts in Bangladesh.

A large proportion of the resources (time, manpower, money) from both public and private sources have already been spent to increase awareness nationwide. Furthermore, most of the research work also confined with detection and identification of water source especially hand tube-well. The majority of the people are now aware about the adverse effects of arsenic contamination in drinking water. Nevertheless, compliance with using arsenic free safe drinking water varies from population to population. Inadequate access to safe drinking water has been one of the main reasons for non-compliance. Additionally, a number of factors influence the compliance including turbidity, bad smells and taste of drinking water, distance and time constraints to fetch water from a distant source, and social conflicts (Milton et al. 2007). It was quite convenient to collect water from tube wells for the rural people. With their long time adaptation to underground water through tube well, many of the people are still finding it difficult to shift towards other water options, which are not as convenient as tube wells (Milton et al. 2007). Exposed individuals with visible arsenical skin lesions are more compliant to use safe drinking water than individuals exposed to arsenic with no visible arsenical skin lesions and individuals not exposed to arsenic through drinking water (Ahamed et al. 2006). Careful analysis of the factors influencing compliance to safe drinking water is essential to success- fully promote safe drinking water options.

Not last but least, economically and technologically, Bangladesh is not in a firm position to solve the arsenic crisis herself. She needs the help of the international community. Environmental experts and funds are desperately needed to save the lives of millions of people affected by deadly arsenic. The international community has the economic resources, environmental experts, and technologies to mitigate the arsenic contamination in groundwater. The support of United Nations, donor countries, donor organizations, agencies, and individuals is essential to save the suffering people from the devastating arsenic disaster.

9.8 Conclusions

For thousands of years, groundwater has served as a unique and reliable source of potable water in developed as well as developing countries. Recent past, groundwater contamination of arsenic in Bangladesh is a serious national problem. To save the lives of millions in Bangladesh, finding the real source and cause(s) of contamination and its mitigation is an urgent need. Then a remedial action plan should be made with participation of experts in relevant fields and community representatives.

In this study, the overall scenario of arsenic causes, impacts, its mitigation processes and future strategies are discussed in Fig. 9.5. But, the main challenges

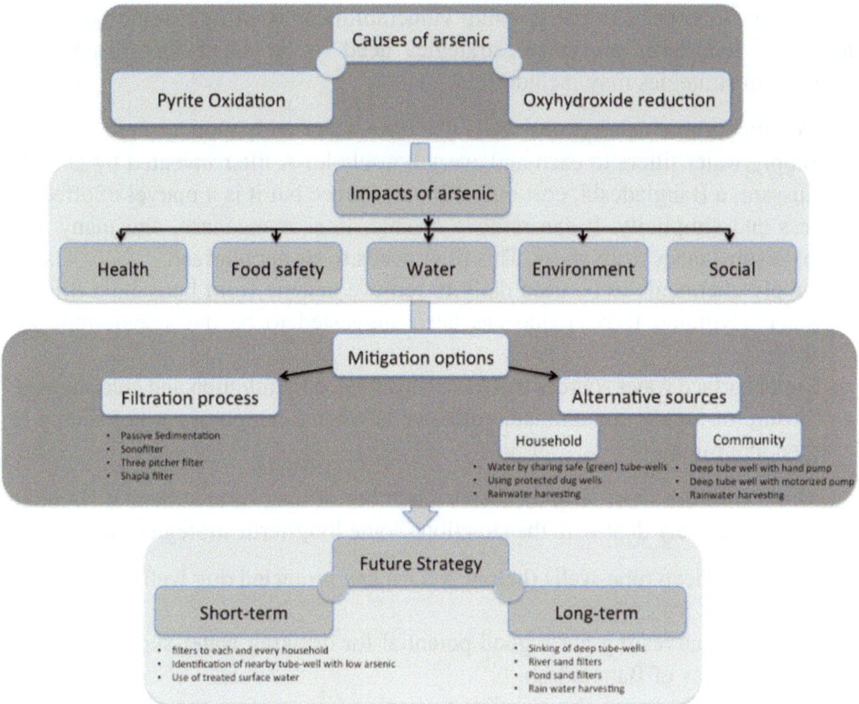

Fig. 9.5 A flow diagram shows arsenic causes, its impact, mitigation and future strategies in Bangladesh

that are ahead for Bangladesh in managing the contaminated groundwater are now (1) locating the sources of the poisoning and closing them to avoid further release of arsenic via the routes of food, soil, and water; (2) creation of mass awareness about the danger of drinking arsenic-contaminated water; (3) provision of arsenic-free, bacteriologically and chemically safe alternative sources of drinking water to the vast majority of the population; and (4) diagnosing all arsenicosis patients and providing effective management of their condition. We strongly believe, however, that mitigation of arsenic should be executed in ways that suit local conditions and requirements with cognizance taken of the health and socioeconomic impacts that are already in effect. Thus, the local community should be fully involved in the planning and development of the water supply system; whether the system is filter-based, uses surface water, harvests rainwater, or takes some other approach, all concerned people should partake in managerial and financial responsibilities. This involvement will satisfy the critical need for social compatibility. The government must cooperate with academic as well as research institutes to assess the causes and impact of arsenic poisoning and take up remedial measures whenever and wherever necessary. Hence, more field surveys eliciting the villagers' response to various facets of health and socioeconomic issues need to be conducted to formulate a long-term community-based management strategy.

In order to cope with the growing vulnerabilities posed by arsenic both short-term and long-term preventive strategies need to be taken for consideration. Short-term strategies may include:

- Identification of nearby tube-wells that have water with a low arsenic content.
- Supply water filters to each and every household. A filter invented by Dr. Abul Hussam, a Bangladeshi, cost only $35 to produce but it is a marvel of effectiveness and simplicity. It can remove arsenic, iron, manganese, and many other toxic substances from water. This filter needs to be popularized.
- Supply chemicals to be used daily to remove arsenic from household drinking water. Hydrated ferric oxide has been suggested to be the arsenic-removing agent.
- Use of surface water sources that have been treated by filtration and chlorination.
- Closure of highly contaminated tube-wells when a temporary water source has been identified.

Along with the short-term measures a number of long-term strategies should be taken to effectively deal with the situation. Some long-term strategies are:

- Sinking of deep tube-wells (below 200 m) and protected dug wells or ring wells (20–30 m).
- Rainwater harvesting has a good potential for drinking water supply in arsenic affected areas of Bangladesh.
- Pond sand filters are the simplest technology of treating the surface water to make it drinkable.
- River sand filters are another way of having arsenic contamination free drinking water.

Acknowledgement The first author gratefully acknowledges support from the Japan Society for the Promotion of Science (JSPS) of Japan in the conduct of this research.

References

Abedin MA (2011) Mitigation of arsenic in groundwater using zero valent iron: removal mechanism and application. PhD Thesis, Kyoto University, Kyoto

Abedin MA, Shaw R (2012) Safe water adaptability for salinity, arsenic and drought risks in southwest of Bangladesh. Environ Hazards J (under review)

Ahamed S, Sengupta MK, Mukherjee SC, Pati S, Mukherjeel A, Rahman MM et al (2006) An eight-year study report on arsenic contamination in groundwater and health effects in Eruani village, Bangladesh and an approach for its mitigation. J Health Popul Nutr 24(1):29–41

Ahmed FM (2001) An overview of arsenic removal technologies in Bangladesh and India, In: Feroze Ahmed M et al (eds) Technologies for arsenic removal from drinking water. A compilation of papers presented at the International Workshop on Technologies for Arsenic Removal from Drinking Water, Bangladesh University of Engineering and Technology, Dhaka, Bangladesh and the United Nations University, Tokyo

Ahmed MF (2003) Treatment of arsenic contaminated water. In: Ahmed MF (ed) Arsenic contamination: Bangladesh perspective. ITN-Bangladesh, Dhaka, pp 354–403

ANL (Argonne National Laboratory) (2003) Review of arsenic removal technologies for contaminated groundwaters, Chemical Engineering Division, Argonne, USA, http://www.ipd.anl.gov/anlpubs/2003/05/46522.pdf

Anonymous (1997) Bangladesh arsenic newsletter 1(1). July-September

BADC (Bangladesh Agricultural Development Corporation) (1992) Final Report of the Deep Tubewell II Project, Vol. 3: Groundwater. Mott MacDonald Ltd. and Hunting Technical Services. Report Produced for Bangladesh Agricultural Development Corporation and the Overseas Development Administration

BGS (British Geological Survey) (1999) Groundwater studies for arsenic contamination in Bangladesh. Main report and Supplemental volumes 1–3. Government of the people republics of Bangladesh, Ministry of Local Government, Rural Development and Cooperatives, Department of Public Health Engineering, Dhaka and Department of International Development (UK) (available at http://www.bgs.ac.uk/arsenic)

BGS report (1999) Ground water studies for contamination in Bangladesh. Phase 1, Final Report. British Geological Survey, Mott. MacDonald Ltd., London

Bhuiyan RH, Islam N (2002) Coping strategy and health seeking behavior of arsenicosis patients of rural Bangladesh: a case study of Ramganj upzilla, Lakshmipur. In: Presented at the international workshop on arsenic mitigation, Dhaka, 14–16 January, 2002

British Geological Survey/Mott MacDonald Ltd. (UK) (BGS/MML) (1999) Arsenic contamination of groundwater in Bangladesh: a review. Prepared for: Department of Public Health Engineering, Government of Bangladesh, Dhaka. S5. 54 p

Chowdhury MAI, Uddin MT, Ahmed MF, Ali MA, Rasul SMA, Hoque MA, Alam R, Sharmin R, Uddin SM, Islam MS (2006) Collapse of socio-economic base of Bangladesh by arsenic contamination in groundwater. Pakistan J Biol Sci 9:1617–1627

Dainichi Consultant, Inc. (1999) http://www.dainichi-consul.co.jp/english/arsenic/arsstat.htm

Dhaka Community Hospital (DCH) (1998) Dhaka Community Hospital. International Conference on arsenic pollution of groundwater in Bangladesh: causes, effects and remedies. Dhaka Community Hospital, Dhaka

DPHE (2000) Groundwater studies for arsenic contamination in Bangladesh. Rapid investigation phase. Final report. Mott MacDonald International Ltd and British Geological Survey. Report prepared for Department of Public Health Engineering (Bangladesh) and Department for International Development, UK

Harvey CF, Swartz CH, Badruzzaman ABM, Keon-Blute N, Yu W, Ali MA, Jay J, Beckie R, Niedan V, Brabander D, Oates PM, Ashfaque KN, Islam S, Hemond HF, Ahmed MF (2002) Arsenic mobility and groundwater extraction in Bangladesh. Science 298: 1602–1606

Hassan MM, Atkins PJ, Dunn CE (2005) Social implications of arsenic poisoning in Bangladesh. Soc Sci Med 61:2201–2211

Huq SMI, Ara QAJ, Islam K, Zaher A, Naidu R (2001) Possible contamination from arsenic through food chain. In: Proceedings of the KTH – Dhaka University Seminar on Groundwater Arsenic Contamination in the Bengal Delta Plains of Bangladesh. Edited by G. Jacks, P. Bhattacharya, and A.A. Khan, KTH Special Publication, Stockholm. TRITA-AMI report 3084

IARC (2004) Some drinking water disinfectants and contaminants, including arsenic, vol 84. International Agency for Research on Cancer, Lyon

Jain CK, Ali I (2000) Arsenic: Occurrence, toxicity and speciation techniques. Water Res 34: 4304–4312

Karim Md M (2000) Arsenic in groundwater and health problem in Bangladesh. Water Res 34(1): 304–310

Khan AH, Rasul SB, Munir AKM, Habibuddowla M, Alauddin M, Newaz SS, Hussam A (2000) Appraisal of a simple arsenic removal method for groundwater of Bangladesh. J Environ Sci Health 35:1021–1041

Khan MMH, Aklimunnessa K, Kabir M, Mori M (2006) Case–control study of arsenicosis in some arsenic contaminated villages of Bangladesh. Sapporo Med J 75:51–61

Klump S, Kipfer R, Cirpka OA, Harvey CF, Brennwald MS, Ashfaque KN, Badruzzaman ABM, Hug SJ, Imboden DM (2006) Groundwater dynamics and arsenic mobilization in Bangladesh assessed using noble gases and tritium. Environ Sci & Technol 40:243–250

Lunde G (1973) Separation and analysis of organic-bound and inorganic arsenic in marine organisms. J Sci Food Agric 24:1021–1027

Milton AH, Smith WT, Dear K, Caldwell B, Sim M, Ng J (2006) Arsenic mitigation: drinking water options in Bangladesh. In: Naidu R, Smith E, Owen G, Bhattacharya P, Nandenbaum P (eds) Managing arsenic in the environment: from soil to human health. CSIRO Publishing, Collingwood, pp 355–362

Milton AH, Smith W, Dear K, Ng J, Sim M, Ranmuthugala G et al (2007) A randomised intervention trial to assess two arsenic mitigation options in Bangladesh. J Environ Sci Health A Tox Hazard Subst Environ Eng 42(1):897–908

NAISU (NGOs Arsenic Information and Support Unit) (2002) An over view of arsenic issues and mitigation initiatives in Bangladesh. NGO Forum for Drinking Water Supply and Sanitation

Nasreen M (2003) Social impacts of arsenicosis. In: Ahmed MF (ed) Arsenic contamination: Bangladesh perspective. ITN-Bangladesh, Dhaka, pp 340–353

Ongley ED (1999) Water quality: an emerging global crisis. In: Trudgill ST, Walling DE, Webb BW (eds) Water quality: processes and policy. Wiley, Chichester, pp 29–31

Rashid MA, Ghani MA, Shah AL (2004) Extent of arsenic contamination in groundwater and its spatial and temporal variation. In: Proceedings of the workshop on arsenic in the food chain, assessment of arsenic in the water-soil-crop system. Shah MAL, Rashid MA, Rashid MH, Mandal MR, Ghani MA (eds) BARI, Gazipur, Dhaka, pp 11–21

Ravenscroft P, Burgess WG, Ahmed KM, Burren M, Perrin J (2005) Arsenic in groundwater of the Bengal Basin, Bangladesh: distribution, field relations, and hydrological setting. Hydrol J 13: 727–751

Selinus O, Alloway B et al (2005) Essentials of medical geology, impacts of the natural environment on public health. Elsevier Academic Press, Amsterdam

Shafiquzzaman M, Azam MS, Mishima I, Nakajima J (2009) Technical and social evaluation of arsenic mitigation in rural Bangladesh. J Health Popul Nutr 27(6):7–83

Shaw R, Thaitakoo D (2010) Water communities: introduction and overview. In: Shaw R, Thaitakoo D (eds) Water communities. Emerald, Bingley, pp 1–13

Smedley PL, Kinniburgh DG (2002) A review of the source, behavior and distribution of arsenic in natural water. Appl Geochem 17:517–568

Smith AH, Lingas EO, Rahman M (2000) Contamination of drinking water by arsenic in Bangladesh: a public health emergency. Bull WHO 78:1093–1103

The New Nation (1996) A Daily Newspaper of Bangladesh, 11 November 1996

Tondel M, Rahaman M, Magnuson A, Akther Chowdhury I, Faruquee HM, Akthar Ahmad SK (1999) The relationship of arsenic levels in drinking water and the prevalence rate of skin lesions in Bangladesh. Environ Health Perspect 107:727–729

US EPA (1982) Arsenic. In: Intermedia priority pollutant guidance documents. U.S. Environmental Protection Agency, Office of Pesticides and Toxic Substances, Washington

Vahidnia A, van der Voet GB, de Wolff FA (2007) Arsenic neurotoxicity – a review. Hum Exp Toxicol 26:823–832

WHO (2001) Arsenic and arsenic compounds. International Programme on Chemical Safety. World Health Organization: Geneva

Chapter 10
Sea-Level Rise Along the Coast of Bangladesh

Md. Golam Mahabub Sarwar

Abstract Sea-level rise along the Bangladesh coast was estimated using tide gauge data. It showed considerable variation in sea-level change in different parts of the coastal zone. The tide gauge station at Hiron Point in the Sundarbans showed a SLR of 3.38 mm/year. SLR at Reyenda and Amtali was observed as 3.64 and 3.16 mm/year, respectively. Sea-level rise along the central coast was observed as 5.73 mm/year at Char Changa station in Hatiya. The station in Companiganj, representing the Noakhali Feni coastal zone, showed a SLR of 2.5 mm/year. Sea-level change along the eastern part of the coast showed very low rise or fall. Sadarghat station in the Chittagong coastal zone showed a sea-level fall of 11.75 mm/year. Sea-level rise calculated from the PSMSL dataset was observed as 1.36 mm/year at the Cox's Bazar station. However, similar calculations from BIWTA in Moheshkhali and Teknaf showed a sea-level fall of 5.59 and 8.33 mm/year, respectively. Additionally, Physical vulnerability of the coast towards sea-level change was assessed revealing that the coastal areas along Patuakhali and Bhola Districts and Hatiya Island are very high vulnerable and the coastal areas along the Sundarbans and Barguna coastal zones are high vulnerable. The Noakhali Feni coastal zone is moderately vulnerable in terms of sea-level change. Vulnerability along the eastern part of the coast is very low.

Keywords Bangladesh • Coastal vulnerability • Coastal zone • Sea-level rise • Tide-gauge data

---

Md.G.M. Sarwar (✉)
Bangladesh Unnayan Parishad (BUP), Dhaka, Bangladesh
e-mail: mgmsarwar@gmail.com

R. Shaw et al. (eds.), *Disaster Risk Reduction Approaches in Bangladesh*,
Disaster Risk Reduction, DOI 10.1007/978-4-431-54252-0_10,

10.1 Introduction

Global climate have changed at a rate of 0.13 °C per decade over the last 50 years (Meehl et al. 2005). It is predicted that global temperature will rise by 1.1–2.9 °C in the twenty-first century (IPCC 2007). One of the significant impacts of global warming is a rise in sea level. Global sea level has been observed to rise at a rate of 1.8 mm/year over the period 1961–2003 and about 3.1 mm/year over the period 1993–2003 as a response to global temperature rise (IPCC 2007). Sea-level rise (SLR) will have a significant effect on global ecosystems (IPCC 2007; Nicholls et al. 2007). It will affect the coastal zone by causing shoreline erosion, saltwater intrusion and increasing threats of flooding. Compounding the effects of sea-level rise, climate change is likely to cause an increase in the number and frequency of extreme weather events globally, including tropical cyclones, floods and storm surges (Malik 1988; Groisman et al. 2004; Emanuel 2005; Webster et al. 2005; Knutson et al. 2010). Consequently, this will affect the coastal communities and populations globally.

As a low-lying country, Bangladesh is highly vulnerable to SLR (Milliman et al. 1989; Ortiz 1994; Warrick and Ahmad 1996; Choudhury et al. 1997; Ali 2000; Sarwar 2005; Sarwar and Khan 2007; Karim and Mimura 2008; Oliver-Smith 2009). The projected SLR of 1 m in the twenty-first century may affect approximately 1,000 km^2 of the coastal land of Bangladesh and its coastal population (Cruz et al. 2007). In spite of great concern about SLR impacts on Bangladesh, few studies have focussed on assessing the actual rates of SLR for this coast. The wide discussion of sea-level rise impacts on Bangladesh has largely arisen from a general perception about the low-lying elevation (Woodroffe et al. 2006) of the coastal zone of the country. Assessing location-specific sea-level rise along different parts of the coast is the main focus of this study.

10.2 Background

10.2.1 The Coastal Zone of Bangladesh

The coast of Bangladesh (Fig. 10.1) is a home to approximately 46 million people. It covers an area of 47,201 km^2 (WARPO 2006). The floodplain of the Bengal delta comprises 2.85 million hectares of cultivable land (Bala and Hossain 2010) and supports 20 % of the rice acreage of Bangladesh (Begum and Fleming 1997). Additionally, the coastal zone contains the world's largest single chunk of mangrove forests; the Sundarbans. From spatial point of view the coast of Bangladesh can be divided into three zones. The southwest coast is the coastal areas of the Ganges Tidal Plain, the southcentral zone is the Meghna Delta Plain and the southeast zone is the Chittagong Coastal belt. Karim and Mimura (2008) have named these zones as western, central and eastern regions. In addition to generalized division of the coastal zone of Bangladesh, Islam and Peterson (2009) have divided it into five

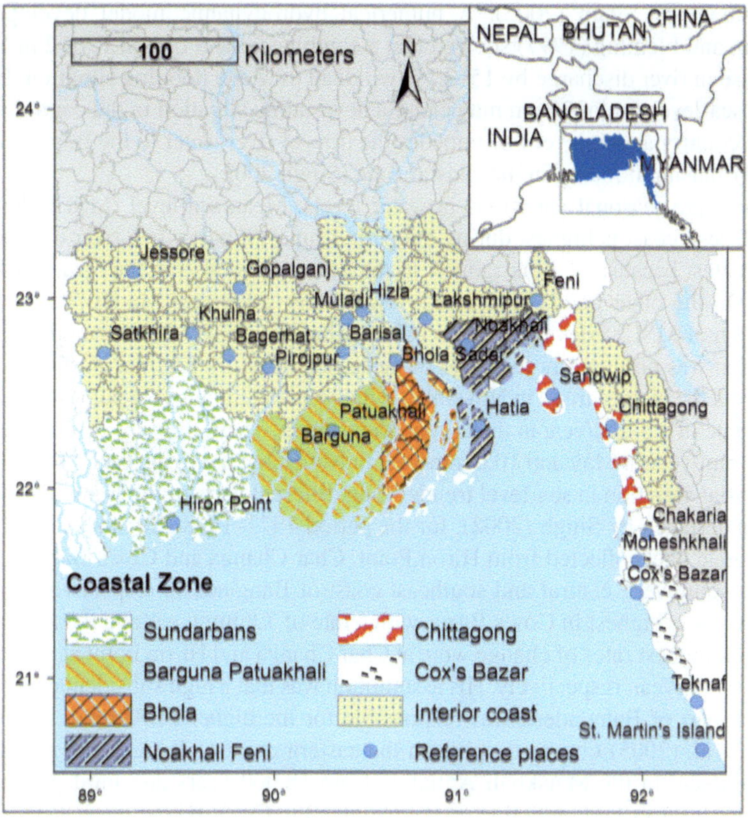

Fig. 10.1 The coast of Bangladesh showing the six coastal zones: The Sundarbans coastal zone, Barguna Patuakhali coastal zone, Bhola coastal zone, Noakhali Feni coastal zone, Chittagong coastal zone and Cox's Bazar coastal zone. Locations referred in the text are also shown

zones on the basis of administrative units. Five coasts are Khulna Coast, Barisal Coast, Noakhali Coast, Chittagong Coast and Cox's Bazar Coast. The present study has divided the coast into six zones: (1) Sundarbans coastal zone, (2) Barguna Patuakhali coastal zone, (3) Bhola coastal zone, (4) Noakhali Feni coastal zone, (5) Chittagong coastal zone and (6) Cox's Bazar coastal zone (Fig. 10.1). Administrative boundary based zonation was done to facilitate mapping and discussion.

10.2.2 Sea-Level Rise Along the Bangladesh Coast

A broad overview of climate change and sea-level rise in Bangladesh was first documented by Warrick and Ahmad (1996). Based on global trends, Choudhury et al. (1997) postulated a sea-level rise of 1.0–1.5 mm/year along the Bangladesh coast. This seems to be an underestimation compared to the range 4–7.8 mm/year in the assessment by Singh (2002). A numerical hydrodynamic model developed by Begum and Fleming (1997) showed that a combined effect of a SLR of 2 m and an increase in river discharge by 15–20 % will lead to worse flooding situation. Even a mean sea-level rise of 20 cm may cause a devastating situation in the central region of the country as sea-level rise impedes the flow of river water to the Bay of Bengal (BOB) causing flooding in the coastal zone (Singh 2002).

There are seasonal variations in water levels. Mean sea level (MSL) along the Bangladesh coast is highest during the period June–August when river discharge is the greatest because of peak rainfall in the country and also in the upstream countries. On the other hand, MSL is the lowest during January–March. The seasonal variation of MSL varies from 0.3 to 0.5 m (Khandker 1997). Furthermore, seasonal variation was observed in the trend of mean tidal level by Khan et al. (2000). They observed the highest trend in November with a rate of 8.5 mm/year but lowest in May at the rate of 2.5 mm/year at the Hiron Point station. They similarly observed a trend of 4.3 mm/year in May and 10.9 mm/year in November along the Cox's Bazar coast.

Spatial variation in sea-level trends along different parts of the Bangladesh coast was undertaken by Singh (2002), for the period 1977–1998. He analysed 22 years tide gauge data collected from Hiron Point, Char Changa and Cox's Bazar stations on the southwest, central and southeast coast of Bangladesh, respectively. Rise in sea level was highest in Cox's Bazar with a rate of 7.8 mm/year and the second and the third highest rates of change were at Char Changa and Hiron Point at rates of 6.0 and 4.0 mm/year, respectively. His assumption was that a high subsidence rate in the eastern part of Bangladesh was responsible for the higher SLR trend. In contrast Khan et al. (2005) observed uplift on the eastern coast of Bangladesh at a rate of 3.6 mm/year in the Maiskhali Anticline from 18,000 years ago until present and 2.86 mm/year in Jaldi anticline on the mainland in Cox's Bazar, from 35,000 years ago until present.

A higher trend of SLR at a rate of 5.22 ± 0.43 mm/year at Diamond Harbour near Kolkata was observed by Unnikrishnan and Shankar (2007) that suggests that a higher SLR rate may be prevailed along the Bangladesh coast. Emery and Aubrey (1989) observed a rise in sea level at Calcutta and Diamond Harbour at a rate of 6.93 and 7.26 mm/year, respectively. The rise seems to be high, but was supported by Unnikrishnan and Shankar (2007) where a rise of 5.74 mm/year was observed in Diamond Harbour by analysing the tide-gauge data of 1948–2004. In contrast, Kidderpore and Sagor Island in the vicinity of Indian Sundarbans experienced a fall in sea level at the rate of 6.06 and 4.2 mm/year respectively (Emery and Aubrey 1989).

10.2.3 Data Used in Global SLR Estimation

In understanding sea-level change, tide gauge data is the most reliable and available data over the past two centuries. Scientists worldwide have been relying on tide-gauge data in researching sea-level rise for long time (Woodworth 1991; Douglas 1997; Haigh et al. 2009; Woodworth et al. 2009). It is expected that tide-gauge data

Table 10.1 Vulnerability of coastal zone on the basis of sea-level rise

Reference	Very low	Low	Moderate	High	Very high
Gornitz (1991)	≤ −1.1	−1.0–0.99	1.0–2.0	2.1–4.0	≥4.1
Gornitz and White (1991)	≤ −1.1	−1.0–0.99	1.0–2.0	2.1–5.0	≥5.1
Gornitz et al. (1994)	< −1.0 (uplift)	−1.0–0.99	1.0–2.0	2.1–4.0 (subsidence)	>4.0 (subsidence)
Dwarakish et al. (2009)	<1.8	1.8–2.5	2.5–3.0	3.0–3.4	>3.4
Abuodha and Woodroffe (2010)	<0.0	0.0–0.9	1.0–2.0	2.1–3.0	>3.1
Ozyurt and Ergin (2010)	<1	1–2	2–5	5–7	7–9 and over
Gornitz and Kanciruk (1989)	≤ −1.0 (land rising)	−1.0–0.99	1.0–2.0	2.1–5.0	≥5.1
Gornitz et al. (1997)	< −1.0 (land rising)	≥ −1 and ≤ +1	>1 and ≤ +2	>2 and ≤4	>4.0 (land sinking)

will remain a strong factor for monitoring sea-level change. The introduction of the Gravity Recovery and Climate Experiment (GRACE) in 1992 has opened another window in researching satellite-based methods of detecting sea-level change (Woodworth et al. 2011). Since 1992, satellite altimetry data has been used for the estimation of SLR (Church and White 2006).

10.2.4 Sea-Level Rise in Coastal Vulnerability Index

Vulnerability is strongly correlated with rate of sea-level rise. Rates of relative sea-level changes are an important input into calculations of a coastal vulnerability index. All studies consider that a lower sea-level rise rate or a fall in the sea level represents the least vulnerable coast (Table 10.1). The studies covering the coastal zone of America place a fall in sea level in the lowest vulnerability class. Negative SLR trends are ignored by Dwarakish et al. (2009), Abuodha and Woodroffe (2010) and Ozyurt and Ergin (2010). Small intervals are used on the coast where the range of sea-level change is small (Dwarakish et al. 2009; Abuodha and Woodroffe 2010) and alternatively, higher intervals are used where there is a high range of sea-level change (Ozyurt and Ergin 2010).

10.3 Seal-Level Change Data

Tide gauge data has been collected from the Bangladesh Water Development Board (BWDB), Bangladesh Inland Water Transport Authority (BIWTA) and the Permanent Service for Mean Sea Level (PSMSL). After receiving data from the provider, an initial screening is done and recorded with PSMSL. This type of data is termed "Metric" data and can be used for different aspects of mean sea-level (MSL) analysis. Metric data is upgraded by verifying monthly and annual MSL against the

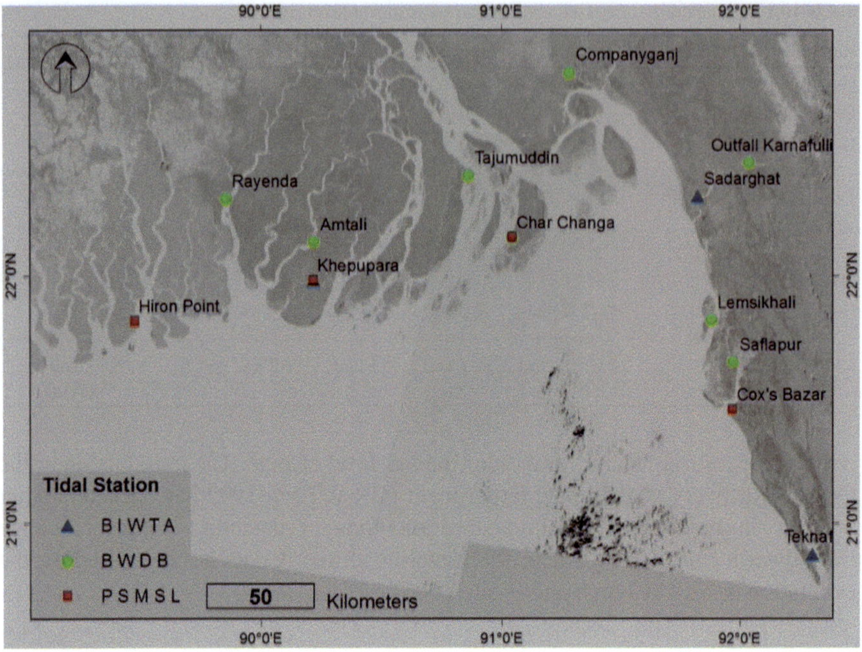

Fig. 10.2 Tide gauge stations along the Bangladesh coast

Tide Gauge Bench Mark (TGBM) of nearby land and is termed the "Revised Local Reference (RLR)" record of the station. Data categorized as RLR are suitable for time-series analysis and can be used for the estimation of long term sea-level change (Woodworth 1991; Woodworth and Player 2003). Tide gauge data from Bangladesh is obtained by PSMSL from BWDB which is then checked for further judgement. Metric data from Bangladesh is available for 11 stations (Fig. 10.2) having a data range from 1 to 33 years, but data for only four stations, Hiron Point, Khepupara, Hatiya Island (Char Changa) and Cox's Bazar, are processed to RLR level (Table 10.2).

Tidal data of Rayenda, Amtali, Tajumuddin, Hatiya, Companiganj, Outfall Karnafully (Chittagong), Lemsikhali and Saflapur stations have been collected from the Bangladesh Water Development Board (BWDB). Data of Khepupara, Sadarghat (Chittagong) and Teknaf stations have been collected from Bangladesh Inland Water Transport Authority (Table 10.2).

Standard PSMSL guidelines have been followed in selecting appropriate data periods. Data for a month is considered if tidal records for a minimum of 15 days of the month are available. Again, data for a year is considered if a minimum of 11 month's data are available. Generally, data from 20 years are analysed for SLR trend estimation (Haigh et al. 2009). However, data for a shorter period are considered in this study for cross-checking a result obtained from one data set, or if longer-term data are unavailable. The time span for the station at Companiganj, Sadarghat and Teknaf are 16, 16 and 15 years, respectively. The data for Khepupara station

Table 10.2 Tidal data for the Bangladesh coast

Station	Data source	Data period	No. of years	Missing data year
Hiron Point	PSMSL	1983–2003	20	0
Khepupara		1979–2000	21	0
Char Changa		1979–2000	21	0
Cox's Bazar		1979–2000	21	0
Rayenda	BWDB	1969–2001	32	1982, 1983
Amtali		1958–2002	44	
Tajumuddin		1969–2002	33	1980–1983
Companyganj		1984–1999	15	1989
Hatiya		1976–2007	31	
Outfall Karnafully		1976–2005	29	
Lemsikhali		1970–2007	37	1972, 1979–1981, 1987–1992
Saflapur		1969–1997	28	0
Khepupara	BIWTA	1986–1999	14	0
Sadarghat (ctg.)		1986–2001	16	0
Teknaf		1986–2000	15	0

obtained from BIWTA is only 14 years. These short-term data were also analysed to check the validity of the high rates of SLR obtained from the PSMSL dataset. Data collected from PSMSL and BIWTA have been provided as mean daily water level. Data collected from BWDB have been provided in the form of mean daily high water level and mean daily low water level. Mean daily water level of BWDB data have been obtained by averaging mean daily high water level and mean daily low water level.

Good quality data is important to have a worthy result. Data quality has been checked beforc analysis. The errors in tidal data have been corrected manually. In case of an overestimation of tidal level, erroneous data are removed from that dataset. Some impractical values, such as same figure for high and low tides, are removed from the data set before obtaining a sea-level rise trend. A total of 31 data points have been identified as erroncous and have had removed manually.

10.4 Methodology

The PSMSL data are provided in the form of mean sea level (MSL) in millimetres (mm), but the water level data from BWDB and are provided as daily high and daily low water levels in metre (m). From the BWDB data, mean water level is calculated by averaging daily high and daily low water in metres, which is then converted to millimetres. The data obtained from BIWTA are in the form of mean daily water level in metres which has also been converted into millimetres. Mean yearly water level has been calculated and plotted in Excel. Finally, trends of mean sea level (MSL) have been calculated by linear regression.

10.5 Results

10.5.1 Sea-Level Change

Sea-level change obtained by analysing tidal data in 13 stations along the Bangladesh coast indicates significant variation in different parts (Fig. 10.3). The SLR at Hiron Point has been observed as 3.38 mm/year. However, SLR at Khepupara and Amtali have been obtained as 14.84 and 3.16 mm/year, respectively, although the distance between the two stations is only 17.5 km. Rayenda station is only 53 km away from Khepupara station but shows a SLR rate of 3.64 mm/year.

Sea-level rise along the central part of the coast shows an extreme increasing trend. SLR at the Char Changa station is observed as 5.73 mm/year but that at Hatiya Island and Tajumuddin stations is observed as 19.81 and 38.82 mm/year, respectively. SLR at the Char Changa station is calculated using data from PSMSL while data from BIWTA are used to calculate SLR at Hatiya Island and Tajumuddin stations. Companiganj station represents the Noakhali Feni coastal zone and shows a rise in sea level at the rate of 2.5 mm/year.

In contrast, sea-level change along the eastern part of the coast shows very low rise or fall but the Outfall Karnafulli station in the Chittagong coastal zone

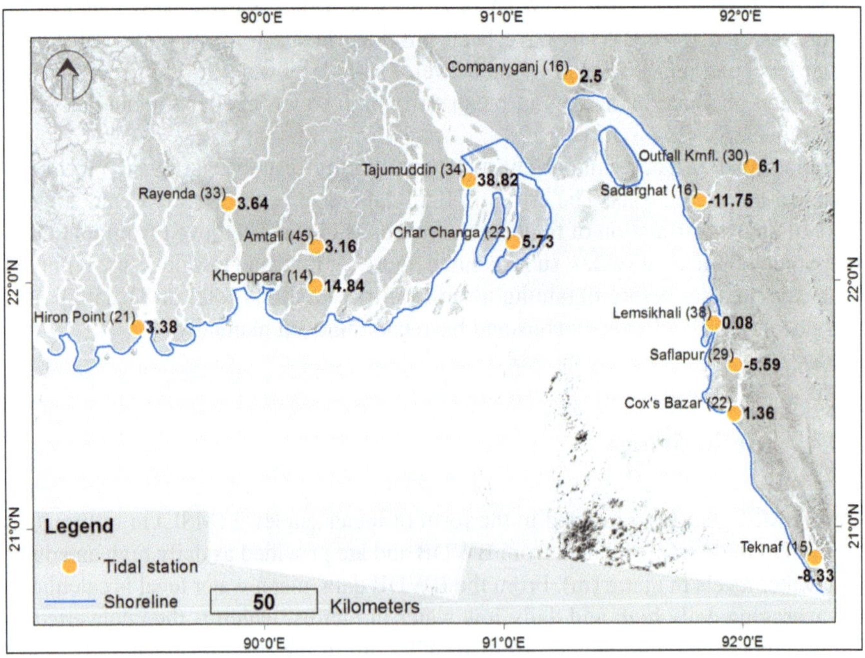

Fig. 10.3 Sea-level change at different tide-gauge stations along the coast of Bangladesh. The numbers in parenthesis are number of years of record used in the SLR calculation (data sources have been shown in Fig. 10.2)

demonstrates a SLR of 6.1 mm/year. However, Sadarghat, another tide-gauge station located in the same zone shows a sea-level fall of 11.75 mm/year. Sadarghat station is located at 15 km inland from the estuary and the Outfall station is located 30 km farther inland from Sadarghat. From the viewpoint of geographic location, the Sadarghat station is closer to the shoreline and represents a more reasonable sea-level change than the Outfall Karnafulli station. On the other hand, the eastern coastal zone of Bangladesh is treated as an active zone where uplifting is observed by Khan et al. (2005). However, a fall in sea level by 11.75 mm/year seems unusual compared to nearby Companiganj station.

Sea-level change along the Cox's Bazar coastal zone calculated from PSMSL data shows a rise of 1.36 mm/year. However, similar calculations from BIWTA data demonstrate a fall of sea level at the rate of 0.08, −5.59 and −8.33 mm/year at the Lemsikhali station on Kutubdia Island, Saflapur station in Moheshkhali Island and Teknaf station, respectively.

10.5.2 Sea-Level Rise Vulnerability of Bangladesh

A huge variation has been observed in calculated sea-level change along the coastal zones of Bangladesh. A wide range of variations between two close stations in the Barguna Patuakhali coastal zone, Bhola coastal zone and in Chittagong coastal zone have been observed. Therefore, logical sense of sea-level vulnerability has been used, instead of using numerical values.

Disregarding a high rate of SLR of 14.84 mm/year at Khepupara station, the southwest coastal zone shows a SLR of 3.16–3.64 mm/year. The Sundarbans coastal zone is believed to be subsiding by 2–4 mm/year (Goodbred and Kuehl 2000; Umitsu 1997). Thus, the relative SLR of the coastal zone becomes 5–8 mm/year because of this high rate of SLR, the southwest coastal zone has been considered as high vulnerable coast (Fig. 10.4).

The central coast demonstrated the highest rate of rise in sea level. The calculation revealed a SLR of 38.82 mm/year at Tajumuddin station in Bhola that seems to be an over estimation of sea-level change. Sea-level at Char Changa station in Hatiya Island has been rising at a rate of 5.73 mm/year. After incorporating local level subsidence, the rise can be more than 8 mm/year. Even if, the SLR estimation in Tajumuddin station demonstrates an overestimation, the coastal zone around Bhola and Hatiya Islands might have experienced the highest rate of SLR. Therefore, this part has been assigned as very high vulnerable coast (Fig. 10.4).

Companiganj station shows a SLR similar to the southwest coast. This station is located offshore with limited connection with the sea and may not represent true sea-level change of the coast. The coastal zone is located near the Meghna River estuary and might show high rates of sedimentation similar to the central coast. However, extreme sedimentation rates in the coastal zone may be compensating the rise of sea-level. Therefore, the extent of sea-level rise in the coastal zone has been treated as moderately vulnerable (Fig. 10.4).

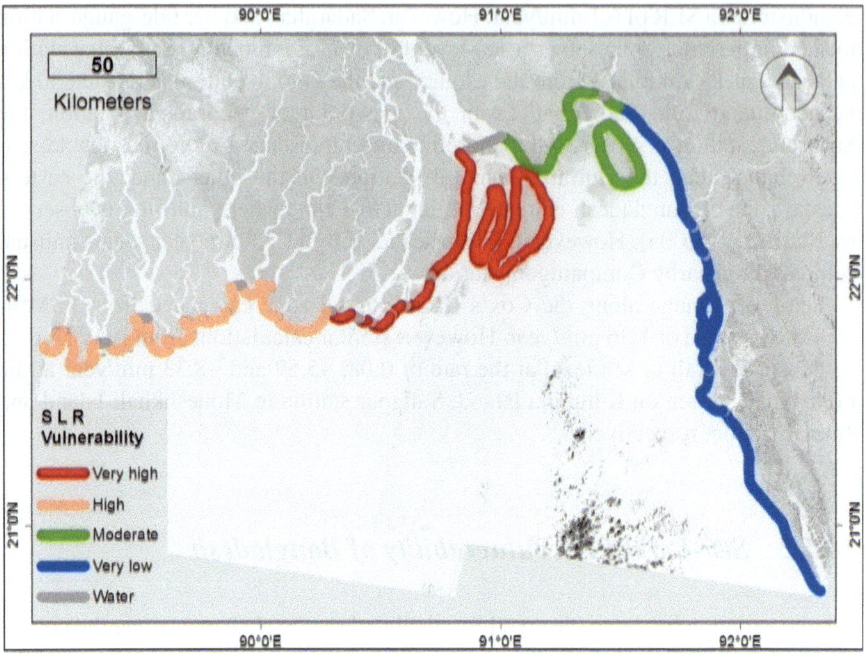

Fig. 10.4 Vulnerability of the Bangladesh coast to sea-level change

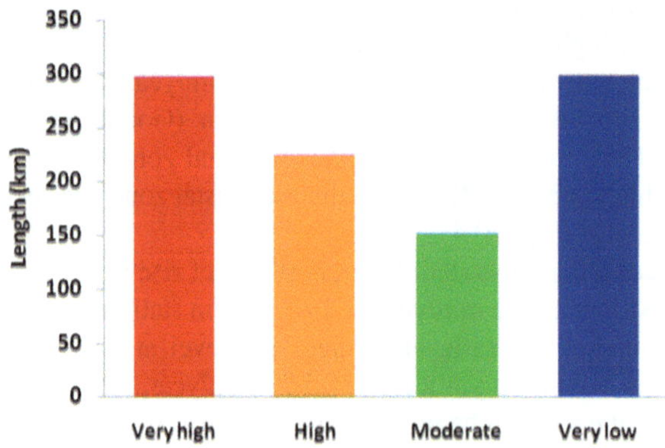

Fig. 10.5 Length of very high, high, moderate and very low vulnerable coast on the basis of sea-level change

Sea-level change along the eastern part of the coast of Bangladesh shows a falling trend, except for the Cox's Bazar station that demonstrates a SLR of 1.63 mm/year. However, this rise is also very low compared to the global SLR rate of 1.8±0.3 mm/year (White et al. 2005). Therefore, the coastal zones of Chittagong and Cox's Bazar have been considered to be a very low vulnerable (Fig. 10.4).

Vulnerability classification has assigned a total of 298 km (Fig. 10.5) shoreline along the Meghna River estuary covering part of coastal zone of Patuakhali district, the whole of Bhola coastal zone and the coasts of Manpura and Hatia Islands as very high vulnerable coast (Fig. 10.4). High vulnerable coast is observed in the Sundarbans coastal zone and part of Barguna Patuakhali coastal zone covering a shoreline length of 226 km. Moderately vulnerable coastal zone has been identified along the shoreline of the mainland in the Noakhali coastal zone, a small section of Chittagong coastal zone and Sandwip Island. The length of moderately vulnerable coastal zone is 253 km and the length of very low vulnerability coast is 297 km which is located in the Chittagong and Cox's Bazar coastal zone. This southeast coastal zone shows a very low rate of sea-level rise.

10.6 Discussion and Conclusion

Bangladesh has got global attention because of potential threat from sea-level rise. Conversely, this study has found it difficult to calculate the rates of sea-level change along the Bangladesh coast. Data availability and data quality are great concerns in estimating sea-level rise. Tide-gauge data for four stations recorded in the RLR list of PSMSL are the most authentic data sources to calculate SLR in the coastal zone, but a rise of 19.14 mm/year in sea level at the Khepupara station seems unrealistic.

The coastal zone of a part of Patuakhali, Bhola, Manpura and Hatiya Islands covering the central coast is very high vulnerable, and the southwest coast is high vulnerable which is consistent with the present coastal setting. Sediment deposition in the central coast is dominated by silt but that in the Sundarbans is mud or mangrove. Therefore, apparent sea-level rise in the Sundarbans should be more than on the central coast because of possible greater compaction. However, the Meghna River estuary area might have experienced a higher SLR because of tectonic activity as well as compaction in the zone.

Tide gauge data from Hiron Point in the Sundarbans and nearby Amtali and Rayenda stations indicate sea-level rise similar to the global average. It has been revealed that the Sundarbans coastal zone was formed in the past few thousand years, and subsidence in the area is suspected by many researchers (Umitsu 1997; Goodbred and Kuehl 2000). Considering the probability of subsidence in the Sundarbans, coastal vulnerability in this zone has been identified as high vulnerable even though SLR in the Sundarbans area has been found to be normal. Because of similar land form in much of the shoreline in the Barguna Patuakhali coastal zone to the Sundarbans coastal area, compared to the mainland and smaller islands covering the rest of BPCZ. Therefore, this part of the shoreline has also been considered as high vulnerable.

Moderate vulnerability has been indicated along the Noakhali Feni Coastal zone. This zone receives the highest volume of sediments along the Bangladesh coast and is located in the vicinity of the Tippera surface which was attributed to uplift by Morgan and McIntire (1959). Geographic location near an uplift zone and receiving a high rate of sediment supply results in a low chance of higher sea-level rise in this zone. Therefore, Noakhali Feni coastal zone has been assigned as moderately vulnerable coast.

Because of low rates of sea-level rise or even the record of sea-level fall Chittagong and Cox's Bazar coastal zones very low vulnerability has been indicated in this eastern part of the Bangladesh coast. Vertical upward movement of land in this zone is causing a fall in sea level. The rate of SLR in the Cox's Bazar coastal zone calculated in this study (1.36 mm/year) is below the global average and indicates the effects of uplift in this area, which is with good agreement with Khan et al. (2005). Despite of a great concern of sea-level rise in southwest and central parts of the coast, eastern coastal zone can be considered as free from such fear.

For Bangladesh, tidal data is the only mode of verification of sea-level change, but quality tidal data are inadequate along the coast. There are frequent tide-gauge stations on the coast but hardly any of them have produced sufficient quality data to allow time-series analysis. Considering potential impacts of SLR on Bangladesh, there is an urgent need to collect and maintain quality data to portray a true picture of SLR along the coast. To collect reliable data, the tide-gauge stations in the country, the data collection methods and data archiving need to be reviewed. Furthermore, there is a great scope for application of sophisticated techniques including altimetry and the conversion of metric data to RLR to augment the meagre tidal records.

There has been much emphasis about the vulnerability of Bangladesh to sea-level rise, and potential sea-level rise impacts on the country are a focus of this study, but variations in the rate of SLR along the coast are still uncertain and incompletely known. Lack of tide-gauge data is one of the main constraints, and may explain why there has been insufficient research on sea-level rise for the Bay of Bengal. Data obtained from three different organizations revealed different sea-level scenarios.

Estimates of sea-level rise along different parts of the coast are inconsistent at the local scale but indicate a general pattern. For the southwest coastal zone a sea-level rise at the global average rate has been indicated, but results obtained by analysing tide-gauge data available for Khepupara station has shown an abnormal rate of about 15 mm/year SLR (Fig. 10.3). The SLR at nearby Hiron Point, Rayenda and Amtali stations have shown rates of 3.16–3.64 mm/year. The SLR obtained at Khepupara has therefore been disregarded. Nonetheless, after considering local subsidence, SLR vulnerability along the southwest coast remains high.

There is a general consideration that the Sundarbans mangrove forest is experiencing sediment deposition and the adjacent coastal zone are experiencing higher rates of SLR because of subsidence and sediment compaction. This has not been detected in the analysis of tide-gauge data. The land area of the Sundarbans zone may have reached sufficient sedimentary maturity over the past 10,000 years (since much of it was deposited), leading to less compaction. On the other hand, sea-level rise along the Meghna River estuary has been found to vary significantly. SLR on Hatiya Island is 5.73 mm/year but at Tajumuddin in Bhola Island is extreme (as high as 38.82 mm/year). In spite of an overestimation of SLR, Bhola Island may be subsiding at a rapid rate possibly in response to big sediment load and local tectonic activities. Sea-level change along the eastern part of the coastal zone has been found to vary but appears very low, or a fall, probably because of uplifting along the coastal zone.

Vulnerability is strongly related to the number of people affected by a hazard and their adaptive capacity. Hence the use of the term "vulnerability" in this study, which mainly assessed physical vulnerability of the coast toward sea-level rise, undermined its true meaning in the disaster literature or development community. The studies by Shaw et al. (1998) and Abuodha and Woodroffe (2010) have designated the term "sensitivity" instead, to explain the susceptibility of a coast. The assessment of vulnerability could be put into practice by integrated sensitivity with affected environment, population and their adaptive response variables.

Sea-level rise imposes the greatest threat to the coast of Bangladesh in its western part of the Meghna River estuary, including the islands of Bhola, Manpura and Hatiya, which is subject to sediment load and compaction induced subsidence. This part of the coastal zone is densely populated and the lands are very fertile. Furthermore, the coastal zone along the Sundarbans has been identified as high vulnerable. Planning to handle the sea-level change impacts in these coastal zones need management attention. These areas must be kept in top priority in dealing with this slow but highly destructive natural threat. In spite of scattered data availability, this study offers a generalized but potential vulnerability assessment of the coast to upcoming sea-level rise. This sea-level rise vulnerability map will assist coastal planners to handle the natural threat in more efficient way.

Acknowledgements I acknowledge guidance by and inputs from Professor Colin Woodroffe. However, mistakes remain with me. Additional advices received from A/Prof. Brian Jones are also acknowledged. Dr. Aminul Islam encouraged submitting the paper to this book. Engineer Ahmed Sayeed (BIWTA) helped in data collection stage. This research was done as part of my PhD research at the University of Wollongong (UoW), Australia; with financial supports from Australian Leadership Award (ALA) Fellowship. Mapping was done using resources at the Spatial Analysis Lab (SAL), UoW, Australia.

References

Abuodha P, Woodroffe CD (2010) Assessing vulnerability to sea-level rise using a coastal sensitivity index: a case study from southeast Australia. J Coast Conservat 14:189–205

Ali A (2000) Vulnerability of Bangladesh coastal region to climate change with adaptation option. Bangladesh Space Research and Remote Sensing Organization (SPARRSO), Dhaka

Bala B, Hossain M (2010) Modeling of food security and ecological footprint of coastal zone of Bangladesh. Environ Dev Sustain 12:511–529

Begum S, Fleming G (1997) Climate change and sea level rise in Bangladesh, part I: numerical simulation. Mar Geodes 20:33–53

Choudhury AM, Haque MA, Quadir DA (1997) Consequences of global warming and sea level rise in Bangladesh. Mar Geodes 20:13–31

Church JA, White NJ (2006) A 20th century acceleration in global sea-level rise. Geophys Res Lett 33:L01602

Cruz RV, Harasawa H, Lal M, Wu S, Anokhin Y, Punsalmaa B, Honda Y, Jafari M, Li C, Ninh NH (2007) Asia. In: Parry ML, Canziani OF, Palutikof JP, von der Linden PJ, Hanson CE (eds) Climate change 2007: impacts, adaptation and vulnerability. Contribution of working group II to the Fourth Assessment Report of the Intergovernmental Panel on Climate Change. Cambridge University Press, Cambridge, pp 469–506

Douglas BC (1997) Global sea rise: a determination. Surv Geophys 18:279–292

Dwarakish GS, Vinay SA, Natesan U, Asano T, Kakinuma T, Venkataramana K, Pai BJ, Babita MK (2009) Coastal vulnerability assessment of the future sea level rise in Udupi coastal zone of Karnataka state, west coast of India. Ocean Coast Manag 52:467–478

Emanuel K (2005) Increasing destructiveness of tropical cyclones over the past 30 years. Nature 436:686–688

Emery KO, Aubrey DG (1989) Tide Gauges of India. J Coast Res 5:489–501

Goodbred SL, Kuehl SA (2000) Enormous Ganges–Brahmaputra sediment discharge during strengthened early Holocene monsoon. Geology 28:1083–1086

Gornitz V (1991) Global coastal hazards from future sea level rise. Palaeogeogr Palaeoclimatol Palaeoecol (Glob Planet Chang Sect) 89:379–398

Gornitz V, Kanciruk P (1989) Assessment of global coastal hazards from sea level rise. In: Sixth symposium on coastal and ocean management, Charleston, ASCE

Gornitz V, White TW (1991) The global coastal hazards data base. In: Future Climate Studies and Radio-Active Waste Disposal, Safety Studies. Norwich, England, 214–224

Gornitz VM, Daniels RC, White TD, Birdwell KR (1994) The development of a coastal risk assessment database: vulnerability to sea-level rise in the U.S. Southeast J Coastal Res (Special Issue No 12):327–338

Gornitz VM, Beaty TW, Daniels RC (1997) A coastal hazards data base for the U S West Coast. Carbon Dioxide Information Analysis Center, Oak Ridge National Laboratory, ORNL/CDIAC-81, NDP-043C, Oak Ridge

Groisman PY, Knight RW, Karl TR, Easterling DR, Sun B, Lawrimore JH (2004) Contemporary changes of the hydrological cycle over the contiguous United States: trends derived from in situ observations. J Hydrometeorol 5:64–85

Haigh I, Nicholls R, Wells N (2009) Mean sea level trends around the English Channel over the 20th century and their wider context. Continent Shelf Res 29:2083–2098

IPCC (2007) Summary for policy makers. In: Climate Change 2007: Impacts, Adaptation and Vulnerability. Contribution of Working Group II to the Fourth Assessment Report of the Intergovernmental Panel on Climate Change, Parry ML, Canziani OF, Palutikof JP, van der Linden PJ, Hanson CE (Eds.), Cambridge University Press, Cambridge, UK, 7–22

Islam T, Peterson RE (2009) Climatology of landfalling tropical cyclones in Bangladesh 1877–2003. Nat Hazards 48:115–135

Karim MF, Mimura N (2008) Impacts of climate change and sea-level rise on cyclonic storm surge floods in Bangladesh. Glob Environ Chang 18:490–500

Khan TMA, Singh OP, Rahman MS (2000) Recent sea level and sea surface temperature trends along the Bangladesh coast in relation to the frequency of intense cyclones. Mar Geodes 23:103–116

Khan MSH, Parkash B, Kumar S (2005) Soil-landform development of a part of the fold belt along the eastern coast of Bangladesh. Geomorphology 71:310–327

Khandker H (1997) Mean sea level in Bangladesh. Mar Geodes 20:69–76

Knutson TR, McBride JL, Chan J, Emanuel K, Holland G, Landsea C, Held I, Kossin JP, Srivastava AK, Sugi M (2010) Tropical cyclones and climate change. Nat Geosci 3:157–163

Malik M (1988) Fear of flooding: global warming could threaten low-lying Asia-Pacific countries. Far East Econ Rev 142:20–21

Meehl GA, Washington WM, Collins WD, Arblaster JM, Hu A, Buja LE, Strand WG, Teng H (2005) How much more global warming and sea level rise? Science 307(5716):1769–1772

Milliman JD, Broadus JM, Gable F (1989) Environmental and economic implications of rising sea level and subsiding deltas: The Nile and Bengal examples. Ambio 18:340–345

Morgan JP, McIntire WG (1959) Quaternary geology of the Bengal Basin: East Pakistan and India. Bull Geol Soc Am 70:319–342

Nicholls RJ, Wong PP, Burkett VR, Codignotto JO, Hay JE, McLean RF, Ragoonaden S, Woodroffe CD (2007) Coastal systems and low-lying areas. In: Parry ML, Canziani OF, Palutikof JP, von der Linden PJ, Hanson CE (eds) Climate change 2007: impacts, adaptation and vulnerability.

Contribution of working group II to the Fourth Assessment Report of the Intergovernmental Panel on Climate Change. Cambridge University Press, Cambridge
Oliver-Smith A (2009) Sea level rise and the vulnerability of coastal peoples: responding to the local challenges of global climate change in the 21st century. UNU Institute for Environment and Human Security (UNU EHS), Bonn
Ortiz CAC (1994) Sea-level rise and its impact on Bangladesh. Ocean Coast Manag 23:249–270
Ozyurt G, Ergin A (2010) Improving coastal vulnerability assessments to sea-level rise: a new indicator-based methodology for decision makers. J Coast Res 26:265–273
Sarwar GM (2005) Impacts of sea level rise on the coastal zone of Bangladesh. MSc Thesis, Lund University, Lund
Sarwar GM, Khan MH (2007) Sea level rise: a threat to the coast of Bangladesh, Internationales Asien Forum. Int Q Asian Stud 38:375–397
Shaw J, Taylor RB, Solomon S, Christian HA, Forbes DL (1998) Potential impacts of global sea-level rise on Canadian coasts. Can Geogr 42:365–379
Singh OP (2002) Spatial variation of sea level trend along the Bangladesh coast. Mar Geodes 25:205–212
Umitsu M (1997) Landforms and floods in the Ganges delta and coastal lowland of Bangladesh. Mar Geodes 20:77–87
Unnikrishnan AS, Shankar D (2007) Are sea-level-rise trends along the coasts of the north Indian Ocean consistent with global estimates? Glob Planet Chang 57:301–307
WARPO (2006) Coastal Development Strategy. Ministry of Water Resources, Government of the People's Republic of Bangladesh, Dhaka
Warrick RA, Ahmad QK (eds) (1996) The implications of climate and sea level change for Bangladesh. Kluwer, Dordrecht, pp 1–415
Webster PJ, Holland GJ, Curry JA, Chang HR (2005) Changes in tropical cyclone number, duration, and intensity in a warming environment. Science 309:1844–1846
White NJ, Church JA, Gregory JM (2005) Coastal and global averaged sea level rise for 1950 to 2000. Geophys Res Lett 32:L01601
Woodroffe CD, Nicholls RJ, Saito Y, Chen Z, Goodbred SL (2006) Landscape variability and the response of Asian Megadeltas to environmental change: the Asia-Pacific region. In: Harvey N (ed) Global change and integrated coastal management. Springer, the Netherlands, pp 277–314
Woodworth PL (1991) The permanent service for mean sea level and the global sea level observing system. J Coast Res 7:699–710
Woodworth PL, Player R (2003) The permanent service for mean sea level: an update to the 21st century. J Coast Res 19:287–295
Woodworth PL, Teferle FN, Bingley RM, Shennan I, Williams SDP (2009) Trends in UK mean sea level revisited. Geophys J Int 176:19–30
Woodworth PL, Gehrels WR, Nerem RS (2011) Nineteenth and twentieth century changes in sea level. Oceanography 24:80–93

Part III
Approaches and Issues

Chapter 11
Urban Risk Reduction Approaches in Bangladesh

Gulsan Ara Parvin, S.M. Reazul Ahsan, and Rajib Shaw

Abstract Asia-Pacific region is one of the most "at risk" parts of the world and eventually its cities are the hub of the risks. The underlying reasons for this high risk are rooted in the rapid urbanization. A majority of Asia's urban growth will be in seven developing countries, which includes Bangladesh as one of the countries having rapid urbanization pace. The rate of urbanization in Bangladesh is over 25 % and the growth rate is more than 3–5 % per year. Similar to the most other bustling cities in Asia cities in Bangladesh are growing along the river-bank, low-lying marshy lands, mining or industrial hubs and steep slopes. Inevitably these cities are emerging as hub of disaster risks. Urban growth and disaster risk are statistically correlated. Disaster risks in urban center are compounded due to unplanned urbanization and unregulated growth in disaster prone marginal areas. Unplanned and uncontrolled urbanization is exacerbating environmental problems and disaster risks in cities of Bangladesh. More than 60 % of urban population lives in four main cities in Bangladesh and Dhaka alone accommodate one-third of urban population and produces 60 % of national GDP. Therefore, disaster in urban areas has severe national consequences in Bangladesh. Eventually, disaster risk reduction from the cities should be the key issue for ensuring overall development of Bangladesh. Unfortunately, like many other countries, in spite of having risk factors, risks and vulnerability to disaster are largely under estimated in urban development of Bangladesh. As an attempt of disaster risk reduction in urban Bangladesh this paper

G.A. Parvin
Khulna, Bangladesh

S.M.R. Ahsan
Urban and Regional Planning, Khulna University, Khulna, Bangladesh

R. Shaw (✉)
Graduate School of Global Environmental Studies, Kyoto University, Kyoto, Japan
e-mail: shaw.rajib.5u@kyoto-u.ac.jp

R. Shaw et al. (eds.), *Disaster Risk Reduction Approaches in Bangladesh*,
Disaster Risk Reduction, DOI 10.1007/978-4-431-54252-0_11, © Springer Japan 2013

intends to diagnose both geological and climate related urban risks in Bangladesh. At the same time it aims to examine the existing risk reduction tools and approaches initiated by Government and civil societies in urban areas of Bangladesh.

Keywords Climate and disaster resilience index • Community participation • Dhaka's risk • Population growth • Urbanization

11.1 Introduction: Urbanization and Disaster Risks

Scholars claim that for the majority of people at risk, loss to disaster are determined more by the processes and experiences of urban development and governance than by the physical processes that shape natural or man-made hazards (UN Habitat 2007). In another way it is also said that urban areas and urban life style are directly or indirectly associated with the precedence and causes of most of the today's global environmental problems (Srinivas et al. 2009). The world is steadily becoming urban, where two out of three people will live by 2030 (Sharma et al. 2011). These people's consumption and lifestyle at individual level are greatly contributing to urban environmental problems and disaster risks. For instance, urban areas accounts for 78 % of carbon emission from human activities. In this urban millennium urban areas are the center of economic activities, industries, infrastructure and services, basic life supports, cultural heritages and amenities and amusements. Therefore, the potential disaster risks and impacts also high in cities. In fact, urban growth and disaster risk are statistically correlated (UN Habitat 2007).

Cities with high population density naturally have high risks of mortality and number of people affected by any disasters, which eventually lead to higher level of economic losses (Slusis and Aalst 2006, cited in Sharma et al. 2011). In contrast, countries having high Human Development Index face low absolute and potential disaster mortality rates (UN Habitat 2007).

Unfortunately in developing countries urbanization is increasing with a rapid pace. It is forecasted that by 2015 there will be 22 cities with populations over ten million and out of these 17 cities will be in developing countries and 13 of them will be in Asia-Pacific regions (UN-ESCAP 2005). This Asia-Pacific region is not only the most densely populated region of the world but also is the fastest growing and the most hazards prone. Therefore, it is one of the most "at risk" parts of the world and eventually its cities are the hub of the risks. The underlying reasons for this high risk are rooted in the rapid urbanization. Asia is recognized as the epicenter of an urban surge. A majority of Asia's urban growth will be in seven developing countries, which includes Bangladesh as one of the countries having rapid urbanization pace (ADB Urban Report 2003).

The percent of urbanization in Bangladesh is over 25 % and the growth rate is more than 3.5 % per year (Fig. 11.1). By 2030 this urban population of Bangladesh will be 40 % of total population (Khan 2008). Rural urban migration cause 40 % of increase in urban population and in some cities of Bangladesh this figure is as high as 70 % (Islam 2006, cited in Khan 2008). After migrating to cities these migrant

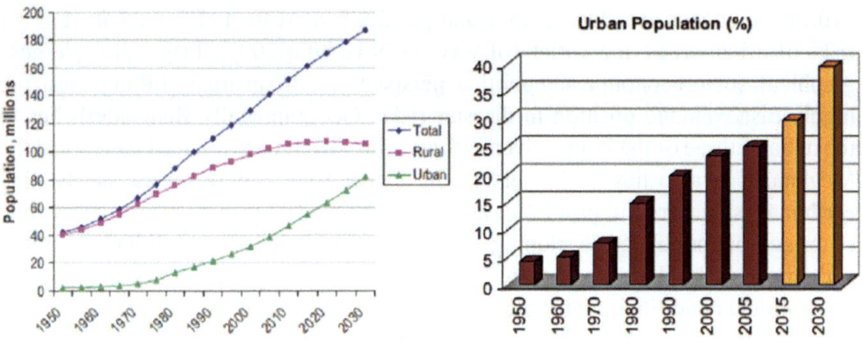

Fig. 11.1 Urbanization in Bangladesh (*Source*: World Bank 2007, Bangladesh Development Series)

people arrange their accommodations in marginal settlements with substandard housing and limited infrastructure and services. Therefore, similar to the most other bustling cities in Asia cities in Bangladesh are growing along the river bank, low-lying marshy lands, mining or industrial hubs and steep slopes. Inevitably these cities are emerging as hub of disaster risks. Unplanned and uncontrolled urbanization is exacerbating environmental problems and disaster risks in cities of Bangladesh. Hyogo Framework for Action 2005–2015, also supported that disaster risk is compounded by increasing vulnerabilities related to various elements including unplanned urbanization. Researchers state that faster growth of cities with insensitive or non-inclusive urban land use planning, urban development and management lead to higher disaster risks (Sharma et al. 2011). Unfortunately, like many other countries, in spite of having risk factors, risks and vulnerability to disaster are largely under estimated in urban development of Bangladesh (UN Habitat 2007). However, HFA priority Action 1 and 4 emphasize incorporation of disaster risk reduction into urban planning. In this context, this paper intends to diagnose the geological and climate related urban risks in Bangladesh. At the same time it aims to examine the existing risk reduction tools and approaches in urban areas of Bangladesh. This paper would contribute to build knowledge base about disaster risks, risk reduction tools and approaches in urban Bangladesh and thus help researchers, academicians, practitioners and policy makers who are working for disaster risk reduction and shaping future cities.

11.2 Geological and Climate-Related Urban Risks in Bangladesh

In order to undertake effective and efficient risk reduction measure in an urban area it is needed to have sound understanding of urban scenario, urbanization trend and risk profile of the region (Sharma et al. 2011). Risk profile and risks level of a region also depends on its individual geographical setting, socio-economic and political condition. It is found that rapid urban growth coupled with geomorphology, hydrology, economics, demography and politics can create and exacerbate disaster risks of urban areas in a variety of ways (UN Habitat 2007). From all these geographical, socio-economic and political perspectives, urban areas of Bangladesh are highly disadvantage position in disaster risks. Geographically Bangladesh lies in the delta of three of the largest river in the world—Brahmaputra, the Ganges and the Meghna. It is also lies in the seismically active zone, where several devastating earthquakes occurred in past (Paul and Bhuiyan 2010). Further, Bangladesh is one of the densely populated developing countries in the world, where political instability and unrest are hindering economic growth. All these geographical, socio-economic and political conditions contributed Bangladesh to be one of the most disaster prone countries in the world and highly vulnerable to disaster risks. Urban areas of this country are more vulnerable to disaster risks, since disaster risks in urban center are compounded due to unplanned urbanization and unregulated growth in disaster prone marginal areas. Urbanization in Bangladesh is mainly concentrated in four largest cities Dhaka, Chittagong, Khulna and Rajshahi. These cities are also the centers of major industrial activities, trade and commerce and service sectors. More than 60 % of urban population lives in these cities and Dhaka alone accommodate one-third of urban population and produces 60 % of national GDP (World Bank 2007). Therefore, disaster in urban areas has severe national consequences in Bangladesh. Eventually, disaster risk reduction from the cities should be the key issue for ensuring overall development of Bangladesh. Before highlighting the risk reduction tools and approaches in urban Bangladesh this section focus on the geological and climate related disaster risks in urban areas of Bangladesh.

11.2.1 Geological Urban Risks in Bangladesh

Researchers have predicted that an earthquake with a million fatalities could occur in Himalayan belt of South Asia and south Asian megacities are the most vulnerable candidates to earthquake (Sharma et al. 2011). The historical seismicity and recent tremors in Bangladesh and its adjoining areas denotes that this country is at high seismic risk. In addition with this present unplanned and unregulated urbanization process and urban trend have increased cities vulnerability to disasters like earthquake. According to the suspicion of experts if a major earthquake occurs in large cities of Bangladesh, there will be huge human tragedy due to structural failure of many buildings (Shah and Murao 2011).

Seismic zoning map divides Bangladesh into three zones, which are zone 1, zone 2 and zone 3 with seismic coefficient 0.075, 0.15 and 0.25 g respectively. Here, zone 1 is the least vulnerable areas to earthquake and zone 3 is the highest vulnerable areas to earthquake (Alam, et al. 2008). Capital city Dhaka and the second largest city Chittagong (city of the largest sea port of the county) are located in Zone 2. Khulna and Rajshahi, which are the third and fourth largest cities, are situated in Zone 3. Mymenshing, Comilla and Sylhet cities are ranked as fifth, sixth and seventh major cities of Bangladesh. Among these cities, Comilla city fall under zone 2 and

Mymenshing and Sylhet cities fall under zone 3. Apart from these, there are about 300 smaller urban centers that account for only 4 % of urban population of Bangladesh (World Bank 2007). These urban centers are situated in different seismic zones.

Analyzing the location of the urban centers in seismic map it can be said that Mymenshing and Sylhet are the highest level earthquake vulnerable cities. But in reality, Dhaka and Chittagong, which are the most important cities of Bangladesh, have the highest level earthquake risks, in spite of their location in zone 2. According to United Nations IDNDR-RADIUS Initiatives, Dhaka and Tehran are the cities with the highest relative earthquake risks in the world (Jahan et al. 2011). The Earthquake Risk Index (EDRI) for Dhaka indicates it as top among the 20 high risk cities in the world (Khan 2008). Further, Munich Re's Natural Hazards Risk Index assessed Dhaka as the city of high risks. Several faults are located in and around the city. Between the years 1885 and 2005, within 200 km radius of Dhaka more that 70 earthquakes with magnitude or above occurred. These frequent earthquakes indicate the possibility of future major earthquakes in the vicinity of Dhaka. Based on the calculation of the recurrence year of 1885 Bengal earthquake Khan et al. (cited in Shah and Murao 2011) predicted that Dhaka would experience an earthquake with a magnitude of about 7.3 around 2017. Dhaka's inherent vulnerability to building infrastructure, dense population, haphazard and unplanned development, poor emergency response and recovery capacity are responsible to create high risk at such a high magnitude earthquake (Al-Hussini 2003, cited in Shah and Murao 2011 and Jahan et al. 2011). Dhaka is pacing with a rapid urbanization where massive real estate development and building construction are going on without following any zoning regulation, building code, proper plan (Fig. 11.2), approved design and appropriate materials (Shah and Murao 2011). Further, in case of relatively older building, owner neither knew earthquake threat not the existence of building code during the construction of their building in past (Shah and Murao 2010). So it is believed that most of the buildings in Dhaka do not have adequate provision of seismic resistance. Bangladesh Government has developed Bangladesh National Building Code (BNBC), which include detail guidelines for earthquake resistant design of concrete and still structures. But it is not effectively enforced (Shah and Murao 2011). All these prevailing circumstances, have contributed Dhaka to be one of the tom most cities with the highest relative earthquake risks in the world.

Chittagong is the second largest among the four metropolitan cities in Bangladesh with a projected population of 2.4 million in year 2011. It is also country's largest port city. Therefore, plays significant role in country's economic development. But unfortunately, this city is located in highly disaster prone area. Along with flood, cyclone, tornados, Chittagong has a long history of earthquakes. Chittagong and its surrounding hill districts are located at a moderately seismic zone. In recent years, since 1996 till to date Chittagong has experienced more than 200 light and moderate earthquakes. Among these in years 1997, 1999 and 2003 there were three earthquakes with a magnitude of 5.2–5.7. These three earthquakes killed more than 30 persons, hundreds of people were injured and hundreds of buildings were damaged. Like Dhaka unplanned and rapid growths have made this city one of the most vulnerable to potential earthquakes. Since 1911 census, Dhaka region emerged as the

a (Zone No.6)

b Filling of water bodies for Housing Project (Basundhara City Project, Zone No. 9)

c Housing in the low-lying flood prone areas (Zone No. 9)

d Zone No. 9

Fig. 11.2 Unplanned and unregulated building construction in Dhaka City, enhancing seismic risks

highest urbanized region and Chittagong region has been following Dhaka since 1941 (Ansary and Alam 2007). Chittagong city is growing so fast that most of the present housing constructions and associated developments are happening in hazardous areas. Hilly areas of Chittagong city are vulnerable to landslide and slope instability; on the other hand flat areas are hazardous because of low bearing capacity of soil deposit that is very much susceptible to liquefaction. Furthermore, a group of people are involved in cutting hill and construction of illegal houses. If there is a moderate tremor, it may cause massive landslide and structures resting at the foothills will become dangerous deathtrap for the dwellers. It is suspected that an earthquake of a larger magnitude ($M > 6.0$) may attack Chittagong region at any time. If so, it will pose great threat to the city. Further, since this city is the most important trade center of Bangladesh the loss would be unimaginable (Ansary and Alam 2007).

It is mentioned earlier that according to the seismic zoning map zone 3 is the most severe zone and Sylhet city, which is the divisional headquarter and one of the important cities of Bangladesh is located at this zone. It is predicted that earthquake in Sylhet region is inevitable. Along with the geographical location, due to rapid and unplanned growths earthquake risks and vulnerability of this city are higher than other urban centers in this region. Presently about 0.6 million people are living in this city. Due to high migration the city is growing very fast without following proper planning. Slums and high-rises both are growing at a rapid pace. New buildings are mostly unplanned and constructed without considering earthquake risks. Beside these, most of the buildings in Sylhet are old and non-engineered. In terms of earthquake occurrence, loss of life and properties Sylher city is considered as one of the most earthquake risks cities in Bangladesh after Dhaka and Chittagong.

Since north and eastern urban centers have higher geological risks, this section has highlighted geological risks of Dhaka, Chittagong and Sylhet cities which are important urban centers of this region. Here geological risks of south and western urban areas are not discussed, as their geological risks are not so high. However, it can be stated that geological risks for whole Bangladesh, especially for all urban centers are high due to its geographical location, high population density and unplanned, unregulated development. North and eastern parts of the country have higher geological risks than other parts. In north of Bangladesh there is a joining point of two plates-Indian plate and Eurasian plate. In east side there is another joining point of two plates-Barmiz plate and Indian plate. There are several fault zones active in this junction area and these are the source of earthquakes in Bangladesh (Alam et al. 2008).

11.2.2 Climate-Related Urban Risks in Bangladesh

Bangladesh is well known to the world for devastating flood and cyclone. Geophysical location of Bangladesh makes it susceptible to climate related disasters, especially flood and tropical cyclone every year. The country is a deltaic one situated at the foothills of Himalayas, which is downstream of about 300 rivers including the Ganges, Bhramaoutra and the Meghna. Therefore, more that 80 % areas of the country arc flood plain. On the other hand, the Bay of Bengal is considered as the breeding ground of tropical cyclone and Bangladesh is the worst victim in terms of fatalities and economic losses. Further, Bangladesh alone is accounting for about 40 % of global fatalities due to storm surges. Out of total 147,570 km^2 territory, Bangladesh has 47,201 km^2 of coastal area, which is one third of the country and highly vulnerable to tropical cyclone and storm surges (MoWR 2006; Islam 2004). Since 80 % areas of the country is floodplain and is one third of the country and highly vulnerable to tropical cyclone and storm surges, it can be said that almost all the urban centers of Bangladesh are in high risks of flood and cyclone. Especially, climate related disaster risks are remarkably high in large urban centers like Dhaka and Chittagong. Unregulated and rapid urbanization in major cities are exacerbating their risks. Comparative analysis of vulnerabilities of 50 different cities in terms of hazard exposure shows that Dhaka's risk level is high in case of earthquake, flood and tropical cyclone. Further it is also estimated that if sea level rise by just 1 m

Dhaka along with many other megacities will be under threat. Not only Dhaka, Chittagong and Khulna which are sea ports and prime cities of Bangladesh are located within 100 km of coastline. So along with these two cities and other small urban centers such as Satkhira, Bagerhat, Noakhali, Comilla etc. are also at high risk of coastal flooding, cyclone and storm surges. In addition to these most of the people in the coastal urban centers are in the high risks of salinity intrusion and scarcity of drinking water that are considered as climate related disasters (WEDO 2008).

For being the hub of administrative, economic, political and cultural hub the capital city Dhaka has the wide national consequence due to its high disaster risks. Therefore, in the following section climate related disaster risks and resilience of Dhaka has been discussed in a broader perspective.

11.3 Climate Related Disaster Risks and Resilience of Dhaka

11.3.1 Encroachment of Natural Drainage and Dhaka's Growth on Flood Prone Areas

The rapid increase of population in Dhaka city is aggravating the acute urban problems and posing tremendous pressure on urban land, infrastructures, utilities and amenities. But the irony is there is severe scarcity of land in Dhaka city (Morshed and Parvin 2009). The central part of Dhaka city is already built up and the expansion of the city is constrained by physical barriers like rivers, canals and low-lying flood prone areas around the city (Roy 2009). But ignoring such barriers Dhaka is expanding towards all directions through rapid development of housing estates (Rahman 2010; Hossain 2008). Since 1990, hundreds of real estate companies emerged in Dhaka city to full-fill the severe shortage of housing supply by the Government. Most of these housing providers have projects in the low-lying areas of Dhaka. During last decade, about 19 housing projects have been initiated in Eastern fringe of Dhaka, which serve as the most important retention area. Furthermore, housing projects of two larger real estate companies have filled up 2,300 m^2 of water bodies in eastern sides and in southern side 3,000 m^2 water bodies have been disappeared by the housing projects of another company. In the western part 91 acres of water bodies have been lost within 1996–2006 due to housing projects of real estate developers (Fig. 11.3). In this context, the most regretting and strange fact is RAJUK, which is the Government Authority for Dhaka's development plan preparation, implementation and development control has used retention ponds for their housing projects. Dhaka Metropolitan Development Plan 1995–2015 indicates eight flood-flow zones and advised to keep those as retention ponds. But many of these areas are already encroached by the real estate developers and even by RAJUK (Rahman 2010).

In addition with numerous environmental problems and challenges, recurrent natural disasters, especially annual flooding is one of the most concerning issues of

Solid waste dumping on the road (Zone No.3)

Filling of water bodies for Housing Project (Basundhara City Project, Zone No. 9)

Housing in the low-lying flood prone areas (Zone No. 9)

Chemical and waste from tannery industry are passing through open drain, which over flow during rainy season (Zone No. 3)

Fig. 11.3 Environmental and disaster issues in Dhaka

Dhaka. The severity of flood and water logging is increasing year on year. Dhaka city faced heavy flood at least 10 times from 1954 to 2007 (1954, 1955, 1970, 1994, 1980, 1987, 1988, 1998, 2004 and 2007) (Alam and Rabbani 2009). Among these last four floods were catastrophic. These catastrophic floods affected more than two-third of city areas and population. The main cause of these floods in Dhaka city was the rise in water levels of the rivers bordering the city during monsoon season (Faisal et al. 2003). Dhaka is surrounded by distributaries of two main rivers, the Brahmaputra and Meghna. All sides of Dhaka city are bounded by rivers and cannels (Hossain 2008). In addition with the rise of river water, internal drainage congestion and uncoordinated operation of flow regulation structures made the flood situation worse. Rapid and unplanned urban growth of Dhaka city causing serious

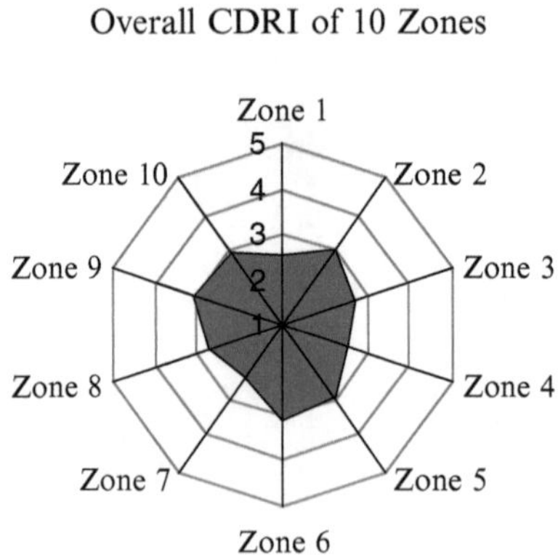

Fig. 11.4 Climate disaster resilience of ten zones from overall perspective (CDRI scores)

encroachment of natural drainage and retentions areas and eventually worsening the natural flow of water and causing serious water logging and flooding almost every year in recent decade. Infrastructures, utilities, livelihood, trade and commerce and public health all sectors are badly affected and vulnerable to recurrent flooding and water logging of Dhaka (Alam and Rabbani 2009).

Scholars claim that beside, floods/drainage congestion, Dhaka will be affected through heat stress due to climate change. Vehicle exhaust emission, industrial activities, increase of built up areas, loss of open spaces and increasing use of air conditioning are contributing to heat generation of Dhaka and this situation will continue in future (DoE 2005; Alam and Rabbani 2009).

Rapid and unplanned urban growth, increase of slum and squatters, air, noise and water pollution, inadequate and poor sewerage and sanitation, inadequate and inefficient waste management system, chaotic transport system, encroachment of natural drainage all are aggravating both the causes and impacts of climate related disasters (especially flood) in Dhaka city.

11.3.2 Climate Disaster Resilience of Different Zones of DCC

Climate disaster resilience of Dhaka City Corporation (DCC) has been assessed by Parvin and Shaw in 2011. This assessment highlights the level of climate disaster resilience of DCC from overall (Fig. 11.4) and from physical, social, economic, institutional and natural perspectives (Fig. 11.5). From Parvin and Shaw's assessment climate disaster resilience scores of each of ten zones have been presented as Climate Disaster Resilience Index (Overall CDRI) in Fig. 11.4. This Overall CDRI is the average of the scores that each micro level area (zone) has obtained in physical, social, economic, institutional and natural dimensions.

Climate Disaster Resilience Index (CDRI) from overall perspective reveals that among the ten zones CDRI scores range from 2.4 to 3.1. Since five is the highest score, these scores of Overall CDRI denote the low/poor to medium level of

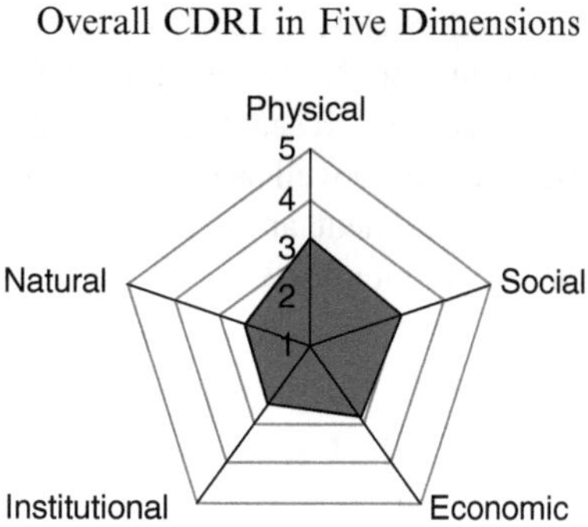

Fig. 11.5 Overall climate disaster resilience of five dimensions

resilience. It is interesting to notice that the planned residential areas (zone no. 6, 9 and 10) have relatively higher level scores in climate disaster resilience index; while, old parts of the city (except zone no. 2) and densely populated low income areas located in the fringe areas have low level of resilience scores than other areas.

In the overall CDRI assessment only three zones (no. 2, 6 and 9) have obtained the scores that are considered as good level of climate disaster resilience. Though these three zones are part of Dhaka City Corporation and many characteristics are similar to other zones there are few features that has made these zones identical. For instance, zone 2 is the smallest zone is DCC (3 km^2) but accommodates the highest number of people per square kilometer. The density of this area is 125,423 km^2, which is extremely high. In spite of being old part of city, having high density, zone 2 has obtained relatively higher score in overall CDRI. In fact this zone is located in high land, free from flooding; it is place of old residents of the city and also place of rich traditional business man of the city. As a result, relatively higher scores in physical, social and economic dimensions have lifted this zone to good position in overall CDRI. On the other hand, it is interesting to notice that zone no. 6, which is the location of National Parliament and residential areas of parliament members and zone no. 9, which is the diplomatic zone and location of all embassies have relatively higher score in physical, economic, institutional and natural dimensions and eventually have obtained higher scores in overall CDRI.

11.4 Risk Reduction Tools and Approaches in Urban Areas

Internationally, the conventional disaster management model on emergency responses is being replaced by a more holistic model. Similar to the international approach, Bangladesh is approaching towards a paradigm shift from reactive responses to a proactive risk reduction culture. This approach emphases to hazard identification, mitigation, community preparedness, capacity building, enhancing community resilience, planning and management and addressing the issues of vulnerabilities. Especially, after the declaration of five priorities of Hyogo Framework for Action (HFA) (2005–2015) different ministries and agencies of Bangladesh Government, Non-Government Organizations, civil societies, community based organizations, researchers, academics and development workers are innovating, introducing as well as revising and updating different tools and approaches of disaster management and thus trying to mainstreaming disaster risk reduction in the development efforts.

11.4.1 Risk Reduction Tools and Approaches Initiated by Government

Government organizations at the national and local level are responsible to assess urban risks due to different disasters and take the initiatives to manage disaster risks. Pro-active planning and action of preparedness and mitigation are the best approach to minimise the urban disaster risks and reduce the future losses. To address risk reduction approaches Government has to take the lead by

- Strengthening policies and institutions
- Identifying, assessing ad monitoring risk and enhancing early warning
- Using knowledge, innovation and education to build a culture of safety
- Reducing underlying risk factors, such as environmental degradation (UNDP 2004).

The planning tools are namely practiced in urban disaster risk reduction in Bangladesh are

- Relocation of settlement at high risk
- Detail land use planning to inform new construction
- Protecting critical structures
- Updating and implementing building codes for disaster resister resistant housing
- Improve early warning system (Pelling and Wisner 2009)

Government initiated risk reduction approaches are mostly policy and planning based where intra and inter organization links are important and a hierarchical link needs to be there to practice appropriate tools in appropriate time to minimize the disaster risks. Under Bangladesh planning and development context regional and local level planning system are responsible for ensuring quality of life through a better urban planning approach. Such planning approaches consider all urban planning challenges associated with the planning components, such has housing, sanitation, and land-use urban utilities. Urban disasters induced by extreme climate events is being added as a new planning challenge which is not clearly understood in the current planning practice both in national, regional, and local levels, therefore adaptive and mitigating planning tools are important to address this challenge with planning and development system especially at the local level (Ahsan et al. 2011).

Following Table 11.1 presents multiple planning tools or activities that are used to address urban disaster risk reduction process in regional and local level planning and the stakeholders associated with those activities.

Table 11.1 Urban disaster risk reduction tools and stakeholders

Tools and stakeholders	Development planning	Development regulation	Risk management	Emergency management
Core tools	Land use planning, development control and planning regulation	Bangladesh National Building code (BNBC), zoning code, infrastructure regulation, environmental and conservation regulation, urban pollution control	Vulnerability and risk assessment, building local resilience, CDMP, CPP, SOD	Early warning, emergency response, environmental monitoring, emergency service, civil defense, disaster management coordination, fire fighter, police, military, red cross/ crescent societies, public and private organization
Stake-holders	Development authorities, urban planners, structure planners, public works department	Environmental conservator, law enforcement authority, developer and contractors	Primary health care facilities, water supply and sanitation, local infrastructure development, waste and drainage management	

Source: Adapted from Pelling and Wisner (2009)

Apart from the National Disaster Management Council, Coordination Committee and Advisory Committee, in recent years Comprehensive Disaster Management Program (CDMP), Cyclone Preparedness Program (CPP), Standing Order on Disaster (SOD, introduced in 1997 and revised in 2010), National Plan for Disaster Management (2010–2015), National Platform for Disaster Risk Reduction (formed in 2009) and Earthquake Preparedness and Awareness Committee (formed in 2009) have been formed to introduce and implement different tools and approaches especially to reduce urban disaster risks. Disaster risk reduction has been incorporated into Interim Poverty Reduction Strategy Paper (I-PRSP) as Annex 9, Disaster Vulnerability and Risk Management.

In addition to the abovementioned disaster management tools and approaches, there are different acts and policies are introduced and enforced by the Government agencies to control urban growth, to regulate land use, population density and building construction. Among different acts and policies noteworthy are Building Construction Act 1952, The Town Improvement Act 1953, Land Development Rules for Private Housing 2004, Bangladesh National Building Code (BNBC) 1993 came into effect with a gazette notification in 2006 and Building Construction Rules 2008. These acts and policies are directly and indirectly aiming to use as effective tools to reduce urban disaster risks by regulating urban growth and building construction.

Central government has to introduce planning and policy tools at the local level. Development control, planning regulations and policy applications are the key instruments to introduce and practice risk reduction tools within the planning and development system. Local Government and urban development authorities as well as planners and developers are the key players to introduce and implement risk

Table 11.2 Organizations and their responsibilities as functional tools to response any natural or manmade disaster

Responsible government authorities/organizations	Responsibilities
Local development authorities/city corporation	Enforce planning and development law and regulation to protect land use in vulnerable areas
	Adopt and enforce building codes
	Develop hazard map based one sites
	Regular monitoring and revise the land use plan and development
	Public awareness campaigning
Health service facilities	Control epidemiological outburst during disaster
	Providing medical facilities to emergency camps and relief centers
	Prepared trained rescue squad for emergency
	Prepared volunteer teams with first aid training
	Reserve emergency medical supply
Fire service and civil defense	Prepared rescue team with all necessary equipment
	Provide security to the disaster affected areas
	Help distribution emergency supplies
	Manage emergency services such as fire-brigade and ambulance
	Regular mock exercise
	Awareness workshop with local communities
Disaster Management Bureau and Armed Force	Develop interactive user friendly geo-hazard maps
	Emergency communication and early warning system
	Contingency plans
	24 h emergency information and service bureau
Public works department	Retrofit the existing structures vulnerable to any kind of disasters
	Renovate and reconstruct public infrastructures
	Published the vulnerable infrastructures list for public awareness

Source: Adapted from Ansary and Alam (2007)

reduction tools for urban areas. In Bangladesh municipalities, city corporations and urban development authorities e.g. Rajdhani Unnayan Katripokkho-RAJUK in Dhaka city, Chittagong Development Authority (CDA) in Chittagong city, Khulna Development Authority (KDA) in Khulna city and Rajshahi Development Authority (RDA) are introducing as well as enforcing disaster risk reduction tools in the respective urban areas.

Responsible authorities in urban local government need to understand the hazards present a risk on a particular place, characteristic, frequency and magnitude of those hazards to design appropriate tools to reduce the risks. Local government and development authorities need to practice different development and planning tools to reduce urban disaster risks. Local government has access to local knowledge and data base as well as national government support and influence, therefore the local government authority is the best to practice emergency response to any hazards. With the local expertise about risk management and considering the socio-economic condition local government authority has to reorganize and design the local level organizations to response the risks. Ministry of Food and Disaster Management (MoFDM) in Bangladesh designed the local government framework to define specific organizations and their responsibilities as functional tools to response any natural or manmade disasters. MoFD (2006) disaster preparedness tools are as follows (Table 11.2).

Among different approaches of DRR, Comprehensive Disaster Management Program (CDMP) and its various tools are recently considered as one of the most effective approaches of DRR in Bangladesh, especially for urban earthquake risk reduction. Through CDMP a comprehensive earthquake program has been developed and earthquake microzonation map has been prepared to identify vulnerable areas. Working with the City Corporations (local government institutions in the largest cities in Bangladesh), a digital database of essential infrastructure was created that now supports sustainable urban planning and research, and underpins the contingency planning process, linking the science, the experts and the community. Second phase of CDMP will also provide training to public and private sector stakeholders highlighting best practices in building construction and demonstrating retrofitting techniques in, for example schools and hospitals, that prepare for, and reduce the impact of, earthquakes. An urban volunteer network has been established and trained on search and rescue and first aid. This force, which will number 60,000, not only represents a significant response capacity, but a clear commitment to one's own community (UNDP 2010). Phase-1 of the CDMP carried out a survey to identify earthquake vulnerable buildings. Around 250,000 buildings in the three major cities of Bangladesh—Dhaka, Chittagong and Sylhet—are extremely vulnerable to earthquakes, according to this survey in 2009.

11.4.2 Risk Reduction Tools and Approaches Initiated by Civil Societies or NGOs

The question of resilience in context of urban areas is not only limited to response and preparedness but also a component to ensure urban sustainability. Rapid urbanization, economic expansion increase disaster as well as climate change risks. Building resilience for urban disasters require political will, active involvement of local government, civil societies, development organization and communities. Disaster risk reduction approach or resilience practice need to be multidimensional rather a top down government action. All who make a city functional from municipal service provider to civil society member need to be committed to design and implement effective disaster reduction tool to ensure urban resilience. For example civil society organizations (CSOs) throughout the world at the regional, and local levels bring attention to development planning. These CSOs include non-governmental organizations (NGOs), faith-based/organizations, indigenous foundations and many others at the local level to ensure quality of life for the poor and vulnerable communities. Those CSOs are key players in reaching the internationally agreed development goals, including the Millennium Development Goals. Therefore, civil society is an essential partner for sustainable urban risk reduction.

The response to disaster risk varies widely at urban level based on the geographical and local environmental context, the size of the city (including population size) and the stage of the city's development (UNSIDR 2012). However around the world five main strategies are considered to design appropriate tools for disaster risk reduction and those are:

1. Consider disaster risk reduction as a part of urban planning regulations, plans development activities;
2. Establishing a planning and development committees dedicated to disaster risk reduction;
3. Design and plan for hazard-resistant infrastructure or improving existing facilities;
4. Establishing education/awareness/training programmes;
5. Organizing multi-stakeholder dialogues

Experiences show that the government either local or national cannot implement those strategies rather community-based approaches offer viable solutions for managing and reducing risks and ensuring sustainable development (Mercer 2010). Therefore it is important to involved CSOs in planning, designing and implementing DRR tools as the CSOs represent the vulnerable communities to design DRR action tools with the support of local knowledge and encourage the engagement of the local communities to minimise the risks.

11.4.2.1 Know the Risk

Building an information base to know the risk and associate vulnerabilities—including drawing on the knowledge of local residents and community organization, as well as experts is one of the key components for DRR tool. Limited available data on natural hazards, risk and vulnerabilities for many developing cities always hold them back to implement appropriate risk reduction and response in time. On the other hand some cities uses modern technology such as GIS base risk map to track the disaster pattern and risk to measure the vulnerabilities. For example Pune in India design comprehensive mapping of flood risk with the support of local communities to design structural and non-structural flood hazard mitigation. The Flood risk map includes detail of city's drainage, number of people living along the drainage network and they level of suffering during monsoon. While preparing the flood risk map experts consider local people's opinion as they are the victim (UNSIDR 2011). Local communities are much aware and well informed about the local disasters and frequencies. The role of NGOS is to combined local knowledge with scientific understanding and advanced technologies to generate a fuller picture of risks. Risk identification and understanding techniques help to build a common understanding of risk (e.g. through creating seasonal calendars or histories of floods/droughts over the last decades). Once the local communities are understood and there is a system for monitoring them, it is easy for them to get prepared to minimise the risk and to get preparation for post disaster effects.

11.4.2.2 Effective Preparedness and Early Warning

Other key components for a city to reduce the disaster impacts due to rapid-onset events are emergency preparedness, early warning and disaster response structures.

Most of the cities are now addressing those issues but yet such practice need to be more effective with the direct involvement of communities and the community based organization that can facilitate different trainings for the communities. Preparedness and early waring process seeks that community members have the knowledge to know and understand the disaster risks and the possible preparedness to face the unevenness. A better approach to ensure public awareness and preparedness is to undertake drills and simulation exercise. For example the Quito city carries out disaster preparedness drills at the institutional level as well as in communities and schools where the local NGOs and community organization take the lead to aware the city residents what to do when it is emergency to minimise the impacts. Disaster preparedness and awareness drills and campaign is also common in Cape Town's small communities and schools who are highly vulnerable to natural disasters.

Members from civil society organizations are the effective working force who can arrange drills and camping to aware the communities, but their role and activities need to be monitored by the local or state government authorities. For example in Cairns community volunteers are responsible for promoting community preparedness, assisting injured persons, protecting community from harm and rescue in emergency situation. Their work is supported and monitored by the Local Disaster Management Group, which clarifies the role and responsibilities between CSOs and local government unit.

11.4.2.3 Education and Public Awareness

Education and public awareness training help to ensure that individual member of the society know what to do in an emergency situation to minimise the risk to themselves, their families and their communities. Education and awareness training program includes campaigns for mass audience, education for the young students in schools and young volunteers as target audience. Education on disaster preparedness is aimed to aware and help people to avoid coming disaster threats and to put plans, resources, and mechanisms in place to minimise the impacts and to ensure that those who are affected receive adequate assistance.

Lack of knowledge on disaster pattern and associated risks such as flood and associated risks or earthquake and risks always affect to build resilience. For example Mumbai, Thimphu and Bhubaneswar are facing the challenges to build urban resilience as they have limited knowledge disaster risk and urban challenges. However Bhubaneswar's Municipal Corporation arrange training for professional such as engineers, architects and planners on hazard-resilience building codes in response to earthquake. Beside train the professionals it is important to educate and aware the community to minimise the risks. For example local NGOs can educate the communities about hazard mapping and aware them to choose the place to live. They can also educate the communities about natural drainage and water channels and environmental mapping to encourage them to protect those to minimise urban flood risks. The key components of community education and public awareness are to build a culture of safety. Safety culture is one of the key components of community based disaster risks reduction tools (UNSIDR 2004b).

11.4.2.4 Capacity Building

Capacity building one of the key components at local level disaster risks reduction process where the community response first. Such capacity building not only covers the physical capacity of infrastructure, but also, institutional, and intellectual capacity, to upgrade the community members with new knowledge, technology and skills to response to any disaster risks. Intellectual capacity of communities knowing how to deal with disasters is as important as physical capacity. Local community based organization along with the local government authorities need to play the critical role for local capacity building. For example local government authority in Bhubaneswar trained 3,000 students from 35 schools and 600 colleges on emergency disaster management and safety tips. At the same time mass public campaigns another technique to change household-level behaviours on disaster risk and early warning system. For example civil societies and local development organizations in Cairns conduct annual campaigns for the local households to build cyclone awareness (UNSIDR 2012). Even in Dhaka Centre for Policy Dialogue (CPD) offers fire safety management for local residents (CPD 2012).

Most of the emergencies are smaller-scale localized events where community members are the first responders, and where the most isolated or underserved communities need greater self-sufficiency in anticipating and responding to them (Benson and Twigg 2007). Local NGOs have access to all level of local communities even the isolated communities; therefore those organizations can play the lead role to build local intellectual capacity to response in case of emergency. Capacity building depends on the need of the community, therefore it can be a physical or intellectual, and the local NGOs based their expertise and community needs can design appropriate tool to addresses disaster risks.

11.4.2.5 Recovery and Rebuild the Communities

After any disaster good recovery practice is important to minimise the risk and to rebuild the society as well as prepared for any future hazards. Recovery plans for future development not only minimise the disaster risk for the city but also minimise economic loss and loss of lives. For example during reconstruction following the earthquake, Santa Tecla city developed long term strategy plan for future development to avoid the risk (Rout nd). Recovery plan requires collaboration with all the local level authorities involved in development process including government, non-government and community-based organizations. For example Cairns, Makassar and all other cities have disaster management and recovery framework where different stakeholders including local communities work together. Cairns' city council has a community support subcommittee that includes the Australian Red, Department of Communities and Homelessness and the Salvation Army to support recovery planning (UNSIDR 2012). Plan for recovery and reconstruction is important tool to put the vulnerable communities in main focus during decision and plan making process, where the civil societies can also raise the voice for them.

Partnership among the supportive organizations working for the disaster management is the most important components for DRR. Partnerships are not only with funding or donor agencies but also with the government to ensure adequate services facilities at all levels to minimise the damage. Government coordination is always top-down in nature, therefore it is always difficult to address all the vulnerable sectors at the community levels. Whereas NGO coordination is usually from local level or bottom up and tend to show great resilience. Therefore partnership consider as a coordination tool that can link all local, nation and international organization working on disaster risks management issues. For example National level large NGOs can provide technical and material support to NGOs already active at grass root level at each district or sub-district level. Local level NGOs are closely connected with the community and have better appreciation of their vulnerabilities, needs and the capacity for revival so as to bring them back to the pre-disaster stage at the earliest. A well-designed partnership and coordination mechanism give a lead role to local NGOs with necessary linkages with national NGOs and District Administration to link local development plan with national policy and strategies (Heather 2011).

Disaster reduction plan always using top-down government and institutional interventions are often considered insufficient as they tend to have a lower understanding of community dynamics, perceptions and needs, and ignore the potential of local knowledge and capacities. Thus, NGOs' involvement in DRR activities has proved beneficial for a number of reasons. NGOs have provided the indispensable link between public authorities and communities in the vulnerable areas before and after disaster crisis. NGOs provide a range of supports including services to vulnerable groups, capacity building, community outreach, social mobilization, advocacy, and awareness rising regarding risk reduction, such as health and hygiene promotion and safety and resilience. NGOs grow out of communities of specific interest, and they are the voice of that community and perform for the community as form of CSO.

Community based disaster risk reduction approach become more popular around the world as the local communities are the prime victims and they have the local knowledge to manage and adopt with those disasters. Therefore, it is important to involve the vulnerable people in disaster risks reduction planning and implement process and ensure partnership with the local and national level organizations involved with process to ensure sustainability.

11.5 Conclusion

It has been mentioned earlier that the geographical location of Bangladesh has made it highly vulnerable to both geological and climate related disasters. Disasters, especially floods and cyclone are most common phenomenon in Bangladesh. Urban Bangladesh is more vulnerable to disaster risks due to rapid, unplanned and unregulated urban growth in marginal disaster prone areas. In last decade three catastrophic disasters (flood 2004, flood 2007 and cyclone SIDR 2007) events in Bangladesh resulted organization of different workshops and attempts, that prioritize disaster risk reduction and emphasis has been given to urban areas, sine these are the hub of economic and administrative activities.

It is thought that through disaster risk reduction approach disaster management can be undertaken in the best way. On the other hand long term reduction in risk and vulnerabilities is possible through adoption of practical and community centered risk mitigation measure within urban planning practice (Sharma et al. 2011). However, like many other countries in Bangladesh risk reduction measures are not considered in local level planning practices. Even though, urban planners are looking for ways to scale down infrastructure, land use and disaster risk reduction planning at local planning and action zones (UN Habitat 2007). Land use planning and building code are treated as the most fundamental tool for risk reduction into urban development process. In Bangladesh development plans prepared for city corporations and municipalities indicate land-uses within the city. At the same time Bangladesh National Building Code (BNBC) is prepared to regulate and maintain standard of building construction. It aims to provide safe and healthy habitat by regulating building planning, design and construction. But with rapid urbanization cities are growing without following land-use planning and zoning regulation. Violation of BNBC is also a common phenomenon. In many cases, building owners do not know about the existence of BNBC and professionals who are involved in planning and design building do not care about BNBC. City development authorities who are responsible to introduce and enforce land-use planning, zoning, BNBC and other planning tools and regulations to control cities unplanned growth, protect environment and thus reduce disaster risks are reluctant to enforce these. Encroachment of natural drainage and Dhaka's growth on flood prone areas denotes this reluctance which enhancing Dhaka's flooding and water logging in every rainy season. Unregulated and uncontrolled urban growth in protected hilly areas of Chittagong city and the consequence of landslide is another example of urban disaster risks due to lack of enforcement of planning regulations.

City Corporation, municipalities and city development authorities need to revise and monitor building construction rules and regulation and land use planning regularly. Hazard mapping for each community and assessment of their disaster resilience level should be a prerequisite component before any development plan preparation and implementation. These risk and resilience assessments can be done by the community participation where city development authorities should be facilitator. Strict enforcement of use and height zoning, set back rules, building codes is needed by urban local government bodies, like City Corporation and municipalities. Rajdhani Unnayan Kotripakkha (RAJUK), Chittagong Development Authority (CDA), Khulna Development Authority (KDA) and Rajshahi Development Authority (RDA), which are development authorities of four principal cities of Bangladesh are needed to strictly control cities unplanned and unregulated growth. Both local government and civil society organizations can play vital role to create public awareness for DRR through following building construction rules and regulation. Disaster Management Bureau, Armed Force, Public Work Department and such other relevant Government bodies are needed to identify, retrofit, renovate and reconstruct emergency and important urban public buildings and services like hospital, school buildings, electricity and water supply facilities. Approaches, actors and actions for urban DRR are given in Fig. 11.6.

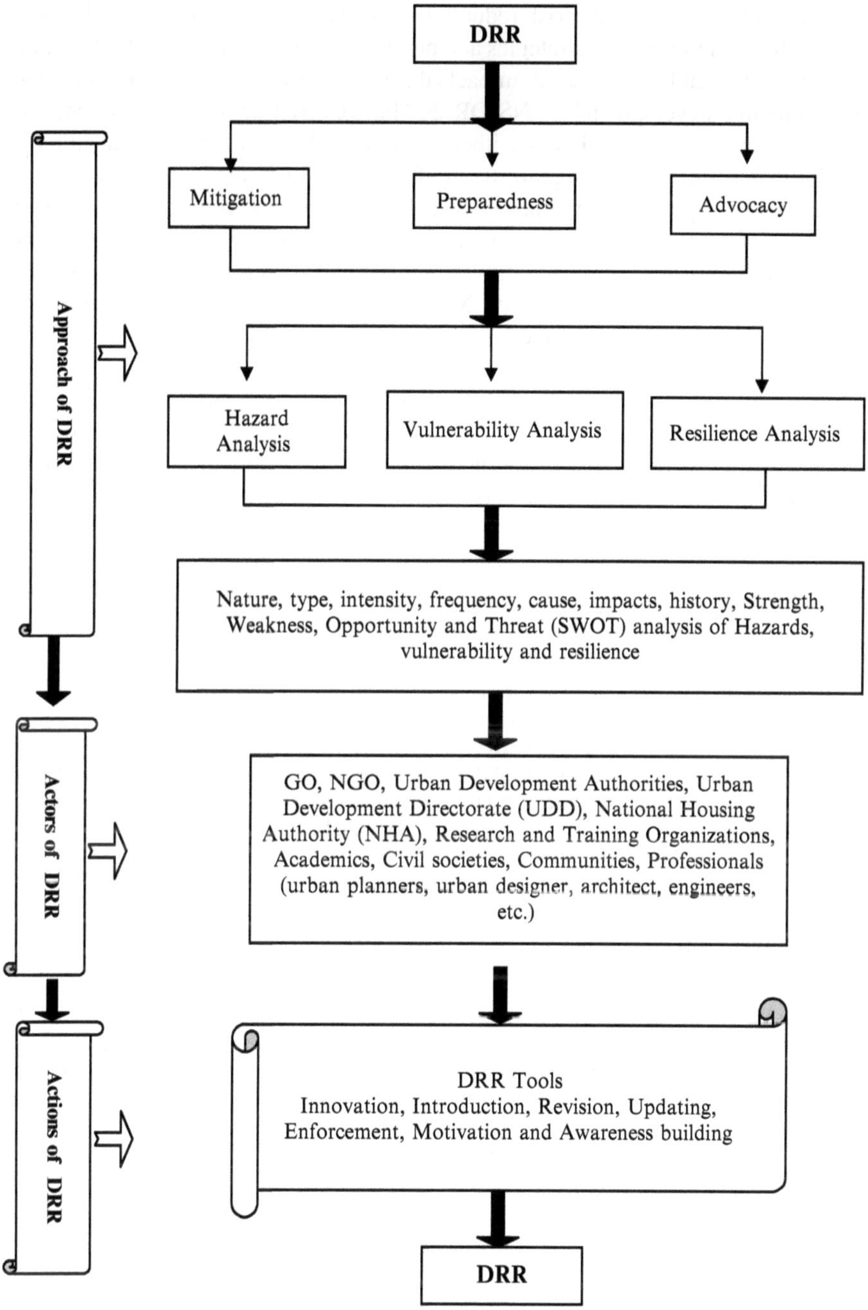

Fig. 11.6 Urban disaster risk reduction approach, actors and actions

UNISDR defined disaster risk reduction (DRR) as "the systematic development and application of policies, strategies and practices to minimize vulnerabilities, hazards and the unfolding of disaster impacts throughout a society, in the broad context of sustainable development" (UNSIDR 2004a, b). DRR involves mitigation, preparedness and advocacy. DRR is neither an individual action nor a responsibility of any individual authority or organization rather it is multi-disciplinary in nature, recognizing the importance of link between hazards and the physical environment (include infrastructure, development and society). Therefore, appropriate DRR tools should link grassroots strategies with top-down strategies designed and introduced by the government authorities, Non-Government Organizations, civil societies, academics, professionals and communities. State and local government has to take the lead and initiatives to incorporate DRR tools with the planning and development system to minimise the disaster risk and to reduce human and economic losses. Successful application of DRR tools help to create a resilient community, whilst ensuring vulnerability is not increased through development efforts or other externally initiated activity in the approach of DRR (UNSIDR 2011).

References

ADB (2003) Moving the poverty reduction agenda forward: priorities and outcomes. Asian Development Bank Annual Report, Singapore

Ahsan R, Karuppannan S, Kellett J (2011) Climate migration and urban planning system: a study of Bangladesh. Environ Justice 4(3):163–170

Alam M, Rabbani GMD (2009) Vulnerabilities and responses to climate change for Dhaka. In: Satterthwaite D, Dodman D, Bicknell J (eds) Adapting cities to climate change-understanding and addressing the development challenges. Earthescan, London, pp 93–110

Alam MJB, Ansary MA, Chowdhuary RA et al (2008) Evaluation of building's vulnerability to earthquake in Old Part of Sylhet and construction safety rules. IUST Int J Eng Sci 19(3):33–43

Ansary MA, Alam MS (2007) Seismic vulnerability of Chittagong City and some mitigation measure. Protibesh 11(1):23–32

Benson C, Twigg J (2007) Tools for mainstreaming disaster risk reduction: guidance notes for development organisations. ProVention Consortium Secretariat Publication, Geneva

CPD (2012) Fire safety management training for local community. Centre for Policy Dialogue, Dhaka

DoE (2005) Dhaka City - State of Environment 2005, this report prepared by Department of Environment (DoE), Ministry of Environment and Forest, Government of the People's Republic of Bangladesh in collaboration with United Nations Environmental Programme (UNEP) and Bangladesh Center for Advance Studies (BCAS), Bangladesh

Faisal IM, Kabir MR, Nishat A (2003) The disastrous flood of 1998 and long term mitigation strategies for Dhaka City. Nat Hazards 28:85–99 (Kluwer Academic Publishers, Netherlands)

Heather HDo (2011) Successful models of non-governmental organizations in consultative status: best practices on disaster risk reduction. Unpublished Master Dissertation, Department of Public Administration in International Development, Tsinghua University, Beijing

Hossain S (2008) Rapid urban growth and poverty in Dhaka City. Bangladesh e-J Sociol 5(1):1–24

Islam MR (ed) (2004) Where land meets the sea: a profile of the coastal zone of Bangladesh. University Press Limited, Dhaka

Jahan I, Ansary MA, Ara S, Islam I (2011) Assessing social vulnerability to earthquake hazard in Old Dhaka, Bangladesh. Asian J Environ Disaster Manag 3(3):285–300 (Research Publishing Services)

Khan H (2008) Challenges for sustainable development: rapid urbanization, poverty and capabilities in Bangladesh. Online at http://mpra.ub.uni-muenchen.de/9290/. Accessed on Mar 5, 2012

Mercer J (2010) Disaster risk reduction or climate change adaptation: are we reinventing the wheel. J Int Dev 22(1):247–264

Morshed M, Parvin GA (2009) Need and consequences: real estate development in Dhaka City, Bangladesh. Presented in International Urban Planning and Environment Association (IUPEA) 8th International Symposium, 23–26 March 2009 Kaiserslautern, p 42. Web link http://www.upe8.uni-kl.de/media/pdf/UPE-compendium-of-abstracts.pdf

MoWR (2006) Coastal Development Strategy, Water Resources Planning Organizations, Ministry of Water Resources, Government of People's Republic of Bangladesh

Parvin GA, Shaw R (2011) Climate disaster resilience of Dhaka City Corporation: an empirical assessment at zone level. Risk Hazards Crisis Public Policy 2(2):1–30

Paul BK, Bhuiyan RH (2010) Urban earthquake hazard: perceived seismic risk and preparedness in Dhaka City, Bangladesh. Disasters 34(2):337–359

Pelling M, Wisner B (2009) Reducing urban disaster risk in Africa. In: Pelling M, Wisner B (eds) Disaster risk reduction: cases from Urban Africa. Earthscan, London, pp 43–59

Rahman AM (2010) Dhaka's peripheral development and vulnerability to earthquake liquefaction effects. Asian Disaster Manage News 16(1):6–7 (Asian Disaster Preparedness Center)

Rout PR (nd) Urban disaster risk reduction a case of Indian State Orissa (Bhubaneswar). Local government network Association, Orissa, India. Online at http://www.preventionweb.net/files

Roy M (2009) Planning for sustainable urbanization in fast growing cities: migration and adaptation issues addressing in Dhaka, Bangladesh. Habitat Int 33(3):276–286 (Elsevier Ltd.)

Shah FMd, Murao O (2010) People's seismic risk recognition in Dhaka and Bangladesh National Building Code (BNBC)1993. In: USMCA symposium, Kobe

Shah FM, Murao O (2011) Understanding seismic risk recognition and intention for safety measures of residents in Dhaka, Bangladesh. Asian J Environ Disaster Manag 3(3):357–372 (Research Publishing Services)

Sharma A, Srujan A, Shaw R (2011) Chapter 1. Overview of urban development and associated risks. In: Shaw R et al (eds) Climate and disaster resilience in cities: community, environment and disaster risk reduction, vol 6. Emerald Group Publishing Limited, Bingley, pp 1–16

Srinivas H, Shaw R, Sharma A (2009) Introduction to urban risk reduction. In: Shaw R, Srinivas H, Sharma A (eds) Urban risk reduction: an Asian perspective. Emerald Group Publishing Limited, Bingley, pp 3–12

UNDP (2004) Reducing disaster risk: a challenge for development, United Nations Development Programme, Bureau for Crisis Prevention and Recovery Publication, New York. Online at http://www.undp.org/bcpr/disred/rdr.htm. Accessed on June 4, 2012

UNDP (2010) Success stories: working for safer cities in Bangladesh available from: http://web.undp.org/eu/successstories/working_for_safer_cities_in_bangladesh. Accessed on Dec 13, 2012

UN-ESCAP (2005) Urban environmental governance for sustainable development in Asia and the Pacific: a regional overview. Economic and Social Commission for Asia and Pacific, Bangkok

UN-Habitat (2007) Disaster Risk: conditions, trends and impacts, Chapter 7, Enhancing urban safety and security, global report on human settlement, United Nations Human Settlement Programme, UN Habitat, Earthscan, London, pp 168–192

UNSIDR (2004a) Terminology: basic terms of disaster risk reduction. United Nations International Strategy for Disaster Reduction Publication, Geneva

UNSIDR (2004b) Disaster risk reduction tools and methods for climate change adaptation, inter-agency task force on climate change and disaster risk reduction. United Nations International Strategy for Disaster Reduction Publication, Geneva

UNSIDR (2011) Revealing risk, redefining development. United Nations International Strategy for Disaster Reduction Publication, Geneva

UNSIDR (2012) How to make cities more resilient – a handbook for mayors and local government leaders. United Nations International Strategy for Disaster Reduction Publication, Geneva

WEDO (2008) Gender, climate change and human security – lesson from Bangladesh. Ghana and Senegal, prepared by The Women's Environment and Development Organization (WEDO) for ELIAMEP, May 2008

World Bank (2007) Urban development and economic growth in Bangladesh, Bangladesh Strategy for Sustain Growth, World Bank Development Series, Paper No. 18, pp 101–118

Chapter 12
Community-Based Disaster Risk Reduction Approaches in Bangladesh

Umma Habiba, Rajib Shaw, and Md. Anwarul Abedin

Abstract A number of disasters are very common in Bangladesh. Among them, flood, tropical cyclone and drought are the most catastrophic disaster, which has substantial effect on livelihood. In every disaster, people within the community suffer most the disaster impacts and they are the first front line responders to survive with disaster. During or after disaster period, it has been often seen that external support comes later to rescue community. In this regard, community-based disaster risk reduction (CBDRR) is a new concept that provides an opportunity to dovetail indigenous knowledge for disaster risk reduction and settle strategy to mainstream risk reduction at the community level. In Bangladesh, the communities have developed a number of community driven approaches with the collaboration with different international as well as governmental organization or NGO. Thus it eventually helps in building the capacity of the local community to prepare and respond with emergency faced by disasters. On the contrary, Government of Bangladesh (GoB) has established Ministry of Food and Disaster Management (MoFDM) through which it operates all disaster management activities within the country. Besides, Government now gives more emphasis on the involvement of community and local government in the disaster management system. Even, GoB established comprehensive disaster management program (CDMP) under MoFDM to reduce the nation's vulnerability to natural hazards and to strengthen disaster management activities. However, CBDRR activities are now institutionalized in local government through the recent national program of CDMP with the assistance of United Nations Development Programme (UNDP).

Keywords Community based risk reduction • Comprehensive disaster management program (CDMP) • Indigenous knowledge • Natural disaster • Resilience

---

U. Habiba (✉) • R. Shaw • Md.A. Abedin
Graduate School of Global Environmental Studies, Kyoto University, Kyoto, Japan
e-mail: shimuagri@yahoo.com

R. Shaw et al. (eds.), *Disaster Risk Reduction Approaches in Bangladesh*,
Disaster Risk Reduction, DOI 10.1007/978-4-431-54252-0_12,

12.1 Introduction

Over more than 100 years ago, before the existence of most of the states, people or communities were taking care of themselves through collective actions during the disasters. After formation of state, government-based disaster risk reduction program started, which failed to serve the needs of the people and communities. For the past 20–30 years, the need of community based disaster risk reduction is taking into account. Because, in the recent years, more research on development has been conducted in various fields that showed the approach to disaster risk reduction became more and more community-based (Blaikie et al. 1994; Quarantelli 1989; Mileti 2001), and much more effort has been put into incorporating disaster management aspects into the holistic development of communities (Shaw and Okazaki 2003; Twigg and Bhatt 1998).

Going back to the traditional approach of risk reduction, it has been seen that community based disaster management (CBDM) is very popular term in later 1980s and 1990s, which gradually evolved to community-based disaster risk management (CBDRM) and then, to community-based disaster risk reduction (CBDRR). CBDRM and CBDRR are often used with similar meaning, with enhanced focus on "risk". However, there still exists a thin line of distinction. According to Shaw (2012), CBDRR focuses more on pre-disaster activities for risk reduction by the communities; while CBDRM focuses a broader perspective of risk-reduction-related activities by communities, both before, during and after the disaster.

However, the concept of CBDRR is relatively new. By definition, CBDRR is a multi-disciplinary agenda for community development; a different paradigm (long-term) that builds on the intrinsic relationship between disasters and development. CBDRR approach brings community together to address common problems which affects everyone. Twigg (1999) further argued that the rationale for community involvement or community-based activities is now well rehearsed. Because community-based activities (and community-based organizations) are deeply rooted in the society and culture of an area, they enable people to express their real needs and priorities, allowing problems to be defined correctly and responsive measures to be designed and implemented. On the other hand, CBDRR strengthens social cohesion and cooperation within the community and society. It builds confidence among individuals, households, and communities for any undertaking including disaster preparedness and mitigation (Shaw 2012).

There are various key players in CBDRR such as governments, UN agencies, international and regional organization, NGOs, CBOs, research institutions, academics, and communities. Across the world, almost all countries CBDRR have been taken up largely through the initiatives of international/national NGOs, UNDP etc. and in some countries in collaboration with the national governments. However, CBDRR has done better in countries where the national and local governments have been proactively involved with the implementation of the programs. In Bangladesh, it has been found that CBDRR first raised after the devastating disaster in 1970 caused by cyclone "Bhola". Aftermath the destructive cyclone, Bangladesh first started CBDRR in South Asia through Cyclone Preparedness Programme (CPP). CPP started their works at the coastal districts of Bangladesh with the help of

International Federation and Red Crescent Societies in 1972. Therefore, it has been observed that the introduction of CPP in coastal districts drastically reduced the causality rate because of its strong contribution for the people in the community (Habiba and Shaw 2012). For instance, 150,000 people lost their lives in 1991 cyclone, but in 2007 cyclone Sidr, only 3,363 people lost their lives. This is achieved only because of the introduction of CBDRR in Bangladesh.

Regarding the development of CBDRR in Bangladesh, the main purpose of this chapter is to illustrate the importance of CBDRR, to provide some best example of CBDRR approaches towards various climatic disasters followed in Bangladesh. Then the focus shifts how to sustain CBDRR in context of Bangladesh and then concluding remarks.

12.2 Importance of CBDRR

Whether a disaster is major or minor, of national or local proportion, it is the people in the community who suffer most its adverse effects. In reality, communities are the frontier to respond this disastrous condition at first sight. They use coping and survival strategies to face and respond to the situation long before help from the government or NGOs arrives (Victoria 2009). The term "community-based" means that disaster management is jointly dealt with by the community. Although the role of the community varies, it is agreed that under this approach, communities are the main actors that develop and implement important policies in relation to disaster management. Adding together, CBDRR takes into consideration the particular context of the community. Therefore, the main aim of CBDRR is to reduce vulnerabilities by strengthening individuals, families and communities. It seeks to address conditions, factors, processes, and causes of vulnerabilities brought by poverty, social inequality, and environmental resource depletion and degradation. It subscribes people-centered development as well as equitable and sustainable development. Therefore, the goal of CBDRR is building safer, disaster-resilient and developed communities. Following are the characteristics of CBDRR that are explained in Table 12.1.

Many developing countries now practice CBDRR. For example, community-based risk reduction projects are running in 500 villages across Aceh as well as in disaster-prone districts of Sri Lanka. In Sri Lanka, village-level disaster teams made up of volunteers who are taking the lead in mapping the hazards as well as learning skills in emergency first aid and spreading awareness amongst the old and young who take part in mock evacuation drills. Further, in Sri Lanka, 400,000 people are benefiting from a grassroots early warning system run by more than 1,000 volunteers who disseminate warnings and help to safely evacuate people during disasters. 3D digital hazard evacuation maps have been developed for all of the communities involved in the project in collaboration with the government's disaster management centre and UN OCHA.

On the other hand, CBDRR is now getting more importance at the development program. Because, by adopting CBDDR, it reduces loss of lives and related risks

Table 12.1 Characteristics of CBDRR

Characteristics of CBDRR	Explanation
1. Participatory process and content	• Community is the key actor and primary beneficiary • Involves all the vulnerable groups
2. Responsive	• Considers the community's perception and prioritization of DRR • Community empowerment through ownership creation
3. Proactive	• Prepares the communities to face disasters beforehand
4. Comprehensive	• Structural mitigation (dam construction, early warning centers) • Non-structural mitigation (education and training, public awareness)
5. Integrated	• Involves all the stakeholders in DRR • Pre-, during and post-disaster measures are planned and implemented as necessary by the community
6. Multi-sectoral and multi-disciplinary	• Combines indigenous/local knowledge with sciences and new technologies • Builds capacity within while bringing resources externally
7. Empowering	• People's options and capacities increased • More access to and control of resources and basic services • Meaningful participation in decision-making
8. Developmental	• Contributes in poverty reduction • Correlated to developmental activities

Source: Facilitator guidebook: Disaster risk reduction training to community volunteer (2010)

by making people self-resilient and helping in strengthening community coping mechanisms through capacity building at the grassroots level. For instance, the CBDRR project in West Bengal has become a success story in India. In 2008, UNICEF, along with Government of West Bengal and 11 key long-term partner NGOs has undertaken the CBDRR project in seven select districts with DIPECHO funds that aimed to build capacities of the most vulnerable communities in flood-prone areas in seven select districts. Due to the accomplishment of the project, the Government of West Bengal has now incorporated CBDRR in its development planning and has budget provisions for the project.

In case of Bangladesh, disasters are on rise because of extreme weather conditions along with climate change that has added more vulnerability to disasters. Since Bangladesh is a developing country that has lack of resources to cope with the disasters, so, there is need to have CBDRR. Because the risk is local and many of the times external support comes later in case of disaster to rescue community. Thus community has to prepare themselves to fight with the disaster to reduce its impact. As a result, CBDRR is inhaling its importance with the increasing threat of disaster in the developing countries.

12.3 Disaster Profile in Bangladesh

During the period of 1991–2000, Bangladesh has suffered from 93 large scale natural disasters which killed 0.2 million people and caused loss of properties valued about US$59 billion dollar in the agricultural and infrastructure sector (Climate Change

Table 12.2 Top 10 natural disasters in Bangladesh for the period 1900–2012

Disaster	Date	No of total affected people
Flood	Jun-1988	45,000,000
Flood	July-1974	38,000,000
Flood	20-June-2004	36,000,000
Flood	May-1984	30,000,000
Flood	22-July-1987	29,700,000
Drought	July-1983	20,000,000
Flood	July-1968	15,889,616
Storm	11-May-1965	15,600,000
Storm	29-April-1991	15,438,849
Flood	5-July-1998	15,000,050

Source: EM-DAT, Centre for Research on Epidemiology of Disasters, Leuven, 2012

Table 12.3 Most vulnerable countries to floods or cyclones (deaths/100 people exposed to floods or cyclones)

Rank	Name of the country	Floods	Rank	Name of the country	Tropical cyclones
1	Venezuela	4.9	1	Bangladesh	32.1
2	Afghanistan	4.3	2	India	20.2
3	Pakistan	2.2	3	Philippines	8.3
4	China	1.4	4	Honduras	7.3
5	India	1.2	5	Vietnam	5.5
6	Bangladesh	1.1	6	China	2.8

Source: UNDP (2004)

Cell 2009). Disasters are being common in Bangladesh and almost every year, this country experiences disaster of one kind or another, such as tropical cyclones, storm surges, coastal erosion, floods, nor'westers, tornadoes and droughts, that causes heavy loss of life and property and jeopardizing development activities (Ali 1996). Among these, flood, cyclone and drought are the most catastrophic one for which a number of people affected. Table 12.2 highlights the top ten natural disasters in Bangladesh from 1900–2012 where it has been seen that more people are affected by flood after that cyclone and drought, respectively. Furthermore, UNDP in 2004 reported that Bangladesh is the most vulnerable country in the world to tropical cyclone and the sixth most vulnerable country to floods that has been shown in Table 12.3.

Hence, after discussing a brief disaster scenario in Bangladesh, this section intends to illustrate its main focus on flood, drought and cyclone, accordingly that has a great impact on live and livelihood of Bangladeshi people.

12.3.1 Flood

Flood is a regular phenomenon in Bangladesh. Mainly four types of floods occur in Bangladesh: flash floods, river floods, rainwater floods and coastal floods induced by storm surges (Ahmad et al. 1994, 2000). Mirza (2002) mentioned that one-fifth

Table 12.4 List of recent severe floods

Year	Inundated area	Deaths	Economic damage
1984	50,000 km^2	n/a	USD 0.38 billion
1987	50,000 km^2	2,055	USD 1.0 billion
1988	89,000 km^2	2,000–6,500	USD 1.2 billion
1998	100,000 km^2	1,100	USD 2.8 billion
2004	56,000 km^2	700	USD 2.0 billion
2007	32,000 km^2	649	USD 1.0 billion

Source: The long road to resilience: Impact and cost-benefit analysis of community-based disaster risk reduction in Bangladesh, 2012)

Table 12.5 Major cyclones that hit the Bangladesh coast

Date	Year	Maximum wind speed (km/h)	Storm surge height (m)	Death toll
11 May	1965	161	3.7–7.6	19,279
15 December	1965	217	2.4–3.6	873
01 October	1966	139	6.0–6.7	850
12 November	1970	224	6.0–10.0	300,000
25 May	1985	154	3.0–4.6	11,069
29 April	1991	225	6.0–7.6	138,882
19 May	1997	232	3.1–4.6	155
15 November (SIDR)	2007	223	–	3,363
25 May (AILA)	2009	92	–	190

Source: Bangladesh Meteorological Department (2007), cited in National Platform for Disaster Management 2010–2015 (2010)

of the country is flooded every year, and even in extreme years, two-thirds of the country can be inundated. Usually, the most severe floods occur during the months of July and August, and severe flooding appears every 4–5 years interval. During last 50 years, at least seven mega floods have occurred, affecting about 35–75 % of the land area. Major flooding recorded in the recent years is in 1988, 1998, 2004 and 2007 among which 1998 was the worst on record (Table 12.4). These devastating floods had an enormous impact on the national economy, agriculture, water and food security, human health, shelter, population displacement, cause hardships for people, and disrupts livelihood systems in urban and rural areas.

12.3.2 Cyclone

Aside from floods, Bangladesh regularly suffers worst disaster that is tropical cyclones, often comes with storm surges. The cyclone of 1965, 1966, 1970, 1985, 1991, 1997, 2007 and 2009 are noteworthy (Table 12.5). Among them, cyclone in 1970 is the deadliest cyclone that hits Bangladesh coastline and took away the lives of over 300,000 people and damaged of about US$2.5 billion equivalent crops and property. In 1991, the catastrophic cyclone killed over 150,000 people and property damages were more than US$ two billion.

Most recently, in 2007 cyclone "Sidr" struck in Bangladesh that killed and injured over 3,363 and 55,282 people, respectively. It is also reported that 563,877 houses were totally destroyed and 955,065 houses were partially damaged. In terms of agricultural sector, it was accounted for fully damaged of 186,883 ha of crop areas and partly damaged of 498,645 ha area (National Plan for Disaster Management 2010). Likewise, in 2009, 64 cyclone "Aila" caused death of 330 lives, made one million people homeless and total damage of US$40.7 million. Estimation depicts that 20 million people of Bangladesh were at risk of post disaster diseases due to Aila.

12.3.3 Drought

Drought is a recurrent phenomenon, afflicting the country at least as frequently as do major floods and cyclones, averaging about once in 2.5 years (Adnan 1993; Ericksen et al. 1993; Hossain 1990). During the last 50 years, Bangladesh suffered from drought conditions about 20 times. The more severe drought-affected years were 1951, 1961, 1973, 1975, 1978, 1979, 1981, 1982, 1992, 1994, 1995, 2000, 2006 and 2009. Among these, the 1973 drought was one of the severest that responsible for the local famine in northern Bangladesh in 1974. Even, in the 1975 drought, 47 % of the country was affected and caused suffering to about 53 % (Adnan 1993; Task Force Report 1992). The consecutive drought of 1978 and 1979 directly affected 42 % of cultivated land and reduced rice production by an estimated 2 million tons (Brammer 1987). The losses due to drought in 1982 were more than double the losses caused by floods in the same year. But the most persistent drought was in 1989 (Kafiluddin 1991). Again, the 1997 drought caused a reduction of around 1 million tons of food grain, of which about 0.6 million tons were transplanted aman. Although droughts were not continuous, they did affect the low rainfall zones of the country, especially in the north–west part of the country.

12.4 CBDRR Evolution in Bangladesh

Bangladesh has pioneered community-based approaches to reduce vulnerability towards climatic disasters. Various types of innovative approaches have been developed by community to survive with these disasters events in collaboration with different international as well as governmental organizations or NGOs (Habiba and Shaw 2012). In Bangladesh, the concept of CBDRR first evolved after the devastating disaster caused by cyclone "Bhola" in 12 November 1970. This cyclone "Bhola" was responsible for the death of 500,000 people that is the most deadliest hurricane in the history of mankind. Nearly 90 % of the marine fisherman suffered heavy losses. Some 9,000 fishing boats were destroyed; the damage to property and crops was colossal. Aftermath the destructive cyclone, the Government of Bangladesh initiated CBDRR through Cyclone Preparedness Programme (CPP) at the coastal districts with the help of International Federation and Red Crescent Societies (IFRCS) in 1972. Among South Asian countries, Bangladesh is the first to start the process of CBDRR. Following the withdrawal of International Federation as a direct partner in the implementation in July 1973, a new joint venture program structure was formulated by Government of Bangladesh (GoB) and Bangladesh Red Crescent Society (BDRCS). An agreement was signed by both parties spelling that the programme is basically continued with Red Crescent and is known as the Cyclone preparedness Programme (CPP) of the Bangladesh. CPP created massive awareness among the communities about the risks and enhanced their capacities to be prepared and respond to disasters, which is reflected in the drastic reduction in causalities in subsequent disasters. Besides these, CPP is also responsible for activities such as the formation of village disaster preparedness committees, development of an extensive awareness-raising campaign, training of the community in disaster preparedness, community first aid, and cyclone warning signals, shelter maintenance, and implementation of disaster preparedness measures such as installation of drinking water and food shortage facilities and construction of poultry sheds (Yodmani 2001). However, in the recent years, it has been observed that CPP carries out their public awareness and capacity building activities for pre-disaster preparedness at family and community level.

At the national level, the Ministry of Food and Disaster Management (MoFDM) serves as the Government's focal point for disaster management and coordination. The mission of MoFDM in Bangladesh is to

> To achieve a paradigm shift in disaster management from conventional response and relief to a more comprehensive risk reduction culture and to promote food security as an important factor in ensuring the resilience of communities to hazards.

Therefore, this country has adopted various legislation and policy to deal with disaster successfully such as National policy on disaster management, National plan for disaster management, disaster management act and standing order (SoD) on disaster etc. According to these policy and act, it has been examined that helping communities to mitigate the potential adverse effects of hazard events, prepare for managing the effects of a disaster event, effectively respond to and recover from a disaster or an emergency situation, and adapt to adverse effects of climate change is the one of the objective of disaster management act. National policy on disaster management gives much emphasis on involvement of community and local governments in implementing disaster management programs. On the other hand, National plan for disaster management (2010–2015) identifies CBDRR as a priority component. This plan has developed by Ministry of Food and Disaster Management where it recognizes community empowerment is operationalzed by CBDRR as one of the six key results areas. It includes the following elements:

- Capacity building of local disaster management committees
- Development of community based early warning system

- Community risk assessment at union level
- Development of risk reduction action plan at union level
- Development of contingency plan at union level
- Bottom up mainstreaming of risk reduction action plans into local development plans
- Local DRR funding mechanism for implementation of community risk reduction actions

Over four decades, the traditional disaster management is mainly focusing on disaster relief and recovery, but aftermath of 1988 devastating floods and the cyclone of 1991, Bangladesh Government has adopted a holistic approach embracing the processes of hazard identification and mitigation, community preparedness and integrated response efforts. Therefore, relief and recovery activities are now planned within all risk management framework seeking enhanced capacities of at risk communities and thereby lowering their vulnerability. In line with the paradigm shift from relief and response to comprehensive disaster management, the Ministry of Relief and Rehabilitation has been changed to the Ministry of Disaster Management and Relief in 1993. It was again renamed as Ministry of Food and Disaster Management (MoFDM). MoFDM coordinates national disaster management efforts throughout all agencies. The Disaster Management Bureau (DMB) under MoFDM is responsible for creating public awareness about the risks associated with natural and human-induced hazards, and to formulate programmes and projects that will better prepare at-risk communities and public officials to mitigate the consequences of disasters. As a technical arm to the MoFDM, DMB overview and coordinate all activities related to disaster management from national to grass-root level.

Furthermore, increasing resilience at all levels, from the national to community level, and to reduce damage and losses from natural disasters and the impacts of climate change, the government of Bangladesh has launched the Comprehensive Disaster Management Program (CDMP) in partnership with DFID and UNDP that started field implementation in 2004. This went beyond relief, recovery and rehabilitation by looking at pre-disaster from the perspective of "risk reduction" and preparedness. It also brought a stronger focus on building coping mechanisms of the communities to manage disasters at the local level. CDMP's community-based disaster risk reduction activities are discussed below:

Community-based disaster preparedness is a major reason for the decrease in the number of deaths in recent years. CDMP Phase I (2004–2009) successfully demonstrated community based disaster risk management. CDMP II (2010–2014) further strengthen the CBDRM initiatives through scale up the coverage from 650 Unions in 19 pilot districts in Phase I to 2,000 Unions in 40 vulnerable districts in Phase II.

Ministry of Disaster Management and Relief is implementing CBDRM under CDMP at local-level by strengthening "Disaster Management Committees" (DMCs) at district, Upazila and Union Parishads, and locally-relevant action through thousands of village-level "Community Risk Assessments (CRA)" and related Risk Reduction Action Plans (RRAP). Now covering almost half of the country subject to disaster risk, these plans are financially supported by the Local Disaster Risk

Reduction Fund (LDRRF), a funding mechanism that provides resources to the most vulnerable groups for local implementation. This is supported under CDMP-II by UNDP in partnership with DFID, EU, SIDA, NORAD and AusAid. Four types of CBDRM interventions are undertaken.

12.4.1 Type-1: Comprehensive Interventions

The comprehensive intervention seeks to expand the small scale and individual interventions approach that was tried out during CDMP Phase I. Instead of stand-alone fragmented interventions the approach promotes the communities and counterparts to consider the different kinds of interventions that, taken together, would address the risk situation from the different sectors in a more inter-related manner. Comprehensive projects are to be planned and/or implemented by local government and or NGOs in a cluster of several geographic or administrative entities addressing the different hazards so that the beneficiaries are able reduce their aggregate risks.

12.4.2 Type-2: Disaster Resilient Habitat

Disaster Resilient Habitat is a new concept being tried out initially in cyclone prone coastal areas. The idea is to provide a most vulnerable village with all the necessary supports to dramatically reduce risks from recurring disaster or climate change hazards. This departs from the analysis that for some villages, the vulnerability is too wholesome and chronic that any single or several interventions, even if provided in a comprehensive manner would not significantly address the vulnerability. And thus the proposition to provide the "A to Z package" involving, for instance, village land use planning, plinth raising, disaster resistant housing, tree plantation, community organizing, income generation activities, etc. CDMP II has already initiated implementing two villages in the coastal areas of Shutarkhali Union of Dacope Upazila in Khulna and in 2012, plans to undertake five Resilient Habitats in areas with major hazards of Flood, Flush Flood, Cyclone and Drought.

12.4.3 Type-3: Interventions by Disaster Management Committees

One of CDMP's major targets is to empower Disaster Management Committees (DMCs) and the LDRRF projects are the ideal avenues to provide substantive role to these very important but otherwise dormant DMCs. DMCs at Upazila and Union levels are invited to submit proposals for LDRRF grant award small scale community-level projects that are identified in the RRAPs. Such schemes may comprise structural and/or nonstructural interventions each not exceeding the value worth US$5,000. DMCs, however, may submit multiple proposals to implement several interventions addressing DRR and CCA against RRAPs. The key feature of this type of intervention is that the local government units are expected to put forward their matching funds and/or inputs to make up the overall interventions. For instance the Upazilla LGED is to provide the infrastructure, the safety net programme to provide cash/good for work for the road, and the LDRRF is to provide the plinth raising and income generating activities. As experienced in phase I, CDMP II also considers engaging NGOs as technical partner to support the DMCs.

12.4.4 Type-4: Specific Interventions

There are disaster and/or climate change major circumstances that require specific interventions. For instance, at the moment, CDMP II is addressing the persistent and widespread saline intrusion to the surface and groundwater that deprives millions of people in the post-Cyclone Aila areas over than 2 years after the disaster occurred. In this case comprehensive intervention, small scale schemes by DMCs, and disaster resilient habitat interventions will not be effective. CDMP II forged collaboration with the GIZ to form a formidable enough input to address the problem in a large scale. In this case the GIZ is to provide the raised solar powered deep tube well in 60 locations and the LDRRF complements with the installation of the pipe distribution networks with at least 600 water points to deliver the safe water to the communities, and the formation of the water users' committees. The LDRRF delivers this intervention through the local DMCs to ensure the participation and ownership. The LDRRD envisions that there are other situations that will require this specific needs-based intervention to reduce disaster and climate change risks.

Aside from governmental activities, several international and national NGOs in Bangladesh are working with community to promote their activities towards disaster. Some of them, like e. g. OXFAM, CARE, CONCERN, and Action Aid as international as well as BRAC (Bangladesh Rural Advancement Committee), BDPC (Bangladesh Disaster Preparedness Center), CNRS (Center for Natural Resource Studies) and other as national NGOs are focusing on pre-disaster awareness and preparedness at household and community level. For example, CARE Bangladesh used food-for-work to help communities with community "flood-proofing" after the 1998 floods. This includes raising of house plinth to a 5-year flood levels, schools and community centers to a 20-year level, raising hand pumps and building foot paths, as well as village level saving schemes that are used as a safety net in a disaster to meet immediate relief needs.

Contrast with these, private sectors such as Grameen Phone and Teletalk have make disaster early warning alerts by sending instant messages to their subscribers particularly during the catastrophe period.

12.5 Community-Based Disaster Risk Reduction Approaches

Local communities are on the frontline of both the immediate impact of a disaster and the initial, emergency response, which is crucial for saving the most lives. Over four decades ago, local people within the community were not so much aware of the catastrophic climatic events occur in Bangladesh and they were not being well prepared about the likelihood of disaster. Currently, community peoples are more awarded about climate change impacts and more prepared to cope with these because of intensified information and increased their capacities to respond to disaster. The evolution of CBDRR helps to give them a platform to deal better with disasters. Within disaster risk reduction field, community based approaches to reduce vulnerability have become increasingly popular over the past years (Allen 2006). It is becoming more common place in development as the development community come to realize the benefits of this approach (Uitto and Shaw 2006) which recognizes and values local culture, conditions, and development issues (Ayers and Huq 2009).

In Bangladesh, a number of community based projects are underway. These are mainly handled by community people with the cooperation of international/national NGOs, UNDP, national governments. Presented below are the three best practices currently taking places, that each integrates disaster risk reduction. CPP, floating garden and rainwater harvest are such type of approach that has been practiced by community people with regard to flood, cyclone and drought in Bangladesh, accordingly.

12.5.1 Floating Garden

12.5.1.1 Evolution of Floating Garden

The southern part of Bangladesh located at the coastal region, is generally flat, and land elevation is low from the mean sea level. Therefore, extended flooding and water logging is very common in this area due to both high tide and also by the river water. As a consequence of flooding and water logging, it affects the agriculture, homestead vegetable cultivation, fish culture, and hampers livelihood activities of this region.

To tackle with this adverse situation, some communities of the southern wetlands practice floating garden. It is also known as *baira*, *boor*, *dhap*, *gathua*, *gatoni*, *geto*, *kandi* and *vasoman chash* and floating agriculture (Islam and Atkins 2007; Irfanullah et al. 2007). Currently, it has generated huge interest in the agricultural field as it has positive contribution in improving livelihood and food security. Therefore, it is not only becoming expanded by different organizations like Practical Action, Strengthening Household Abilities for Responding to Development Opportunities (SHOUHARDO) program of CARE Bangladesh in northeastern Bangladesh but also Research on Innovative and Strategic Policy Options (RIPSO) of APEIS in southern Bangladesh. Moreover, the National Action Program for Adaptation (NAPA) has listed floating garden is one of the 15 major projects.

Fig. 12.1 Floating garden practice in Southern Bangladesh (*Source*: Aarjan Dixit 2011)

12.5.1.2 Approach

Community people in southern Bangladesh has developed and established a sustainable community based floating garden that helps to protect them from the devastating effects of floodwater and allow farmers to grow crops in a floating platform. Usually people of this region use bamboo to make the floating base which help to avoid damage due to wave action or drifting (Fig. 12.1). Various local materials are used to construct the floating layer.

Most commonly used material is water hyacinth that is piled and then covered with soil and cow dung to form a seed bed that helps to plant different types of crops. The main crops cultivated in the floating gardens are okra, cucumber, bitter gourd, tomato, cauliflower, chili, amaranth, taro, pumpkin, turmeric etc. The community people practiced this floating cultivation during the winter season as well as the monsoon season (mainly June to August).

12.5.1.3 Lesson Learned

- Floating garden provides as an alternative choice to grow vegetables on the floating bed by turning the flooded and water logged area into productive lands
- It generates income for the rural poor, and also leads to significant and substantial increase of food production in that rural area
- It provides a proper gender balance by involving men and women to maintain this activities

12.5.2 CPP (Cyclone Preparedness Program)

12.5.2.1 Evolution of CPP (Cyclone Preparedness Program)

A massive cyclone with wind speed 62 m/s accompanied by a storm surge 6–10 m in height struck Bangladesh on 12 and 13 November in 1970, claiming an estimated 300,000 people, making millions of homeless and totally destitute. Following the cyclone of 1970, the United Nations General Assembly requested the League of Red Cross (now known as International Federation) to take the lead role in establishing and improving the pre-disaster planning program in Bangladesh. The League of Red Cross with the collaboration of the Red Crescent societies undertook an extensive program named CPP in 1972 consisting of 20,310 volunteers in 204 unions of 24 thanas (sub-district) and a transceiver telecommunication system (Wireless transceiver radio in 22 stations). In 1973 the League of Red Cross and the Red Crescent societies withdrew the field level program. But, considering the vulnerability of coastal people, the Government of Bangladesh came forward and took the responsibility. Afterwards, the program appears as the joint program of Government of Bangladesh and Bangladesh Red Crescent Society (BDRCS). The goal of the CPP is "to minimize loss of lives and properties in cyclonic disaster by strengthening the capacity in disaster management of the costal people of Bangladesh".

12.5.2.2 Approach

CPP is a mechanism which relies on technical skills and volunteers commitment for ensuring all potential victims of an approaching cyclone are given sufficient warning to 11 million coastal people residing in the 710 km long coastal areas. So that they can move to safe sites including the cyclone shelters and building after receiving the warning signals from volunteer of CPP. The designated warning signal for approaching cyclone comes from Bangladesh Metrological Department (BMD) and transmitted this information from zonal office to Upazila level offices through high frequency (HF) radios. The Upazila office further passes this information to village unions and lower levels through very high frequency (VHF) radios. At present, CPP extended their activities through 49,365 volunteers in 37 Upazilas (sub-district) in 321 unions (village level) divided into 3,281 units.

Volunteers are the core of CPP who come from the community and remain in the community. They play a crucial role in the dissemination of cyclone warning in the coastal districts. They are responsible for alerting people by megaphones and mikes, house to house contact, raising danger signal flag, rescue of survivors, first aid to the wounded, post-cyclone security measures, distribution of relief materials, and surveying damages caused by cyclones and reporting these to their local headquarters. They do these types of activities with the help of warning equipments like transistor radio, megaphone, siren, signal light, first aid kit, etc. provided through CPP (Fig. 12.2).

Besides these, CPP implements different public awareness program in various ways in the cyclone prone coastal areas such as public awareness program through

Fig. 12.2 Volunteer's alarts people using megaphone during cyclone period (*Source*: Bangladesh Red Crescent Society)

volunteer; cyclonic drills and demonstration; publicity campaign; radio and television program; poster, leaflets and booklets; staging of dramas for rural communities that helps to enhance their better preparedness activities toward disaster risk reduction.

12.5.2.3 Lesson Learned

- CPP is an effective; grass-root oriented, disciplined and tightly knots organization which is remarkable and successful example of CBDRR in this sub-continent.
- It shows the dedication of protecting the population along with community capacity build up activities.
- Volunteer is the backbone of CPP where they exhibit a high level of commitment for their program and readiness to meet the community requirements for better disaster preparedness.
- CPP involves community with full participation, increase awareness of the community to have a practical approach of reducing risks and losses.
- Achievements of cyclone preparedness program are significant and impressive and it can be replicated elsewhere.

12.5.3 Rain Water Harvesting

12.5.3.1 Evolution of Rainwater Harvesting

The coastal belt of Bangladesh is suffering from scarcity of drinking water because of its salty ground water as well as drought that causes a major water problem in this area.

On the other hand, most of the hand tube-wells in this area are inoperative because of presence of high levels of salt, arsenic and iron. To cope with this safe drinking water, rain water harvesting practice was first carried out at community level in Dacope Upazila of Khulna districts with the help of Bangladesh Agricultural University in 1988. After that, with the support from Department of Public Health Engineering, UNICEF extended this rain water harvesting practices in Chittagong hill tract and then extended it in coastal belt areas. But, due to contamination of well with arsenic, it was looking for alternative option for drinking water. WHO (World Health Organization) and Swiss technology transfer agent, SKAT agreed to identify the feasibility of this technology in 1998. At the same time, International Development Enterprises (IDE) offered locally acceptable, affordable technologies for the community people. Furthermore, OXFAM, CARE and Bangladesh University of Engineering and Technology (BUET) were also involved in this rainwater harvesting systems.

12.5.3.2 Approach

Rain water harvesting refers to the collection and storage of rain water in situ or within the vicinity of rainfall. In Bangladesh, it is mainly used for drinking and cooking purposes. It provides a viable option than the other technologies. Because, annual rainfall of Bangladesh is 2,350 mm that does not uniformly distribute in all areas of the country. For instance, the amount of annual rainfall in north–east part is about 5,500 mm whereas, in south–west part, it is around 1,200 mm. Again, most of the rainfall occurs from July to October. Depending upon the scarcity of drinking water, various type of rain water harvesting model is used at house-hold and community level. Among them, cement/mortar jar, ferro-cement tank, R.C.C. ring tank is used by the household level and constructed underground rain water harvesting system is used for the community level (Fig. 12.3). Besides these, there is an indigenous process of rain water harvesting system which is practicing in different part of Bangladesh named construction of do-it yourself model (motka). While using these models, it should be kept in mind about hazardous aspects like contamination of water, mosquito breeding etc.

12.5.3.3 Lesson Learned

- The rain water harvesting in rural areas gives as an alternate water source especially in the arsenic, salinity and drought affected area
- The number of rain water harvesting is rapidly increased but it needs time to maintain the pH value and checking of water quality
- During dry season, the situation goes to worse, because, most of the water tanks start to go empty. This situation bound the local people to collect drinking water from pond using the filtering technology
- Rain water harvesting and its judicious utilization for agriculture and municipal use would be essential in our future life

Fig. 12.3 Rain water harvesting in Bangladesh (*Source*: Author)

From the above discussion, it is proven that community people are the key actor for doing any types of activities to reduce the disaster risk. They can utilize their local resources and share indigenous knowledge to resolve the disaster impacts. Therefore, these best practices can be used as successful example for the vulnerable communities in any disaster affected area that is prone to climate change as well as disaster effects.

12.6 Sustainability of Community-Based Approaches

It has been observed that CBDRR has mostly been taken up on a project mode and has not been integrated within the existing governance and development programs for their long term sustainability. However, over the last two decades, there has been a growing realization that disaster risk reduction is most effective at the community level where the specific local needs, resources, and capacities are met. From the sustainability point of view, it is very important to involve the community from the very beginning of any development as well as disaster risk reduction initiatives. On the other hand, Shaw and Okazaki (2003) mentions that community involvement often faces the problem of sustainability over a longer period of time. Shaw (2012) further highlighted that CBDRR faces two major challenges that include: (1) sustainability of the efforts at the community level and (2) incorporation of the CBDRR issues at the policy level.

Sustainability is the end goal for every community. It provides an infrastructure to identify and address community challenges in the long run and helps a community make the best use of the resources and makes it possible to leverage resources for future activity. In Bangladesh context, various stakeholders have taken different approaches to reduce the disaster risk. Despite many setbacks progress has been achieved in disaster risk reduction, but in terms of community based approach, improvements are still beyond the requirement. Therefore, it is necessary to identify and impart essential skills of community that can translate risk awareness into concrete practices of sustained risk management. Moreover, Shaw (2012) suggested a list of factors that helps to enhance the sustainability of CBDRR. These are as follows:

- Promote and strengthen a "culture of coping with crisis"
- Enhance people's perception on vulnerability
- Recognize motivation of community initiatives
- Increases community participation and empowerment through institutionalization
- Focus on need-based training approaches
- Involve diverse stakeholders based on the needs and objectives in both formal and/or informal ways
- Promote tangible and intangible accumulation of physical, technological and economic assets as the project outputs
- Promote the integration of community initiatives into regular development planning and budgeting to ensure sustainability

On the other hand, institutionalization of CBDRR must need legislation that support CBDRR activities and must be incorporated into government framework and plan of action of all development stakeholders. Government must need to mention the role of NGOs and other organizations for CBDRR and recognize the need for involving communities and community groups in disaster risk reduction work into their development plans and policies. In addition, local governments need to institutionalize CBDRR in investment programming and project design, budgeting and revenue generation. At the same time, it needs to take pragmatic steps to address the lack of activities between national, sub-national level and community level. Therefore, institutionalization of CBDRR requires the following as its precondition:

- A guideline for disaster risk reduction
- Integrate into National Disaster Management Framework
- Strong policy support
- Institutionalized budget
- Definite role of stakeholder
- Transparency and accountability of the framework

12.7 Conclusions

From the study, it is clear that the people within the community of Bangladesh frequently experiences climate change impacts in a form of various climatic disasters. At the same time, experiences from community based organization show that involvement of community is proven to be useful in disaster risk reduction processes. Because, local people in the community are the reservoir of time tested knowledge on disaster risk reduction. But, communities require encouraging reviving; adapting and using such knowledge, to the extent it is possible, in the community based disaster risk reduction (CBDRR). Moreover, top-down disaster programs sometimes fail to address the specific local needs of vulnerable communities, overlook the potential of local resources and capacities, and may even increase people's vulnerability (Bhatt 2002).

On the other hand, CBDRR shifts from a top-down to a bottom-up approach. The community-based approach corrects the defects of the top-down approach in development planning and disaster risk reduction which failed to address the local needs, ignored the potential of indigenous resources and capacities, and may have been increased people's vulnerabilities. In this regard, community-based activities will be effectively reduced the disaster risk, as they can emphasize their needs and prioritize their responsive measures toward disaster.

In Bangladesh, it has been seen that most of the community based practiced developed in the coastal area that are innovative in nature and runs successfully in those areas. Therefore, this proven approach should be scaled up by different organizations and also by governmental institution to adapt it in the country level whenever it will be applicable. Similarly, to organize disaster resilient community, there is a need of for integrating the principles and process of disaster risk reduction into local government institutions, community-based organizations and NGOs. Local level institutions bear the responsibility for several services including health, safety and welfare. In order to ensure effective and efficient response, community based disaster risk reduction activities will be implemented to build safer communities in Bangladesh. The activities will benefit from and supplement the regional initiative "Building Safer Communities in South Asia" through the sharing of best practices and knowledge and co ordination on a regular basis. Disaster risk reduction initiatives in Bangladesh will also contribute to the government's national disaster risk reduction framework.

Communities are central in disaster management, so, they need more awareness information, encouragement and support during disaster events from government and other actors, particularly from NGOs. Moreover, community participation and control is essential for any successful implementation, orientation and maintenance of any disaster risk reduction project. Although, Government of Bangladesh carries out several disaster management programs to reduce the vulnerability, but, it is important to involve people in decision making on policies and strategies that should be followed for their development in the community (Kafle and Murshed 2006; Shaw and Okazaki 2004). Aside from the above, it can be concluded that CBDRR will serve as a bridge among communities, governments, donors and other stakeholders in any successful disaster risk reduction program. Because, community driven disaster risk reduction approach will be more reliable and effective as community people directly involve in this process that will be sustained for a long time.

Acknowledgements The first author gratefully acknowledges the support from GCOE-ARS postdoctoral program of Kyoto University. The third author gratefully acknowledges the support from the Japan Society for the Promotion of Science (JSPS) of Japan for conducting this research.

References

Aarjan Dixit (2011) Adapting to climate change in Bangladesh. Retrieved from http://www.wri.org/stories/2011/04/adapting-climate-change-bangladesh

Adnan S (1993) Living without floods: lessons from the drought of 1992. Research and Advisory Services, Dhaka

Ahmad QK, Ahmad N, Rasheed KBS (eds) (1994) Resources, environment and development in Bangladesh with particular reference to the Ganges, Brahmaputra and Meghna basins. Academic Publishers, Dhaka, 134 p

Ahmad QK, Chowdhury AKA, Imam SH, Sarker M (eds) (2000) Perspectives on flood 1998. The University Press Limited (UPL), Dhaka, 142 p

Ali A (1996) Vulnerability of Bangladesh to climate change and sea level rise through tropical cyclones and storm surges. Water Air Soil Pollut 94(d):171–179

Allen KM (2006) Community-based disaster preparedness and climate adaptation: local capacity building in the Philippines. Disasters 30:81–101

Ayers J, Huq S (2009) Community based adaptation to climate change: an update. IIED, London

Bhatt RM (2002) Corporate social responsibility and natural disaster reduction: local overview of Gujarat. Disaster Mitigation Institute, Ahmedabad (67 pp)

Blaikie P, Cannon T, Davis I, Wisner B (1994) At risk: natural hazards, people's vulnerability, and disaster. Routledge, London

Brammer H (1987) Drought in Bangladesh: lessons for planners and administrators. Disasters 11(1)

Climate Change Cell (2009) Department of Environment, Ministry of Environment and Forests, Dhaka

Ericksen NJ, Ahmad QK, Chowdhury AR (1993) Socio-economic implications of climate change for Bangladesh. Bangladesh Unnayan Parishad, Dhaka

Facilitator guidebook: Disaster risk reduction training to community volunteer (2010) Retrieved from http://www.preventionweb.net/english/professional/publications/v.php?id=16788. Accessed 10 Jan 2013

Habiba U, Shaw R (2012) Bangladesh experiences of community based disaster risk reduction. In: Shaw R (ed) Community, environment and disaster risk management, vol 10. Emerald Group Publishing Limited, Bingley, pp 91–111

Hossain M (1990) Natural calamities, instability in production and food policy in Bangladesh. Bangladesh Dev Stud 18(1):33–54

Irfanullah HM, Adrika A, Ghani A, Khan ZA (2007) Introduction of floating gardening in the north-eastern Wetlands of Bangladesh for nutritional security and sustainable livelihood. Renew Agric Food Syst 23(2):89–96

Islam T, Atkins P (2007) Indigenous floating cultivation: a sustainable agricultural practice in the Wetlands of Bangladesh. Dev Pract 4(1):130–136

Kafiluddin AKM (1991) Disaster preparedness for Bangladesh floods and other natural calamities. Padma Printers & Colour Limited, Dhaka, p 251

Kafle SK, Murshed Z (2006) Community-based disaster risk management for local authorities. Participant's Workbook, Asian Disaster Preparedness Centre (ADPC), Pathumthani

Mileti DS (2001) Disasters by design. Joseph Henry Press, Washington

Mirza MMQ (2002) Global warming and changes in the probability of occurrence of floods in Bangladesh and implications. Glob Environ Chang 12:127–138

National Plan for Disaster Management (2010) Disaster Management Bureau, Disaster Management & Relief Division, Government of the People's Republic of Bangladesh

Quarantelli EL (1989) Conceptualizing disasters from a sociological perspective. Int J Mass Emerg Disast 7:243–251

Shaw R (2012) Overview of community-based disaster risk reduction. In: Shaw R (ed) Community, environment and disaster risk management, vol 10. Emerald Group Publishing Limited, Bingley, pp 3–17

Shaw R, Okazaki K (2003) Sustainability in grass-roots initiatives: focus on community based disaster management. UNCRD Publication, Kobe (99 pp)
Shaw R, Okazaki K (2004) Sustainable community based disaster management practices in Asia: A user's guide. Kobe: UNCRD
Task Force Report (1992) Report of the task forces on Bangladesh development strategies, vol 4. UPL, Dhaka
Twigg J (1999) The age of accountability? Future community involvement in disaster reduction. Aust J Emerg Manag 14(4):51–58
Twigg J, Bhatt M (1998) Understanding vulnerability: South Asian perspectives. ITDG, London
Uitto JI, Shaw R (2006) Adaptation to climate change: promoting community-based approaches in the developing countries. Sansai 1:93–108
UNDP (2004) A Global Report: Reducing Disaster Risk: A challenge for Development cited in Bangladesh Climate Change Strategy and Action Plan 2009. Retrieved from http://www.undp.org/bcpr. Accessed 6 Jan 2013
Victoria L (2009) Community capacity and disaster resilience. In: Shaw R, Krishnamurty R (eds) Disaster management: global challenges and local solutions. University Press, Hyderabad, pp 338–351
Yodmani S (2001) Disaster risk management and vulnerability reduction: protecting the poor. Social Protection Workshop 6: Protecting communities – Social funds and disaster management. Under the Asia and Pacific Forum on Poverty, Manila

Chapter 13
Crop Insurance as Risk Management Strategy in Bangladesh

Umma Habiba and Rajib Shaw

Abstract In recent years, natural disasters, particularly climate-related disasters have increased both in terms of frequency and magnitude. As a consequence of climatic disasters, agriculture and agricultural dependent livelihood of Bangladesh is facing its substantial impacts. Structural measures for disaster risk management are often found less effective. In this regard, non-structural measures such as crop insurance is suggested as a risk management strategy. All over the world, especially USA, Canada, Australia, Japan and India, practices crop insurance through Multiple Peril Crop Insurance (MPCI), weather index-based crop insurance etc. To some extent, crop insurance has to be found successful or failure. In Bangladesh case, it has been seen that crop insurance has been introduced in 1977 in corporation with the Sadharan Bima Corporation (SBC) as a pilot project. But, due to lack of policy support and partnership, expertise and monitoring and methodological problem, this project did not sustain. Therefore, this chapter briefly discusses the types and products of crop insurance, the worldwide experiences of crop insurance and which types of crop insurance is important in context of Bangladesh. This chapter also gives emphasis on the implementation of crop insurance because of improving the welfare of risk averse farmers, particularly small and marginal farmers. In conclusion, this chapter recommends the integration of a solid partnership among relevant stakeholders at local, national and international level.

Keywords Agricultural livelihoods • Climatic disaster • Crop insurance • Partnership • Sadharan Bima Corporation

U. Habiba (✉) • R. Shaw
Graduate School of Global Environmental Studies, Kyoto University, Kyoto, Japan
e-mail: shimuagri@yahoo.com

R. Shaw et al. (eds.), *Disaster Risk Reduction Approaches in Bangladesh*, 281
Disaster Risk Reduction, DOI 10.1007/978-4-431-54252-0_13, © Springer Japan 2013

13.1 Introduction

Bangladesh is one of the most climate vulnerable countries in the world and facing many of harmful impacts of climate change due to its geography, high population density and poverty (MoEF 2008). As a result of climate change, various types of climate-related disaster such as floods, cyclones, storm surges and droughts are likely to become more frequent and severe in Bangladesh in the recent years. Besides, this situation will be aggravated in the warmer climate resulting in severe droughts and increasing floods. Moreover, it is projected that by the year of 2030, an additional 14.3 % area of Bangladesh would become extremely vulnerable to floods, while the existing flood-prone areas will face higher levels of flooding (Alam and Rabbani 2007). On the other hand, it is also projected that by 2030 sea level could rise by 30 cm (World Bank 2000).

As a consequence of climate change impacts, agriculture, the main economic driver of Bangladesh is always vulnerable to its unfavorable weather events and climate conditions. Every year, large scale crop failure occur in one part of the country or the other due to various natural calamities such as flood, drought, cyclone etc. and damaging the crops in wide spread areas and making agriculture as the most risky business. Although, lives and livelihoods of Bangladesh depend mainly on agriculture, that engaged 48 % of population and contributes about 23 % GDP. Therefore, any consequence that happens through the climatic disasters that directly affects agriculture and agriculture dependent livelihood experiences its substantial effects. Especially the poor and the marginalized farmer of the country are the worst victims to these disasters (POVCC 2003) and they have least capacity to adapt.

In this regards, non-structural measures like micro or crop insurance have been suggested as complementary risk management strategies (NAPA 2005). The UN Climate Convention (Article 4.8) and the Kyoto Protocol (Article 3.14) have included the provision of insurance as a mechanism to address the risks from climate change and discussions to implement them have started in earnest (FCCC/SB/1999/9). The Bali Action Plan adopted at COP13 in 2007 further calls upon governments to consider insurance part of an adaptation strategy. Since crop insurance is a valuable risk management tool that allows growers to insure against losses due to adverse weather conditions, fire, insects, disease, and wildlife. Geographically these crop insurance premiums are concentrated in developed regions, i.e. in USA and Canada (55.0 %) followed by Europe (20.1 %), Oceania (0.9 %), Latin America (4.0 %), Asia (19.5 %) and Africa just (0.5 %) (Mahul and Stutley 2010). Many of these countries are continuing this business, but in some others, it has stopped functioning because of incurring heavy loss. In Bangladesh case, crop insurance is not new. It was launched by Sadharan (General) Bima (Insurance) Corporation (SBC) in 1977 which was extended to the Aus, Aman and Boro rice, wheat, jute and sugarcane and ran until 1995. A total of 15,420 farmers were covered by crop insurance, but failed to attract many farmers, partly because farmers may not be aware of the benefits of the scheme, but also because the premiums are high and are not properly adjusted according to the different risks in producing particular crops.

Based on the above discussion, the main aim of this study is to highlight the utilization of crop insurance across the world, its conceptual framework and how it can help the vulnerable farmers to cope with the climatic disaster in future.

13.2 Rationale of Crop Insurance and Its Increasing Demand

Agriculture has always been risky business. The risk and uncertainties in the agriculture sector might be derived from a number of causes. Among them, climate change is one of the major reasons that cause the agriculture sector more risky. Because, in the recent years, climate related disaster have increased in frequency and magnitude (IPCC 2007) resulting in economic losses rising manifolds. Furthermore, the trend is increasing day by day and compared to developed countries developing countries seem to be more prone to such calamities because of their low ability to adjust.

Contrasting with the climate change, insurance fundamentally removes or reduces the risk of experiencing an uncertain financial loss. Because, it is widely recognized that insurance can also play a role in physical risk reduction and adaptation. IPCC'SREX also concluded that "risk sharing through formal insurance, micro-insurance, crop insurance can be a tool for risk reduction and for recovering livelihood" (Cutter et al. 2012). Aside from this, it is known that modern insurance sector can play a major role to solve the problems and considerably strengthen the financial security of farmers. In this case, agricultural Insurance is a more efficient instrument and an effective institutionalized mechanism for dealing with the problem. It helps to streamline the relief efforts and reduces the direct and indirect costs on the national economy (Jain 2004). For example, new models of crop insurance like weather index-based crop insurance has been introduced in several developing countries, such as India, Malawi and Ethiopia (World Bank 2005). Initial experience in these countries indicates a positive outcome (Hellmuth et al. 2009). On the other hand, a number of reasons demand for crop insurance is increasing day by day, which can be grouped as:

- Evidence is accumulating of connections between climate change, and the increasing incidence of crop damaging weather events of extreme severity.
- Farming is becoming steadily more commercialized, with greater financial investment. Farmers/investors and their banks frequently examine the feasibility of using a financial mechanism i.e. insurance, in order to address part of the risk.
- The World Trade Organization (WTO) regulations generally forbid governments from subsidizing agriculture directly; however, they permit the subsidization of agricultural insurance premiums. In the case, demand for crop insurance will increase in those economies that wish to implement a policy of permitted subsidization of their farmers.
- Insurance can also assist in managing the on-farm production risks consequent to changes in pest management practices. Such changes are increasingly required in order to address environmental protection and food safety concerns. This is that

any insurance arrangement will involve not only the farmer and the insurer, but also important third parties. Consideration is now given to these changes to the business of farming.

13.3 Products of Crop Insurance

Recently Mahul and Stutley (2010) summarized the traditional and new index-based crop insurance products. Based on their view, the details of these insurance products are discussed under the following sub-headings.

13.3.1 Traditional Crop Insurance

13.3.1.1 Damage-Based Indemnity Insurance (Named Peril Crop Insurance)

In this type of crop insurance the insurance claim is calculated by measuring the percentage damage in the field, soon after the damage occurs. The percentage damage measured in the field, less a deductible expressed as a percentage, is applied to the pre-agreed sum insured. The sum insured may be based on production costs, or on the expected crop revenue. Where damage cannot be measured accurately immediately after the loss, the assessment may be deferred until later in the crop season. Damage-based indemnity insurance is best known for hail, but is also used for other named peril insurance products (e.g. frost, excessive rainfall, wind).

13.3.1.2 Yield-Based Crop Insurance (Multiple Peril Crop Insurance, MPCI)

In this case, an insured yield (e.g. tons/ha) is established, as a percentage of the historical average yield of the insured farmer. The insured yield is typically between 50 % and 70 % of the average yield on the farm. If the realized yield is less than the insured yield, an indemnity is paid equal to the difference between the actual yield and the insured yield, multiplied by a pre-agreed value of sum insured per unit of yield. Yield-based crop insurance typically protects against multiple perils, meaning that it covers many different causes of yield loss. This is because it is generally difficult to determine the exact cause of the loss.

13.3.1.3 Crop Revenue Insurance

This product combines conventional loss crop yield based MPCI insurance with protection against loss of market price at the time of sale of the crop. Currently, this product is only marketed on a commercial basis in the USA for grains and oilseeds that are quoted on commodity markets (Chicago Board of Trade) and where future price contracts can be combined into the revenue policy.

13.3.2 Index-Based Crop Insurance

13.3.2.1 Area-Yield Index Insurance

Area-yield index insurance is insurance where the indemnity is based on the realized (harvested) average yield of an area such as a county or district. The insured yield is established as a percentage of the average yield for the area and typically ranges from 50 % to a maximum of 90 % of the area average yield. An indemnity is paid if the realized average yield for the area is less than the insured yield regardless of the actual yield on a policyholder's farm. This type of index insurance requires historical area yield data on which basis one can establish the normal average yield and insured yield.

13.3.2.2 Weather Index Insurance

This is insurance where the indemnity is based on realizations of a specific weather parameter measured over a pre-specified period of time at a particular weather station. The insurance can be structured to protect against index realizations that are either so high or so low that they are expected to cause crop losses. For example, the insurance can be structured to protect against either too much rainfall or too little. An indemnity is paid whenever the realized value of the index exceeds a pre-specified threshold (e.g. when protecting against too much rainfall) or when the index is less than the threshold (e.g. when protecting against too little rainfall). The indemnity is calculated based on a pre-agreed sum insured per unit of the index (e. g. US$/mm of rainfall).

13.3.2.3 NDVI/Satellite Insurance

This refers to indexes constructed using time-series remote sensing imagery, for example applications of false colour infrared waveband to pasture index insurance where the payout is based on a NDVI (normalized dry vegetative index), which relates moisture deficit to pasture degradation. Research is currently being conducted into applications of SAR (synthetic aperture radar) to crop flood insurance.

13.4 Crop Insurance: Conceptual Framework

Discussing the crop insurance product in previous section, this part states the conceptualization of crop insurance. According to the Climate Change Cell of Ministry of Environment and Forest's (2009) report, the processes of crop insurance is general and different components in it. A conceptual framework of crop insurance program has been discussed as the first step. Based on the agricultural situation, dominant socioeconomic conditions and available administrative infrastructure,

Table 13.1 Conceptual framework for crop insurance program (adapted from Jain 2004)

Determination of basic structure	→	Critical elements	→	Perils to be covered, public/private involvement Individual/area approach, voluntary/compulsory participation
Super structure of program	→	Key elements	→	Coverage of farmers, coverage of crops, determination of sum insured and loss assessment, determination of premium Loss adjustment mechanism, organization structure Financing of the scheme, communication with farmers Reinsurance arrangements
Operational sustainability of the program	→	Other requirements	→	Availability of adequate data, availability of trained the program personnel, evaluation and monitoring

crop insurance program varies among different countries. But in general, the broader perspectives of the insurance scheme should be covered. There are some critical elements which determine the basic structure and some key elements which shape the ultimate scheme. In addition to these elements, there are some essential requirements which create the operational viability and sustainability of the scheme. The elements and requirements of the crop insurance scheme are presented in Table 13.1.

13.4.1 Critical Elements for Crop Insurance

Critical elements determine the basic structure and lay foundation of the scheme. The elements that are critical for structuring the scheme are:

13.4.1.1 Perils to Be Covered

This fundamental issue determines whether to cover all or certain specified risks. In place of yield insurance, where insured farmer gets the indemnity if the yield is below some specified level; insurance coverage against crop losses caused by specific perils (e.g. drought, storm, flood etc.) are more effective. Insurance scheme based on perils can be of specific perils (e.g. Mauritius, Cyprus) or multi-perils (e.g. USA, India, Sri Lanka).

13.4.1.2 Involvement of Public or Private Sector

Most of the crop insurance schemes are developed in the public sector. The basic advantage of public sector insurance is that it can access government budget and support from other public institutions and banks. On the other hand, private crop insurance is in place mostly in developed countries in specific weather risk management (e.g. hail insurance in USA, Canada, and Europe).

13.4.1.3 Individual or Area Approach

To determine the indemnity of crop insurance, losses of individual farmer are assessed separately under individual approach. On the contrary, indemnity of a group of farmers on the basis of average loss experienced by a specified homogeneous area (e.g. district, village) is covered under area based approach. As both of the methods have pros and cons, selection of the method for the target insurance scheme should be based on: target farmers, farm size, crop insured and communication facilities.

13.4.2 Key Elements

Key elements shape the structure and influence the working of a crop insurance scheme is discussed below:

13.4.2.1 Coverage of Farms and Farmers

Farms with specialized activities (horticultural, aquaculture, poultry) adopt specialized technology and have access to institutional finance. Private sector insurance has already shown interest in these specialized firms. Medium and large size firms are commercially viable and the risks are insurable. The semi-commercial and emerging sectors refers to firms are in a state of transition from traditional to commercial agriculture. Both private and public sector insurers have scope of participation in this sector. Farmers with small holdings are in the traditional and subsistence sector. They are most vulnerable to agricultural risks and need insurance the most. In developing countries, addressing this sector is of greatest challenge.

13.4.2.2 Coverage of Crops

Though the objective of crop insurance is to stabilize farmer's income, it is not feasible to cover all crops grown by a farmer. In the beginning, the coverage can be limited and extended gradually to achieve the desired goal.

13.4.2.3 Determination of Sum Insured and Loss Assessment

Sum insured coverage is usually based on cost of production, part of the value yield and amount of crop loan. In most cases, sum insured is based on cost of production due to its simplicity to assess and availability of information. But this assessment faces some theoretical and practical problems (e.g., variable cost vs. fixed cost, imputed value of family labor). The assessment of losses of agricultural insurance is more difficult than other insurance schemes (e.g., fire, property). Crop insurance relates to something yet to come into existence. The deductible level and its nature and application in relation to the risks insured are also important for determining the loss.

13.4.2.4 Determination of the Premium

For a viable crop insurance scheme, the premium rate needs to cover the following: pure risks, administration cost, contribution to catastrophe reserve and a reasonable return. The insurance premium may be on a net or gross basis. Net premium covers only the average loss over a period and possible an additional amount to accumulate a small reserve. Gross premium involves some "loading" to include cost of administration and some return or profit.

13.4.2.5 Loss Adjustment Mechanisms

The loss adjustment mechanism should be effective enough to minimize spurious claims and at the same time fair to the insured. The procedure will depend on the approach of the scheme (area or individual). In case of the area approach, indemnity will be based on the average crop yield. In case of individual approach, field inspections are necessary to estimate yield through eye estimation and crop cutting procedure.

13.4.2.6 Organizational Structure

There is diversity of organizational structure across countries. It may be a private organization (China), private organization with government support (Japan), parastatal organization with minimal government control (Mauritius) or a public sector organization (India). The administrative structure chosen by a country depends on socioeconomic infrastructure, type of insurance schemes, target farmers and crops, comprehensiveness of its coverage and size of operation.

13.4.2.7 Financing of the Schemes

Financing is important to insurance programs as in times of disaster the requirement of funds are very large. A scheme will be self financing if the premium rats are correct and the loss adjustment mechanisms are properly structured. In practice, there may be at times imbalance between the premium income and fund required to pay the indemnity (e.g., catastrophic events). Thus, building a reserve in required during the early years of a program. The size of the reserve should be based on realistic estimate of the maximum possible loss.

13.4.2.8 Reinsurance Arrangements

Reinsurance provides access to larger reserves by spreading the risk wider. It can take various forms like: reinsurance support from the government (Japan, Canada), loans funds from the government (Japan, Canada), budgetary fund from the government (India) and private insurance in the international market (Mauritius).

13.4.2.9 Communication with Farmers

In case of mandatory participation in the scheme, the farmers must be convinced that the program is of their interest to avoid dissatisfaction. In case of voluntary participation, the coverage will depend on how and to what extent farmers perceive it is beneficial to them. Farmers should believe that the terms and conditions are fair and have confidence on claim settlement issues. Communication with farmers can be done through media, education programs and group interactions.

13.4.3 Other Requirements

Other requirements (adequate data, trained personnel and monitoring and evaluation) will make the scheme operationally sound.

13.4.3.1 Adequate Data-Base

To work out the financial implications of a possible crop insurance scheme, availability of data is crucial. Data is required to form the basis for determining the premium, guaranteed yield, indemnity etc. Moreover, information on climatic conditions (e.g., frequency of flood, drought), land tenure, land record systems, cropping pattern, availability of agricultural inputs including credit and other infrastructure in an area are essential. This information facilitates realistic assessment of exposure of various crops to the perils to be covered.

13.4.3.2 Availability of Trained Personnel

Trained personnel are required to operate an insurance scheme both at headquarters and filed level. Due to complexity of agricultural schemes, specialized training on theory, techniques and practice is a must.

13.4.3.3 Monitoring and Evaluation

Monitoring and evaluation in a regular basis is required to take remedial measures on time, if necessary.

13.5 Crop Insurance: Worldwide Experiences

In USA, crop insurance is offered through the Federal Crop Insurance Program (FCIP), a public–private partnership between the federal government and a number of private sector insurance companies, created in 1938. The program officially aims to improve the social welfare of farmers as well as deliver insurance products in an actuarially sound manner. The government provides subsidies to farmers to pay the premium. In addition, the government reimburses administrative and operating expense for private insurance companies that sell and service FCIP policies. The reimbursement is approximately 22 % of total premiums. Lastly, the government provides reinsurance to the private insurance companies at an estimated subsidy rate equivalent to 14 % of total premiums. In total, the government is subsidizing 70 % of the total cost for the FCIP. Net government costs for administering the program (total costs less premium paid by producers) have ranged from $1.1 billion in 1995 to $1.7 billion in 2000 (EU 2001). Recent estimates put the cost of crop insurance subsidies (2004–2005) at approximately $3 billion.

In Canada, the federal government started to provide disaster assistance to grain producers on the prairies in 1939. Since then, a tripartite system has evolved that consist of three separate programs: Crop Insurance (CI), the Net Income Stabilization Account (NISA), and the Agricultural Income Disaster Assistance (AIDA). The programs are administered at the level of provincial governments and no private insurance companies are involved. In addition, programs are highly participatory with farmers, provincial governments, and federal government participating in discussing surrounding product designs, rate setting, and performance feedback. The Federal and provincial governments each pay 25 % of the total premium and 50 % of the administrative costs. The combined cost to the Federal and provincial governments has been trending upward sharply, rising 34 % from US$338 million in 1995 to US$454 in 1999 (EU 2001).

In Australia, there are no premium subsidies, but crop insurance has been available for many years and the demand for agricultural insurance is very high. In Australia market penetration rates are expressed as the actual gross written premium as a percentage of the potential market gross written premium for each product line/class of business. Named-peril broad acre crop insurance is extremely popular with farmers and in 2009 it is estimated that 75 % of the potential market for this class of business was underwritten.

Among Asian countries, it has been observed from Japan that a highly subsidized cooperative crop insurance program covers approximately 45 % of all cultivated cropped area and possibly as high as 90 % of the cultivated area of paddy rice, wheat and other cereals, which are insured on a compulsory basis.

In India, Crop insurance began in 1972 on a pilot basis. India has implemented subsidized public-sector area-yield index multiple peril crop insurance (MPCI) since 1979. In 1985 the comprehensive crop insurance scheme (CCIS) was introduced in 16 states and 2 union territories by the General Insurance Corporation of India (GIC). The CCIS was provided only to recipients of crop credit (loanees) on a compulsory basis. The CCIS was replaced by the National Agricultural Insurance Scheme (NAIS) in the *rabi* season 1999/2000. The NAIS was closely modeled on the CCIS area-based approach. In 2009/2010, the Agriculture Insurance Company (AIC) implemented NAIS is marketed in 25 states and 2 union territories and insured a total of about 25 million farmers and an insured area of about 27 million hectares with Total Sum Insured (TSI) of US$9.55 billion. In addition, in 2009/2010 AIC also underwrote its Weather Based Crop Insurance Scheme (WBCIS) in 139 districts in 13 states with an additional two million farmers, with an insured area of approximately 2.7 million ha, TSI of US$900 million and a premium of about US$80 million.

According to the above, it has been seen that crop insurance is practiced in a number of countries across the world. But there are some extents to which the success and failure of crop insurance depend. For example, in America, crop insurance is more successful because, it is considered as an important risk management tool. Crop insurance remains an integral part of farming, and most producers consider crop insurance to be the centerpiece of their risk management planning. The fact remains that the crop insurance program is vital to and is working for the vast majority of America's farmers, and most of them would not think of risking their livelihood and future without the coverage provided by crop insurance. Therefore, it is reported by Rain and Hail Insurance Society (RHIS) in 2012 that in 2011, more than 265 million acres were insured, among the highest in the history of the program. Farmers are electing higher coverage's (92 % of crop insurance contracts are at buy-up levels) and electing revenue coverage (almost 84 % of the protection in force provides coverage for both price and production losses) to protect their production and revenue exposures. In a time of rapid changes in agriculture, including increasing price volatility and production costs, farmers and their lenders are relying on the coverage provided by crop insurance as they work to maximize opportunities and minimize risks. For a number of reasons demand for crop insurance is increasing day by day, which can be grouped as;

- Evidence is accumulating of connections between climate change, and the increasing incidence of crop damaging weather events of extreme severity
- Farming is becoming steadily more commercialized, with greater financial investment
- Farmer/investors and their banks frequently examine the feasibility of using a financial mechanism i.e. insurance, in order to address part of the risk
- The World Trade Organization (WTO) regulations generally forbid governments from subsidizing agriculture directly; however, they permit the subsidization of agricultural insurance premiums. In the case, demand for crop insurance will increase in those economies that wish to implement a policy of permitted subsidization of their farmers
- Insurance can also assist in managing the on-farm production risks consequent to changes in pest management practices. Such changes are increasingly required in order to address environmental protection and food safety concerns. This is that any insurance arrangement will involve not only the farmer and the insurer, but also important third parties. Consideration is now given to these changes to the business of farming.

On the other hand, it has been seen that India has taken various schemes through crop insurance such as first individual approach scheme (1972–1978), Pilot Crop Insurance Scheme (1979–1984), Comprehensive Crop Insurance Scheme (1985–1999), Experimental Crop Insurance Scheme and National Agricultural Insurance Scheme (1999–2000 onwards). But these schemes are failed to meet the expected results due to the low policy implications, unawareness of farmers, unsatisfied performance of implementing agencies as well as limited State and central Government (Mahajan and Bobade 2012). They also identified that all crops are not covered under National Agricultural Insurance Scheme. National Agricultural Insurance scheme has been introduced by Government of India from *rabi* season (1999–2000). But it also failed to influence the farmers as well as work effectively.

13.6 Crop Insurance: Bangladesh Experiences

In Bangladesh, crop insurance was introduced in 1977 through the state-owned insurance company Sadharan Bima Corporation (SBC) as a pilot project. The main aim of the program is firstly to indemnify farmers against crop loss, stabilize farm income and promote agricultural growth; and secondly, to undertake research necessary for promotion and development of such crop insurance program. The scheme started with two sub-districts, and later expanded up to 56 sub-districts in 1981. Overall SBC received premium of BDT 39, 62, 337.50 and provide cover of BDT 11, 05, 20, 277.40 (about US$150 million at current exchange rate) against an insured area of 23, 794.43 acres. The scheme was interrupted twice, and after an operation of around 19 years. The scheme was closed in 1996, registering a loss experience of about 500 %. There are several reasons responsible for the failure of crop insurance project in Bangladesh. These are described below:

13.6.1 Lack of Policy Support and Partnership

- Considering the CI delivery model adopted, it was a full service model, administrated by the SBC, without any partnership with public insurers, banks or MFIs. Neither it had the provision of community involvement. Obviously it provided scope for inefficiency and corruption. So, there was no sense of ownership among policymakers.
- The CI project was not integrated with the mainstream agriculture development or disaster management policy; rather it was a discrete effort by SBC as a standalone insurance scheme.
- Expansion of the program was initiated without developing any workable model and fine-tuning of the delivery mechanism.

13.6.2 Methodological Problem

- The program was made voluntary and based on individual approach. This leads to adverse selection, i.e. only the more risky lands were preferred for insurance. Uniform premium rate for all types of land farther aggravated the problem. The approach was totally contrary to the principle of risk pooling, where the farmers should be selected from diverse agro-ecological zones so that not all the insured suffer from disaster at once.
- Peril covered for the program was too many: Flood, Cyclone, Hailstorm, Windstorm, Drought, Plant Disease, Pest and Insects. At the initial stage of introducing crop insurance in Bangladesh, the peril covered should be limited to one or two.
- The sum insured as 80 % of the average yield was too high. A common figure in individual based programs in the region during 1980s was of around 50 %.
- Weak and unscientific ocular or eye estimation method of damage calculation made ample scope of moral hazards, inaccuracies and anomalies among different assessors.

13.6.3 Lack of Expertise and Monitoring

- The program was introduced hastily without adequate preparation like a clear policy and well defined structure and proper training of the SBC staffs and other relevant people.
- There was no grassroots level monitoring of the program at all, which is a must for either microfinance or micro insurance. Rather, the SBC head office controlled the program, which was totally irrational.

13.7 Farmer's Impression About Crop Insurance

To know the farmer's view on climate change as well as crop insurance, a survey was undertaken by climate change cell of Department of Environment in flood affected area, drought and manga affected area, and cyclonic storm surge and salinity intrusion affected area namely Tahirpur upazila of Sunamganj district, Hatiabandha Upazila of Lalmonirhat district and Mathbaria upazila of Pirozpur district, accordingly (Fig. 13.1). They conducted questionnaire survey with 150 farmers of those areas and also conducted three FGD. They also did key interviews and collected secondary information like climatic data, disaster data etc. from that area.

The findings from the survey show that farmers in these three areas were quite aware of the fact that climate is changing and they mentioned that the existing disaster management mostly includes relief distribution by government, not necessarily for agri-damage. But for overall losses and the distribution of relief is done by the

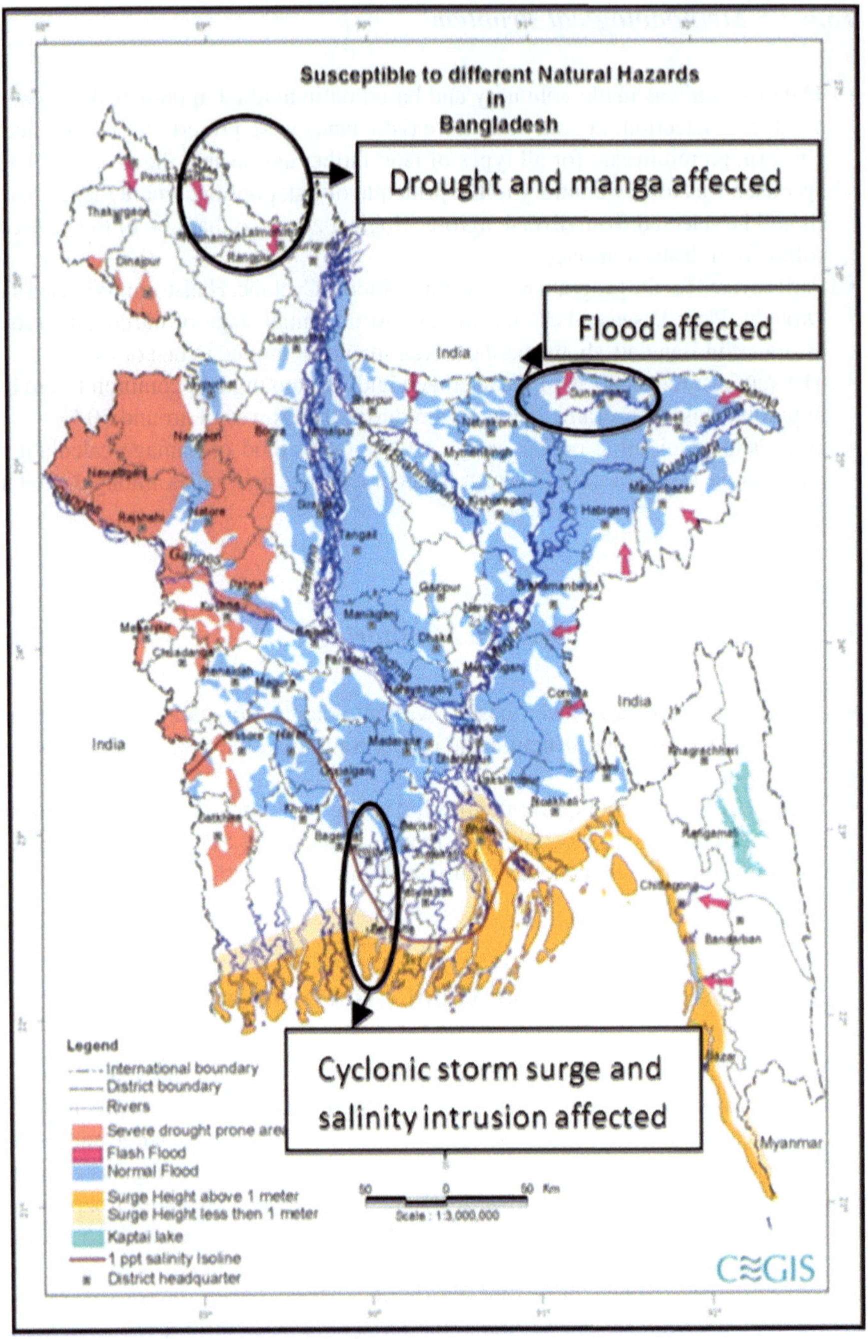

Fig. 13.1 Map showing the survey area in different parts of Bangladesh

local government bodies, in some places by local NGOs, but no inclusion of farmer's representative. Moreover, in terms of crop insurance, farmers were not well aware about crop insurance, but they had the idea what insurance means. They wanted to be part of crop insurance scheme and eager to know details of the scheme. On an average of three areas the willingness to premium payment rate is around 3–6 % of yield value per season-mentioned by farmers of those areas. In case of Sunamganj and Pirojpur district, farmers preferred NGO as the most suitable for crop insurance program. On the other hand, farmers from Lalmonirhat preferred bank. They further mentioned multi peril is desirable because it protects from all sorts of disasters.

13.8 Weather Index Based Crop Insurance: Demonstration of a Model

Based on experiences elsewhere, and in view of the social conditions in Bangladesh, it is suggested that area based index, especially weather index-based crop insurance might be the best approach. The land distribution pattern, tenant farming, and the inherent problems of insurance, such as moral hazard, adverse selection and huge administrative costs favor such an index-based CI. While administrative cost is minimized in weather index method, it however, requires a large network of hydro-climatic data stations with data of good quality or accuracy. Along with this, detailed land classification and elevation map is also required, especially for assessment of flooding and salinity intrusion. So, proper technical expertise is required.

A successful weather risk management and transfer program for agriculture involves four essential steps that are:

- Identifying significant exposure of an agricultural grower/producer to weather
- Quantifying the impact of adverse weather on their revenues
- Structuring a contract that pays out when adverse weather occurs; and
- Executing the contract in optimal form to transfer the risk to the international weather market.

Each of the steps is outlined in the following four subsections:

13.8.1 Identifying the Risk

Identifying weather risk for an agricultural grower or producer involves three steps:

- Identifying the regions at risk from weather and the weather stations that reflect that risk
- Identifying the time period during which risk is prevalent
- Identifying the weather index providing the best proxy for the weather exposure

This last step is the most critical in designing an index-based weather risk management strategy. Rather than measuring the actual impact on crop yields—or related fluctuations in demand, supply, or profitability—the index acts as a proxy for the loss experienced due to weather and is constructed from actual observations of weather at one or more specific weather stations.

13.8.2 Location and Duration

All weather contracts are based on the actual observations of weather variables at one or more specific weather stations. Transactions can be based on observations from a single station or a basket of several stations or on a weighted combination of readings from multiple stations like:

- If an individual farmer is interested in purchasing weather protection for his particular crop, the index-based weather contract must be written on the weather station nearest the farmer's land to provide the best possible coverage for the farmer client.
- A larger grower, with several production regions, may be more interested in purchasing a weather contract based on several weather stations to reflect the weather conditions in all areas covered by the business. The grower's risk management strategy can be either to purchase a weather contract on each of the identified weather stations or to purchase a single contract on a weighted average of several stations.

All contracts have a defined start and end date to limit the period over which the underlying index is calculated. This calculation period describes the effective dates of the risk protection period during which relevant weather parameters are measured at the specified weather stations. For agricultural end users, the duration of the weather contracts will be determined by the specific requirements of their business. Contract duration has the flexibility to address individual end-user business exposures; which can be weekly, monthly, seasonal, and even multi-annual. Final settlement of the weather contracts typically occurs up to 40 days after the end of the calculation period, once the collected weather data have been cross-checked and quality controlled by the relevant data-collecting body, usually the National Meteorological Service.

13.8.3 Underlying Indexes

A weather index can be constructed using any combination of measurable weather variables and any number of weather stations that best represent the risk of the agricultural end user. Common variables include temperature and rainfall, although transactions on snowfall, wind, sunshine hours, river flow, relative humidity, and storm/hurricane location and strength are also possible but used to be more complex. The normal process for designing an index-based weather insurance contract for an agricultural grower, for example, involves identifying a measurable weather index strongly correlated to crop yield. After gathering the weather data, an index can be designed by

- Looking at how the weather variables have or have not influenced yield over time
- Discussing key weather factors with experts, such as agro-meteorologists and farmers; and/or
- Referring to crop growth models using weather variables as inputs for yield estimates or phenology models illustrating how weather variations relate to pest development.

A good index must account for the susceptibility of crops to weather factors during different stages of development, the biological and physiological characteristics of the crop, and the properties of the soil. If a sufficient degree of correlation is established between the weather index and crop yield or quality, a farmer or an agricultural producer can insure his production or quality risk by purchasing a contract that pays if a specified undesirable weather event occurs or a specified desirable weather fails to occur.

13.8.4 Weather Index: Daily Average Temperature

Critical temperatures causing crop damage may vary depending on the length of time that temperatures deviate from expected level as well as on the variety, health, and development stage of a plant. Likewise, for a particular crop the critical temperature required for different growing period can be fixed and deviation from it can be correlated to yield loss. The required temperature at different months can be used as Weather Index for that crop. Temperature data can be collected from the nearby weather stations. Crop damage can also be the result of specific or critical temperature events that can be detrimental to yield or quality. Freezing conditions, for instance, were reported to have caused more than US$600 million in damage to the U.S. citrus crop in a single week of December 1998. In 2005, in Bangladesh, due to a prolonged cold spell along Brahman Baria district a large area of Boro rice fields were affected, where there was no rice inside the husk of rice. During the flowering and maturation time of rice the temperature was not at required level so that no rice could be formed there. During field survey at Sunamgonj almost same incidents were heard and according to the farmers such incidents are occurring more frequently nowadays.

13.8.5 Weather Index: Deficit Rainfall and Drought

Meteorological drought is usually defined in terms of deviation of precipitation from normal levels and duration of a region's dry periods. Agricultural drought refers to situations in which soil moisture content no longer meets crop growing needs in an area due to insufficient rainfall. Crops, particularly rain-fed crops, often have a minimum overall threshold of cumulative rainfall necessary for successful and healthy plant development. These water requirements must be met by natural rainfall, stored soil moisture from precipitation prior to the growing season, or supplemental irrigation. Therefore, a deficit of rainfall below these levels, in the absence of irrigation, can cause plant moisture stress that affects development and reduces yields.

A simple cumulative rainfall index can be developed to suit a grower's specific insurance requirements with regard to such decreases in rainfall and yield. Looking at historical yield rainfall and yield. The distribution of rainfall during the growing season or at specific stages of a plant's development is often more important than total rainfall, however, and customized indexes must be developed to capture this risk. Such indexes may also include other weather parameters, such as temperature and relative humidity. Other factors like availability of irrigation water can be embedded into a more complex index. Crop growth models or historical yield data can be used to infer the empirical relationship between rainfall amounts and yield/quality for specific soil and crop types. A number of alternative rainfall contracts is possible to develop. One of them, the proportional contract simply pays in percentage terms for levels of rainfall below a well specified strike or threshold. For example, if the median rainfall in a given area is 300 mm from November to March, one might begin payments anytime rainfall is below 250 mm. These payments would be based on the level below the strike of 250 mm. The percentage calculation would be performed as follows:

$$\text{If rainfall from November to March} < 250\ \text{mm, then}$$
$$\text{Payment percentage} = (250 - \text{actual rain})/250$$

For example, if the rainfall is 200 mm, the payment percentage would be 50/250 or 20 %. Those at risk (farmers, agribusinesses, farmer organizations, banks, etc.) would purchase contracts at some specific value, say 1,000 MAD. If the payment rate is 20 % and the insured purchased 10 units of the 1,000 MAD, the actual payment would equal. $20 \times 10 \times 1{,}000 = 2{,}000$ MAD. The contract could also simply be sold in any MAD unit value. The principles are the same:

$$\text{Indemnity} = \text{payment percentage} \times \text{total MAD value or liability.}$$

13.8.6 Weather Index—Flooding: Without and with Embankment

Excess water that causes inundation destroys crops. In case of Bangladesh flooding is the most serious culprit for crop damage. Even though it is not much practiced elsewhere but it is still possible to include flooding effect in weather insurance contract. If the croplands are situated in the floodplain of a particular river, the recorded water level of the river at a particular gauge station can be compared with topography or land elevation and determine the inundation height of an area. For simplicity of calculation, for a particular gauge station the inundation height for crop damage can be fixed so that any crop land within the coverage area of the station will get compensation if the water level is above a fixed level.

For an area protected by embankment, if the recorded water level is above the embankment height, full damage of crop can be assumed. If water level is below the embankment height no indemnity will be paid. It is because there are possibilities that even though embankment did not fail but artificially it among farmers to protect those embankments by themselves as well along with govt. officials. Especially, for submergible embankments in the Haor area this might be the most appropriate as because those submergible embankments are mostly made and maintained by farmers. So they will have better sense of belonging as because if those embankments fail for water level below desired height, they will lose their crop and will not get any indemnity. Because of low flood risk within the embankment, there should be provision that if one declares himself within embankment, his premium will be lower.

13.8.7 Extreme Weather Event

For extreme weather event like cyclone, Tornado, Hailstorm, etc. the methodology might be almost same as that of traditional insurance scheme. However, to avoid actuarial investigation or damage assessment, which is the main source of moral hazard, it is still possible to take help from weather stations or any other technical measures like satellite picture. For cyclone the damage assessment is possible either through observing the magnitude or category or by the height of storm surges. In fact cyclone and storm surge used to affect a large area and in that case the coastal gauge stations to measure water level can collect the inundation height. However, for disaster occurring at a smaller spatial scale like tornado and hailstorm, it might not always happen that the weather station can record it, especially for the case of Bangladesh where weather stations are not densely located. In that case if possible, satellite pictures can be used. A simplified actuarial analysis can be made there as well to confirm the region affected by such event instead of assessing each and every land unit.

13.8.8 Combined Index

In some cases, variation in crop yield might be due to a combination of several climatic factors like temperature, water availability, groundwater level, relative humidity, etc. For example, drought possibility increases with both the increase in temperature and decrease in rainfall. Lowering of groundwater will further aggravate it. In the same case, simply rainfall deficit might not a good approximation, provided the location of the site whether close to a river or not so that irrigation facilities are abundant. Once again, the river flow in the river not necessarily always correlated with the rainfall recorded in the nearby station as because water might come from the upstream segment of the catchment. In that case the contribution off low in the river from upstream should be added with the rainfall occurring in the vicinity.

Methodology to develop an index in that case is possible by providing a weight factor for each of the hydro-climatic variables as follows:

$$\text{Index} = (\text{Rainfall deficit} / \text{Target rainfall}) * (\text{Runoff deficit} / \text{Target runoff})$$
$$\text{Or,} = 0.3 * (\text{Rainfall deficit} / \text{Target rainfall}) + 0.25 * (\text{Runoff deficit} / \text{Target runoff}) + 0.25\,(\text{GW level fall} / \text{Target GW level})$$

13.8.9 Quantifying the Loss

Once the index has been identified, it must be calibrated to capture the financial impact of the specified weather exposure as measured by the index. Two approaches are possible at this stage: identifying the financial exposure per unit of the defined index, and/or establishing the limit, the total financial protection, required per risk period, that is, the maximum payout necessary in a worst-case scenario. The approach chosen depends on the nature of the underlying index and weather event. If the weather exposure is event driven, for example, such as a Category 5 hurricane hitting a particular location or a cold winterkill event destroying an entire wheat crop, the latter approach is more appropriate. If the weather exposure is of a cumulative nature, such as drought the former approach should be chosen.

13.8.9.1 Unit Exposure

In designing an index, expert scientific agrometeorological assessments, either in conjunction with crop model output or with verification using historical yields, used to be employed that best proxies the weather sensitivity of the crop in question. After developing weather indexes to capture the impact of adverse weather conditions on a specific crop's yield, it is straightforward to calculate the financial impact of these events for producers. Having identified the index, the crop yield, *Y*, or volume, *V*, variability per unit of the defined index, *I*, can be defined, as follows:

$$\Delta Y = \Delta V / H = a(I)\Delta I$$

where, $a(I)$ is some function of I that relates the index to the yield Y, and H is the planting area of the crop. In order to calibrate an appropriate weather contract, the variation in crop yield must now be converted into a financial equivalent that mirrors the producer's exposure. This can be done, for example, by considering a producer's production and input costs per hectare planted or by considering his expected revenue from the sale of the crop at harvest.

Producers with fixed-price delivery contracts or those using price risk management instruments to protect themselves from market fluctuations in the price of their crop at harvest time know the financial value of each kilogram or metric ton they produce and hence can quantify the financial cost of a shortfall in production. If a grain producer, for example, knows he will receive \$*X* per metric ton of crop, the following relationship must hold for his change in revenue:

$$\begin{aligned} \Delta Revenue &= X \times (Actual\ Yield - Expected\ Yield) \times H \\ &= X \times \Delta V = \pm X \times H \times \mathrm{a}(I) \times \Delta I \end{aligned} \tag{13.1}$$

A good weather hedge must offset the negative $\Delta Revenue$ fluctuation in the event of a drop in yield from budgeted levels if a producer is to protect his earnings. In order to perfectly replicate his position, the farmer could enter into a weather contract with the following incremental payout *P* per unit index:

$$\Delta P = X \times H \times a(I) \times \Delta I$$

Therefore, his overall position would be:

$$\Delta Revenue + AP = -X \times \Delta Y \times H + X \times H \times a(I) \times \Delta I = 0$$

Producers may have contractual obligations to deliver a predefined amount of their farmed product to a buyer at harvest time, with associated penalties if these obligations are not met. In such a situation, it would be straightforward to quantify and structure a hedging product to protect producers from these contractual costs in the event of weather related shortfalls in production.

13.8.10 The Limit

Most weather contracts have a limit, which corresponds to the maximum financial payout or recovery from the contract in a worst-case scenario, such as a complete crop failure. The maximum payout can be set by either considering the value-at-risk for the producer in the event of a total crop failure or by looking at historical index, production, and sales data to find the worse-case scenario historically in order to establish a limit. Alternatively, a producer may simply want to insure his production and input costs in order to recover these outlays if the crop fails. If a producer's production costs are \$*Z* per hectare farmed, \$*Z* will therefore correspond to the maximum payout, the limit of the weather contract, for each hectare the producer wishes to insure. The unit exposure *P* will therefore be as follows:

$$\Delta P = \left(-\Delta Y / Expected\ Yield\right) \times Z = (a(I)\Delta I / \text{Expected Yield}) \times Z,\ for \Delta \text{Y} < 0$$

13.8.11 Structuring the Product

Once the index has been identified and calibrated, the next step is to structure a contract that pays when the specified adverse weather occurs, thus performing a hedging or risk smoothing function for the agricultural grower or producer. The approach for offering such a contract can be either of the following:

In general, most weather contracts are defined by a set of standard specifications including:

- The reference index, *I*, and weather station(s): complete specification of the index and data used to construct it;
- The term, *T*: the risk protection period of the contract, including the start and end date of the contract;
- A strike, *K*: also known as an attachment level, the level at which the weather protection begins;
- The payout rate, *X*: the financial compensation per unit index deviation above or below the strike at maturity, defined as the unit exposure in the previous section; and
- The limit, *M*: the maximum payout per risk protection period.

13.8.12 Risk Retention and Premium

It is clear that an important aspect to consider when structuring an index-based solution is the retention of risk by the party seeking protection. This means defining the index trigger level at which the weather protection begins. The strike determines the insured party's level of risk retention and is the key to pricing and success in transferring the risk. A strike very close to the mean of the index indicates a low level of risk retention by the end user and a high probability that the contract will pay out. This implicitly means a large premium, as well as the possibility of inspiring little interest in the weather market if the location or nature of the risk is outside the main liquid trading hubs or variables. A strike farther away from the mean reduces the probability of a payout and hence the premium of the contract, as the entity is retaining the more frequent, near-the-mean risk internally and transferring less to the market.

The level of risk retention will depend on the risk appetite and business imperatives of the end user and the sensitivity to the premium associated with entering into a contract. By retaining more risk, all things being equal, the producer would reduce the premium of the contract.

13.9 Limitations of Crop Insurance in a Climate Change Context

Key limitations of insurance include (a) it does not prevent or reduce the likelihood of direct damage and fatalities from extreme weather events; and (b) it is not always the most appropriate option to manage risks (for example, in terms of cost-effectiveness or affordability). These same limitations are potentially aggravated in a climate change context (i.e. more frequent and intense extreme events). Climate change also poses additional challenges for insurance, a point that further underscores the vital importance of disaster risk reduction. Two of these issues include:

- *Potential un-insurability associated with increasing frequency and magnitude of extreme weather events*
  The United Nations Environment Programme's Finance Initiative (UNEPFI) reports that by 2025, insurers may withdraw from some markets as the risks become too high for the pool of premium available. This has happened periodically in the United States. CERES, a United States-based NGO, has identified a growing move by insurers to reduce coverage in coastal areas. In this context, it would be beneficial to further explore the use of alternative risk transfer products such as catastrophe bonds (cat bonds), which pass the risk on to investors in the capital markets rather than to reinsurers. At the very least, maintaining affordability will be challenging as climate risk impacts increase in frequency and magnitude, becoming less insurable. Given increased levels of uncertainty coming with climate change, higher risks to insurers ultimately mean higher premiums for clients unless significant risk reduction measures are in place.
- *Unsuitability of traditional insurance for longer-term foreseeable hazards like sea-level rise and desertification*
  Two preconditions for insurability of disasters are the unpredictability of a specific event, which means that losses occur suddenly and cannot be foreseen; and the ability to spread the risk over time, regions and between individuals/entities. For two of the already ongoing changes caused by global warming, that is, sea-level rise and desertification, the insurability criteria cannot be fulfilled. Both processes are slow and continuous changes that potentially affect the population of one or more countries. They can lead to a deterioration of living conditions in developing or poor countries and, in the long term, could threaten the survival of human populations in affected regions. Further, only rapid and significant reduction of greenhouse gas emissions that lead to global warming could effectively prevent these risks in the long run. For this reason, insurance alone is hard-pressed to address some of the dire effects of climate change.

13.10 Conclusions

In conclusion, this chapter emphasizes on the implementation of crop insurance in Bangladesh. Because, agriculture is an uncertain business for this country, partly due to its high dependence on the weather, living 14.40 million farm household vulnerable to serious hardship. Moreover, due to the global climate change, agriculture is at great risk (Agarwala et al. 2003). Especially climatic disaster flood, drought and cyclonic storm have the most severe impact and cause maximum amount of losses. To reduce the loss of agricultural production in Bangladesh, crop insurance the most popular risk management tool for farmers, is the key to their financial stability, enabling them to supply food and fiber to our country despite severe weather and other challenges that impact their business. Furthermore, it can directly improve the welfare of risk averse farmers, particularly the 80 % of small and marginal farmer households operating less than 2 ha. Perhaps even more importantly, affordable agricultural insurance can in effect act as collateral against loans, increasing the creditworthiness of farmers and allowing them the opportunity to invest in appropriate inputs to increase agricultural productivity (Hazell 1992).

However, the provision of crop insurance is challenging, particularly in developing countries. Multiple Peril Crop Insurance programs, where each policyholder is indemnified against their own crop loss, were fraught with moral hazard, fraud and adverse selection, leading to high costs (Hazell 1992; Skees et al. 1999). In view of literature, it can be said that the application of new model of weather index-based crop insurance in some developing countries for the last few years, the Weather based index based weather insurance has the potential for a greater resilience to moral hazard, fraud and adverse selection than the former and lower basis risk, the risk of a mismatch between incurred losses and indexes claim payments, than the latter (Carter et al. 2007).

Although there is currently no crop insurance schemes in Bangladesh, but they can revise the previous crop insurance program handled by SBC. Moreover, Bangladesh can follow the India's most successful crop insurance program. Nevertheless, crop insurance is a highly specialized line of business that requires intensive institutional capacity building. Therefore, it is necessitated to work together in an integrated manner to support crop insurance. In this case, public–private-NGO partnership is what may make the CI a viable and sustainable venture. However, the huge moral hazard in the CI process could be substantially reduced if the damage assessment authority was given to a committee composed of insurance expert, agriculture specialist, economist, climate expert, local government officials and farmers group together instead of just one person or only insurance company officials. However, successful implementation of this scheme, a solid partnership among relevant stakeholders at local, national and international level is nascent. Moreover, in order to sustain such a partnership, financial resources and capacity building at different levels and scales hold the key.

Acknowledgements The first author gratefully acknowledges the support from GCOE-ARS program and the Graduate School of Global Environmental Studies of Kyoto University.

References

Agarwala S, Ota T, Ahmed AU, Smith J, van Aalst M (2003) Development and climate change in Bangladesh: focus on coastal flooding and the Sundarbans. OECOD, Paris

Alam M, Rabbani MG (2007) Vulnerabilities and responses to climate change for Dhaka. Environ Urban 19(1):81–97 (London: Sage)

Carter MR, Galarza F, Boucher S (2007) Underwriting area-based yield insurance to crowd-in credit supply and demand. In: Giannini Foundation of Agricultural Economics working paper, UC Davis 07-003

Climate Change Cell (2009) Crop insurance as a risk management strategy in Bangladesh. Department of Environment. Ministry of Environment and Forests. Government of the People's Republic of Bangladesh, Dhaka
Cutter S, Osman-Elasha B, Campbell J, Cheong S-M, McCormick S, Pulwarty R, Supratid S, Ziervogel G (2012) Managing the risks from climate extremes at the local level. In: Field C, Barros V, Stocker T, Qin D, Dokken D, Ebi K, Mastrandrea M, Mach K, Plattner G, Allen S, Tignor M, Midgley P (eds) Managing the risks of extreme events and disasters to advance climate change adaptation a special report of working groups I and II of the Intergovernmental Panel on Climate Change (IPCC). Cambridge University Press, Cambridge, pp 291–338
EU (European Commission) (2001) Risk management tools for EU agriculture. Brussels
FCCC/SB/1999/9 (1999) UNFCC subsidiary body of implementation: annotations to the provisional agenda, Bonn, 25 October to 5 November 1999
Hazell PBR (1992) The appropriate role of agricultural insurance in developing countries. J Int Dev 4:567–581
Hellmuth ME et al (eds) (2009) Index insurance and climate risk: prospects for development and disaster management, Climate and Society No. 2. International Research Institute for Climate and Society, Colombia University, New York
IPCC AR4 (2007) Intergovernmental Panel on Climate Change 4th Assessment Report. Oxford University Press, London
Jain SRCA (2004) Challenges in implementing agriculture insurance and re-insurance in developing countries. The Journal, India
Mahajan SS, Bobade AH (2012) Growth of NAIS: a study of crop insurance in India. BAUDDHIK 3(1)
Mahul O, Stutley C (2010) Government support to agricultural insurance: challenges and options for developing countries. World Bank, Washington
MoEF (Ministry of Environment and Forests) (2008) Bangladesh climate change strategy and action plan 2008. Ministry of Environment and Forests, Government of the People's Republic of Bangladesh, Dhaka
National Adaptation Programme of Action (NAPA) (2005) Final report. Ministry of Environment and Forest, Government of the People's Republic of Bangladesh, Dhaka
POVCC (2003) Poverty and climate change: reducing the vulnerability of the poor through adaptation. DFID, UNDP, London
Skees JR, Hazell PBR, Miranda M (1999) New approach to public/private crop yield insurance in developing countries. Environment and Production Technology Division (EPTD) Discussion Paper 55, International Food Policy Research Institute (IFPRI)
World Bank (2000) Bangladesh: Climate Change & Sustainable Development. Report prepared by the South Asia Development Team, Report No. 21104 BD
World Bank (2005) Managing agricultural production risk: innovations in developing countries, agriculture and rural development department. World Bank, Washington

Chapter 14
Disaster Education in Bangladesh: Opportunities and Challenges

Umma Habiba, Md. Anwarul Abedin, and Rajib Shaw

Abstract During the last couple of decades, there has been an increase in the frequency and severity of disasters in Bangladesh, particularly those of hydro meteorological origin such as flood, cyclone and drought. To deal with disasters, Bangladesh has achieved significant progress in disaster preparedness and taken substantial steps forward in terms of risk reduction. However, to create awareness towards disaster, Hyogo Framework for Action (HFA) 2005–2015, priority for action-3 gives emphasis on the role of knowledge and education, and highlights formal and non-formal education, and awareness-raising as important components for disaster risk reduction. In this regards, this book chapter intends to highlight the ongoing disaster education at different levels in Bangladesh. It has been seen that Bangladesh is carried out disaster education through course curricula developed by National Curriculum and Textbook Board (NCTB) in 1994/1995. In 2006 Disaster management Bureau (DMB) and Comprehensive Disaster Management Program (CDMP) had undertaken a review of the books of NCTB that mainly focused on the measures necessary to be undertaken at the family and community levels and emphasized on the need for risk reduction. This chapter further point outs key gaps and challenges to implement disaster education in context of Bangladesh. Besides, this chapter gives emphasis on non-formal and informal education, and addresses ESD (Education for Sustainable Development), CCE (Climate Change Education) and mentions community's role for achieving sustainable development that helps to develop attitudes, knowledge to make decisions for the benefit of students, communities, and others, now and in future.

Keywords Curriculum • Disaster education • Risk reduction • Student's action • Teacher's training

U. Habiba (✉) • Md.A. Abedin • R. Shaw
Graduate School of Global Environmental Studies, Kyoto University, Kyoto, Japan
e-mail: shimuagri@yahoo.com

R. Shaw et al. (eds.), *Disaster Risk Reduction Approaches in Bangladesh*, Disaster Risk Reduction, DOI 10.1007/978-4-431-54252-0_14,

14.1 Introduction

Due to a number of hydro-geographic and geological features like geographical location, flat deltaic topography with sea-facing low elevation, extreme tropical climate variability, unstable tropical atmospheric condition and regional setting, Bangladesh becomes more prone to natural disasters. Adding together, some socio-economic conditions like overpopulation with high density, extreme poverty, less infrastructural development, dependency of majority of population on crop agriculture, continue to take heavy causality on the people. Moreover, it has been seen from IPCC's (2006) assessment on climatic disaster risks throughout the world, Bangladesh ranked second among the ten extremely disaster vulnerable countries. In Bangladesh, the natural disasters which usually occur include flood, drought, tropical cyclone and storm surge, etc. Yet, the total number of natural disasters has quadrupled in the last two decades—mostly due to floods, storms, cyclone, and tidal surge. Statistics show that over the last 200 years at least 70 major cyclones have hit the coastal region.

To deal with these types of natural disasters, earlier disaster management was mainly focused on post disaster management. That is, rescue, relief, rehabilitation, and reconstruction were main issues of consideration. Recently, disaster management goes beyond post disaster management. It also gives emphasize on pre-disaster management issues that include planning and preparedness activities, organizational planning, training, information management, public relations and many other fields. It means building awareness and undertaking preparedness towards disaster risks reduction. Here, disaster awareness (having knowledge of disaster) is not only the precondition but also the first step of today's disaster management. In this regard, disaster education is considered as the main device of disaster awareness building.

In 1990s, significant public education efforts are emerged by many nations through the International Decade of Natural Disaster Reduction (IDNDR). Therefore, various educational materials for school children and the general public have been produced. Agenda 21 articulated on chapter-36 about "Promoting education, public awareness and training" where it mentioned education, including formal education, public awareness and training should be recognized as a process by which human beings and societies can reach their fullest potential (UNEP 1992). After that, the Hyogo Framework for Action (HFA) 2005–2015, priority for action 3, emphasizes the role of knowledge and education and highlights formal and non-formal education and awareness-raising as important components for disaster risk reduction. According to the UN/ISDR (2005), "DRR education is an interactive process of mutual learning among people and institutions. It encompasses far more than formal education at schools and universities, and affects all aspects of life through the concerted efforts to overcome universal barriers of ignorance, apathy, individual interests and lack of political will present in communities. It also involves the recognition and use of traditional wisdom and local knowledge for protection from natural hazards. Education is conveyed through experience, established learning arrangements, information technology, staff training, electronic and print media and other means that facilitate the sharing of information and knowledge to citizens, professionals, organizations and policymakers, among a range of other community stakeholders". Moreover, in 2006–2007, UN/ISDR campaigned on "Disaster Risk Reduction begins at School" addressed the emphasis of integrating disaster risk reduction into education.

In context of Bangladesh, Government of Bangladesh (GoB) raised awareness among students through the incorporation of various hazards and disaster management tasks in the educational curricula from class V to XII in 1994/1995 by National Curriculum and Textbook Board (NCTB). They included one chapter of disaster management in each class. The GoB has also made to hold a compulsory session of at least 2 h on disaster management in the curricula of all types of Training Institutes that train officials and non officials. Since disaster is more frequent in the recent years, therefore, the main purpose of this study is to disclose (1) how disaster education is exercised at different level of ongoing educational system in Bangladesh through course curricula and (2) to identify the gaps and challenges for successful implementation of disaster education in context of the country.

14.2 Disaster Education: Key Prospects

It is widely acknowledged that school disaster education programs have the potential to build community resilience to natural hazards as:

- If properly planned, built, maintained and managed, schools provide a haven to the children who spend most of their waking hours at school and safe facilities for public shelter in the aftermath of a disaster.
- Strong leadership of teachers has been proven to be very effective in dealing with emergency situations in disaster-prone countries (UNCRD 2008).
- Incorporating risk awareness and risk reduction strategies into school curricula contributes to institutionalize the reach of these messages to the broader public as what learnt in childhood is incorporated into collective knowledge and is carried out into future decision-making. In most countries, children are very influential and effective communicators and knowledge disseminators at the family level (OECD 2009). However, it has to be considered that students learn from a broad range of sources, unless there is a prescribed activity (e.g. homework task to develop a family home emergency plan) that students may not take home hazard-related learning and messages (Dufty 2009).

However, the aim of disaster education is to accelerate the progress of societies toward disaster resilience and at the same time, increase awareness and develop proper knowledge and skills among individuals. This can be achieved through appropriate and effective tools. In the 2006 Review of the role of education and knowledge in disaster risk reduction, Professor Ben Wisner commented, "Education, knowledge and awareness are critical to build the ability to reduce losses from natural hazards, as well as the capacity to respond and to recover efficiently from extreme natural events when they do, inevitably, occur" (Wisner 2006).

It has seen from a survey of school children in New Zealand that students who received disaster education may have raised risk awareness, their ability to differentiate and respond appropriately to disaster was weak (Ronan et al. 2001). Furthermore, another study carried out in Nepal where it shows that the student who received disaster education in school, only can raise risk perception, but cannot enable students to understand the importance of pre-disaster measures and to take actual action for disaster reduction (Shiwaku et al. 2007). However, the importance of disaster education at school is increasing because of the following reasons: (1) children are one of the most vulnerable sections of the society during a disaster, (2) they represent the future, (3) school serves as a community's central location for meetings and group activities, and (4) effects of education can be transferred to parents and community (Shiwaku 2009; UNSIDR 2006; OECD 2008). On the contrary, Pandey and Okazaki (2005) emphasized the importance of education on community by ensuring dissemination of best practices in disaster risk management at the community level and formulation of integrated programs for sustainable development. Moreover, Paton (2005) mentioned that disaster education should be integrated with community development initiatives to increase resilience and facilitate self-help capacities so as to reduce reliance on external response and recovery resources. Petal (2008) further suggested that the goal of developing "disaster-resilient communities" is widely understood to depend heavily on the success of disaster risk reduction and further mentioned that the integration of both formal and non-formal/informal education through school is one way of ensuring these messages reach into every family and community, so that the learning can be sustained for the future generations (Disaster risk in public education system in Bangladesh).

14.3 Various Streams/Tracks of Education Followed in Bangladesh

In Bangladesh, there are three main tracks of education: General, Madrasa, and Technical. Each of these three main systems is divided into four levels:

1. Primary level (from grades 1–5)
2. Secondary level (from grades 6–10)
3. Higher secondary level (from grades 11–12)
4. Tertiary/higher level (from grades 13 to above)

14.4 Educational System of Bangladesh

14.4.1 General Education System

14.4.1.1 At Primary Level

In Bangladesh, primary education is the first stage of four levels of education. It consists of class 1–5 and is free for all children. The primary education compulsory act passed in 1990 that made primary education free and compulsory for all children up to grade 5. During the past decade, Bangladesh has made great progress in improving the primary education situation making the country one of the largest primary education systems in the world. It has been seen that the number of enrolled students increased from 12 million in 1990 to over 16 million in 2008 and net enrolment rate was boosted from 60 % in 1990 to 90.8 % in 2008 (Unicef 2012). At present, Bangladesh has 80,397 primary schools; among them 47 % is government school (MoE 2012). The rest includes registered non-government primary school, non-registered non-government primary school, experimental school and private school. The government of Bangladesh distributes free books and education kits to students in all primary schools. The primary curriculum is competency-based developed by the National Curriculum and Textbook Board (NCTB) and is controlled by the Ministry of Primary and Mass Education. Recently government has started nationwide examination at the end of the fifth year named PSC (Primary School Certificate) examination. The Education policy 2010 envisages the introduction of primary (basic) education from grade 1 to grade 8 (e.g. incorporating junior secondary education).

Table 14.1 represents the educational structure of Bangladesh where it shows the tracks of and levels of educational system now carried in Bangladesh. At all levels of schooling, students can choose to receive their education in English or Bengali (Bangla). Private schools tend to make use of English-based study media while government-sponsored schools use Bengali (Bangla).

14.4.1.2 At Secondary Level

Secondary education consists of two stages: junior secondary (grades 6–8) and secondary (grades 9–10). Starting from grade 9, in general secondary education, there are three streams of course—humanities, science and business education. Students are free to choose their course out of these streams. At the end of grade 10, Secondary Education Certificate (SSC) examination is held that helps students seeking to move to the higher secondary education. The Boards of Intermediate and Secondary Education (BISE) are responsible for conducting Junior School Certificate (JSC), Secondary School Certificate (SSC), in addition to granting recognition to non-government secondary schools. There are eight boards in different places in Bangladesh namely: Dhaka, Rajshahi, Jessore, Comilla, Chittagong, Sylhet, Dinajpur and Barisal. The Ministry of Education is responsible for secondary education in Bangladesh. On the other hand, books for all the levels are being produced by the lone National Curriculum and Text Book Board (NCTB).

14.4.1.3 At Higher Secondary Level

The secondary education is designed to prepare the students to enter into the higher secondary stage. The higher secondary consists of grade 11 and 12 which is being offered by intermediate colleges or by intermediate section of degree or master colleges. The same streams such as humanities, science and business education are

Table 14.1 The educational structure of Bangladesh

Level of education	Class	Age	Grade	General education					Technical education			Madrasa education
Tertiary	XX	25+		PhD	PhD (Eng. Med.)							
	XIX	24+	M. Phil			Post MBBS						
	XVIII	23+				M. Phil (Med.)	MBA					
	XVII	22+	MA/MSc/ Mcom/MSS		MSc (Agri.) MSc. (Eng.)	MBBS BDS	BBS	LLM				
	XVI	21+	Bachelor degree		BSc. (Agri.)			LLB (Hons)			D. in eng./ agri./ Tex. tec	Kamil
	XV	20+			BSc (Eng.)							
	XIV	19+			BSc (Text)							Fazil
	XIII	18+			BSc (Leat.) BSc (Tec edu)							
Higher secondary	XII	17+	Higher secondary education		H. S. C				H. S. C (Voca)	H. S. C. (BM)		Alim
	XI	16+										
Secondary	X	15+	Secondary education		S. S. C				S. S. C (vocational)			Dakhil
	IX	14+										
	VIII	13+	Junior secondary education		J. S. C				Junior secondary education			
	VII	12+										
	VI	11+										
Primary	V	10+	Primary education		P. S. C				Primary education			Ebtedayee
	IV	09+										
	III	08+										
	II	07+										
	1	06+										
Pre-primary education		05+										
		04+										
		03+										

followed in higher secondary. At the end of grade 12 students sit the Higher Secondary School Certificate (HSC) examination. The boards under BISE are responsible for conducting Higher Secondary Certificate (HSC) examination.

14.4.1.4 At Tertiary/Higher Level

Higher education is offered in the universities (both public and private) and post-HSC level colleges and institutes of diversified studies in professional, technical, technological and other special types of education. HSC holders are qualified to enroll in 3-year degree pass programs while for honours, they enroll in 4-year bachelor' degree (honours) programs in degree colleges or universities (bachelor's degree program in the field of architecture, agriculture, dentistry, medicine and traditional medicine last 5 years). After successful completion of a pass/honours bachelor's degree, students can pursue a master's degree. Programs leading to the award of a master's are of 1 year duration for bachelor's degree (honours) holders and 2 years for pass bachelor's degree holders. For those students pursuing a M. Phil and Ph. D in selected disciplines or areas of specialization, the duration is of 2 years for M. Phil and 3–4 years of the doctoral degree.

There are 34 government universities and approximately 59 private universities in Bangladesh. Specialized universities are Bangladesh University of Engineering and Technology (BUET), Bangladesh Agricultural University and Bangabandhu Sheikh Mujib Medical University. The number of government and non-government medical colleges stand at 13 and 5 respectively. There are 4 engineering colleges and 2,845 colleges and institutes. In addition to the normal universities, non-campus distance education programs are offered by the Bangladesh Open University (BOU)—especially in the field of teacher education. There also is a National University which works as an affiliating university to supervise college level education.

14.4.2 *Madrasa Education System*

Madrasa education system has several levels like the general education. Their Primary level is called "Ebtedayee", while their "Dakhil" is equivalent to secondary level which spans grades 6–10. At the end of Dakhil education, students appear in a public examination which is organized by the Bangladesh Madrasa Education Board. The Dakhil certificate is equivalent to the Secondary School Certificate (SSC) obtained after successful completion of secondary education from mainstream schools. The minimum requirement for admission to the higher level of Madrasa education is the Alim (equivalent to HSC) certificate. Alim pass students are qualified to enroll in 2-year Fazil education. This level of education is imparted in Fazil Madrasa and fazil level of Kamil Madrasa. After the successful completion of Fazil degree students can enroll in 2-year Kamil level education. There are four streams of courses in Kamil level education (hadis, tafsir, fiqh and adab). Fazil and Kamil are under the Islamic University. The rest are under the Bangladesh Madrasa Education Board that administers the examinations and award certificate.

Madrasa Board has their own curriculum. Along with Quran and Hadith and Arabic, there are English, Bengali, Mathematics, Islamic History, and Social Science subjects in their curriculum. For the Dakhil Science group there are Physics and Chemistry.

14.4.3 Technical Education System

Technical education starts at class 9/10: SSC Vocational for general education, and Dakhil Vocational for Madrasa education. At class 9/10 the students study Social Science, and the text books of NCTB are followed. At the higher secondary level, it is H.S.C BM (Business Management). Then there are Diplomas on many technical subjects, e.g. Diploma in Engineering, Textiles, and Agriculture etc. For the Diplomas the students have to study 4 years right after completing SSC Vocational. In the Technical Education System, after obtaining Diploma-in-Engineering degree (4 years long curriculum), students can further pursue their educational carrier for obtaining a Bachelor degree from Engineering & Technology Universities, and normally it takes two and half or three years long courses for students with a Diploma-in-Engineering degree, to obtain a Bachelor degree, but often in some cases these students take more than 3 years to complete their bachelor degree (undergraduate degree) (16th Grade) in Engineering. Then they can enroll into post-graduate studies.

There are some 36 subjects on trade at the SSC level, and the students have to choose one. H.S.C BM has four semesters, each consisting of 6 months. First and third semesters are assessed at the colleges, while second and fourth semesters are assessed in board final exams. This ensures more intensive education and requires skilled and qualified teachers.

There are some 4,500 schools under the lone Technical Education Board in Dhaka: 1,800 SSC Vocational schools; 1,350 HSC. Business Management; 180 schools providing degree on diploma in Engineering; 103 provide Diploma in Agriculture; and 31 schools Diploma in Textiles. Over the last few years there has been quite an influx of students in technical education. In 2007 there were some 457,000 students in SSC Vocational, which was a 35 % increase.

For the Madrasa, the technical education of class 9 and 10 is being introduced from 2008 in 100 Madrasa with IDB funding.

14.5 Disaster Management Issues in the Existing Course Curriculum of Bangladesh Education

14.5.1 In General Education

In Bangladesh, National Curriculum and Textbook Board (NCTB) is responsible for the development of curriculum, production and distribution of textbooks at primary, secondary and higher secondary levels. The NCTB has introduced the issues of

Table 14.2 Disaster related course at primary level

Class	Text	Name of the chapter	Number of the chapter in the text	Contents of the chapter
V	Bangla	Cyclone		
	English	Improved burner		
	General Science	Environmental pollution and natural disasters	16th	– The reason and impacts of environmental pollution – Natural and manmade disaster

natural and human induced hazards, vulnerability, risk and disaster preparedness in different chapters of the subjects on disaster management in different books e. g. social science, general science, bangle, English. They have done it in 1994–1995 with the help of Disaster Management Bureau (DMB). However, it has been seen from the below mentioned Tables 14.2–14.4 that how disaster related issues is being taught through various course curricula at different level of Bangladesh education system.

14.5.1.1 Disaster Related Course at Tertiary/Higher Level

Over the last few years higher education in Disaster Management has been turning into a popular discipline for students to opt for at the University level. In Bangladesh, **BRAC University** has only the full-fledged higher education program granting as master's degree in disaster management. They provide Master's program on Disaster Management since 2004. Besides, Institute of Governance Studies at BRAC University has been running a Masters program for the last 2 years on "Environmental Management and Sustainable Development" as an elective course, targeting the mid level government officials. **North South University** (**NSU**) established the department of Environmental Science and Management since 1994, beginning Environmental Studies at the under graduate since then. With the assistance from the University of Manitoba, Canada, NSU is currently drawing up a Masters program on "Resource and Environmental Management", where there is scope to integrate climate change and risk reduction issues. Besides, **Independent University Bangladesh** (**IUB**) has carried out research, and post graduate Certificate course in leadership and management, and incorporated risk reduction issues in other degree program curricula, courses and in course contents through Department of Environmental Science.

Some of the public Universities offer one course or short courses on this topic. At **Dhaka University** the department of Geography and Environmental Science offers a course on disaster management. Similarly the Department of Urban and Regional Planning of **Jahangir Nagar University** offer a course on disaster management in the third year: Disaster Management and Planning having three credits, with three classes a week. In the fourth year many students are doing thesis on disaster management issues. Apart from this there are two more courses: (1)

Table 14.3 Disaster related course at Secondary level

Class	Text	Name of the chapter	Number of the chapter in the text	Contents of the chapter
VI	Social Science	The natural disasters of Bangladesh	14th chapter	• Definition and classifications of disasters • Natural disaster in Bangladesh Cyclone and tidal wave, flood, river erosion drought, tornado, earth-quake, cold wave • Plan and program to face disaster
VII	General Science	Flood, river bank erosion and drought in Bangladesh	24th chapter	• A brief discussion about Bangladesh Geographical location, physical features, climate • Flood Definition, types and causes, devastation, preparedness and mitigation • River bank erosion Overview, causes, affected areas, mitigation measures • Drought Definition, nature and types, effect, mitigation measure
VIII	General Science	Natural disaster: cyclone and tidal wave	25th chapter	• A brief discussion about Bangladesh Geographical location, various types of disaster • Cyclone Formation, area and, loss • Tidal wave Areas, time period and loss, causes of tidal wave formation • Regional cyclone and tidal wave • Timing of natural disaster • Diseases following disaster and their prevention • Possible measure to combat the natural disaster Pre disaster management, Post disaster management, Inter-disaster management warring signals with meaning for the maritime ports
IX and X	General Science	Disaster management and Bangladesh	21st chapter	• Objectives of disaster management • The cycle of disaster management • Disaster management system at national to union level

Environmental Planning and Management, (2) Geography, where disaster management issues are discussed. They are currently reviewing the syllabus of the course: Disaster Management and Planning to contextualize it further and capture the latest thinking in the field of disaster reduction. The existing curriculum is tilted more on

Table 14.4 Disaster related course at Higher Secondary level

Class	Name of the subject	Name of the chapter	Contents of the chapter
XI	Bengali	Disaster Prone earth: Bangladesh and the world	• Different types of natural and man-made disaster Description of manmade peppercress, contribution and initiatives of United Nations in disaster management • Post disaster management • Different steps for disaster mitigation
XI	Geography (second part)	Flood control, irrigation	• Detail water management plan for agriculture and living with flood can be included
XI	Geography (first part)	Rivers (third chapter)	• Flood Causes of flood, influence of flood, measures of flood control • Govt. measures for flood control • Problems of flood control
XI	Commercial Geography	Flood control and drainage	• Definition of flood, floods in Bangladesh, effects of flood, flood controlling system in Bangladesh • Problems of flood, control in Bangladesh, initiatives taken by the Govt. of Bangladesh • Drainage. initiatives, taken for drainage, salinity

environment focusing mainly on geographic and scientific causes and effects of different types of disasters, and preparedness.

The Social Science Department of **Dhaka University** has been teaching Sociology of Environment from early 1990s. Since 1996, sixth chapter deals with "Environmental Hazards and Disaster", and seventh chapter on "Management of Environmental Hazards and Disaster". Since 1999 Sociology of Disaster has been included in the curriculum for third year. Although it's not a main course, many students are taking this optional course, and at the Masters level students are doing research and writing thesis on this issue. At the second year basic concepts of disasters are taught, while in third year, in Sociology of Disasters topics taught are: Origin, scope, and importance of this topic; major disasters and learning; causes and consequences of disasters; Gender and disaster, National Disaster Policy; GO/NGO Collaboration; Disaster planning and management, national and international policies; empirical evidences, coping mechanism, best practices etc. are taught too. The Urban & Regional Planning discipline of **Khulna University** offers courses on disaster management.

Some of the technical University such as **Bangladesh University of Engineering and Technology (BUET)** offers disaster management courses at both Bachelors and Masters Degree programs through the Department of Urban and Regional Planning under the Faculty of Architecture and Planning. Besides, **Bangladesh Agricultural University (BAU), Mawlana Bhashani Science and**

Technology University (MBSTU), **Shahjalal University of Science and Technology (SUST)**, **Chittagong University of Engineering and Technology (CUET)**, **Patukhali Science and Technology University (PSTU)** carried out research and offers Professional Certificate course in Disaster Management. Even, **PSTU** commenced Bachelor of Science in DM and Masters in DM and to incorporate risk reduction issues in other degree program curricula, courses and in course contents through the Department of Environmental Science and Disaster Management (DESDM).

The National University is carrying out one course on Disaster Studies and Management under "Geography and Environmental Science". At the Degree and Honors levels disaster management is studied as part of a course. In Sociology there is one course on "Sociology of Environment", where the issue of disaster management is included.

14.5.2 In Madrasa Education

In Social Science group, the text books of class 6 and 9/10 have an entire chapter on disaster management, which is reproduction of the chapters from the social science and general science books of the NCTB for general education of the corresponding classes.

14.5.3 In Technical Education

At the higher secondary level, that is H.S.C BM, they have the book of Commercial Geography of NCTB, where chapter 9 of second part is on "Flood Control and Drainage". At the Diploma level, Environmental Management is a core course. Curriculum of this course might be revised to incorporate risk reduction.

14.6 Curriculum Review Process at Different Levels

14.6.1 General Education

The last time NCTB had produced the curriculum was 1994/1995. Subsequently in 2002 NCTB reviewed the curriculum of class 1 and 2, of class 3 in 2003, class 4 in 2004, and class 5 in 2005. In 2008, NCTB reviewed the curriculum of class 6–8. Recommendations for review or any revised materials should have to be submitted to the Ministry of Education (MoED) by the relevant Ministry, in case of Disaster Risk Reduction, by the Ministry of Food and Disaster Management. The Ministry of Education would then send the materials to NCTB who review the submitted the materials or documents and take the necessary action. The curriculum experts will review both the revised texts and/or the curriculum. Once the materials are cleared/reviewed/revised by the curriculum experts committee, it is then sent back to the Ministry of Education where the National Curriculum and Coordination Committee (NCCC), chaired by the Secretary MoED gives the final approval to the revised curriculum, and NCTB then incorporate it into the text books.

On the other hand, subject or course teacher of Dhaka University has to initiate the curriculum review process. The teacher revises the curriculum and places it to the Academic Committee, comprised of all teachers of the University. After that it goes to the Syllabus Review Committee consisting of a few teachers.

14.6.2 Madrasa Education

Madrasa Education Board has its own Curriculum Committee headed by the Chairman of Madrasa Education Board. Under that they form subject committees to review and revise curriculum. These consist of curriculum specialists, who are usually nominated from NCTB, subject specialist and class teacher. Once the curriculum is revised, the "Curriculum and Courses of Studies Committee" review and clear the revised curriculum, which is then placed to the Academic Committee chaired by the Madrasa Board Chairman for the final approval.

14.6.3 Technical Education

The Technical Education Board has their own curriculum review committee headed by the Director, Curriculum. Experts committee, consisting of three members, prepares the curriculum which is then assessed by the Courses Studies Committee (which is considered the executive level approval body). The Board has the final authority to approve revised curriculum. The curriculum can be reviewed anytime the teachers feel the need; don't need any approval from Ministry or any other authority. The board has the final say on curriculum.

14.7 Proposed Amendments for All System of Education Up to Higher Secondary Level: In Disaster Risk Reduction Point of View

In Bangladesh, the existing text books mainly focus on hazards and disaster management, nothing is there about risk reduction. Therefore, DMB under CDMP had undertaken a review of the books of the National Text Book Board in 2006, and made the necessary amendments and if required reproduce the relevant chapters on disaster management to address Disaster Risk Reduction (DRR) (Haider 2008). The proposed revisions are focused on the measures necessary to be undertaken at the family and community levels and emphasizes on the need for risk reduction, rather than giving theoretical explanations of the hazards themselves. In view of that, the chapters of the Social Studies books (Poribesh Porichiti, Poribesh Bigyan) of class 5–11 were rewritten. According to their revision, following are the analysis of the existing text books (point a), and proposed amendments (point b):

i. a. Class Five: chapter 16 of the book "Poribesh Porichiti, Bigyan" (Introduction to Environment, Science) book is on "Environment Pollution". While describing the reasons and impact of the environment pollution, natural and manmade disasters are being mentioned.

i. b. Porisbesh Porichiti (Bigyan): sixteenth chapter should be titled "Environment Pollution and Natural Disasters". After the first part of the chapter which tells about Environment and it's elements, it should have the section on "hazards" explaining that due to environment pollution natural and manmade hazards occur which further deteriorates our environment: due to improper river and land management we have flood, and floods also can destroy arable land with sand deposit. The idea of ecological imbalance causing disasters, and that our lifestyle and behavior are responsible for that, could be introduced at this level: using air conditioner, refrigerator, air conditioner in car contributing to increasing CFC emissions resulting in green house effect and ecological imbalance might be introduced in very non technical "story telling" way. Felling of trees exacerbating greenhouse effect should also be told at this stage. This would help children to be careful about their behavior towards nature. The exercise section should have assignments for children to explore ways how they can contribute in minimizing the ecological imbalance and therefore decrease the possibility of disasters.

In the existing chapter, section on "reasons of pollution" focuses on forest fire, which does not occur in our country that frequently. Instead flood or cyclone might be brought in. Reference should be made to SIDR, 2004/2007 flood, Catrina, tsunami etc. In the existing book, there is a drawing of the aftermath of cyclone and tidal surge showing dead bodies and carcass. It's better not to present such images to the children. There could be different drawings, or better pictures, showing drowned houses and trees, swelled sea, swaying trees etc.

ii. a. Class Six: chapter 14 of "Shamajik Bigyan" (Social Science) book of class 6 is solely on "Natural Disaster in Bangladesh", discussing the types of disasters in details, and finally a paragraph on the future plan to face disasters, which is mainly about post disaster response. The language in this chapter seemed a bit high compared to the level of students of class 6 and the introductions to the types of disasters seemed too detail.

ii. b. Shamajik Bigyan (Social Science)—should avoid giving definitions, instead should start with the differences between natural and manmade disasters, and the reasons for those (geographic, natural, and economic reasons). Children should be introduced with the main types of disasters in Bangladesh: flood, cyclone, storm surge, tornado, river erosion, earthquake, drought, etc. However, it should not focus much on scientific explanations on why and how the hazards occur. Rather it should dwell on why and how hazards turn into disasters: social, economic, livelihood patterns, etc.; and what we could and should do to prevent hazards from turning into disasters. It should also refrain from dishing out misleading information like comparing 1998 cyclone which had hit only Cox's Bazar and St. Martins Island with those of 1970 or 1991 and claim wrongly that the reason of less damage to life and property was timely intervention of government. The reality was that it took place at a time and in an area where the damage could not be bigger or even close to the mega cyclones of 1991 or 1970. The information given to the children needs to be very objective and authentic. DRR experts should be consulted in such cases.

iii. a. Class Seven: chapter 20 fourth of the "Shadharon Bigyan" (General Science) book of class even is titled "Flood, River Erosion and Drought in Bangladesh". The chapter begins with description of the geographical characteristics of the country and its climatic condition. After that it gives details of flood: what it is, why it happens, some of the devastation floods in the country—1987, 1988, and 1998. Then it talks about the short and medium term preparedness measures to mitigate the impact of floods at the pre, during and post flood period. Then it goes over to river erosion: what it is, the reasons for river erosion and the usual spot of this disaster, ways to stop river erosion. Finally the chapter goes on to drought: what it is, reason and nature, the repercussions of drought, what to do to stop drought.

iii. b. Shadharon Bigyan (General Science)—the existing chapter deals with disasters induced by the river system: flood, river erosion, and drought. It could remain the same: focus on flood, river erosion, and drought, but the approach needs to be changed. Instead of citing the latitude, longitude, and other related geographical realities making the country disaster prone, the chapter should deal with the social and economic reasons, in detail, as well as impacts of these disasters on lives and livelihoods. It should discuss types of these disasters, areas prone to these disasters, risk management at the family and community levels; including what the students can do to mitigate these disasters. While introducing floods, it should mention that flood is not only a normal phenomenon in a country like Bangladesh, it is also necessary for the land and agriculture. Therefore flood is not a disaster on its own. It is when the "limit" is crossed that it turns into a disaster. There could be small stories on coping with disasters like flood and drought as boxed items. Key issues might be boxed too as bullet points.

iv. a. Class Eight: the existing chapter of 25, in the "Shadharon Bigyan" (General Science) of class 8, is titled: "Natural Disaster: Cyclone and Tidal Surge". It begins with discussion on cyclone: the climatic and geographical phenomena which induce the information of a cyclone; tidal surge, and why tidal surge hit the coastal belt of Bangladesh; vulnerable locations/areas for cyclone and tidal surge; the usual periods of natural calamities; the nature of devastation of cyclone and tidal surge; tidal surge induced diseases and their prevention; four stages of coping with cyclone and tidal surge: pre, post, during, and long term mitigation measures. Finally the cyclone warning signals are being explained.

iv. b. Shadharon Bigyan (General Science)—the existing chapter is on disasters of coastal belt: cyclone and tidal surge, and it can remain the same with amendments in its approach in line with the proposed class 7 amendments. There needs to be some corrections made with regards to some information and data about past events. Under the section on "disaster related diseases and prevention", it says due to natural disasters ecological balance is lost causing diseases which is not the right perspective. It also says that because of different diseases tablets and salines should be distributed, while the appropriate recommendation should be health care assistance not distribution of tablets as such. Besides, it also says that as extra caution even when tube well water is available, water purifying tablets (WPTS) should be distributed. Actually water surge can contaminate tube-wells with saline water only, not with germs, and rehabilitation of tube-well is very cheap, and tube-well water does not need WPTs. Thus the text/content needs to be carefully edited and issues inserted with the help of the DRR experts.

v.a. Class Nine/Ten: chapter 20 first of "Shadharon Bigyan" (General Science) book of class 9 and 10 is titled as "Disaster management and Bangladesh". This chapter discusses the objectives of disaster management (focusing on disaster response): the cycle of disaster management is described accordingly. After that disaster management structure beginning from National Disaster Management Council down to Union Disaster Management Committee is being discussed.

v. b. "Shadharon Bigyan" (General Science)—the chapter should discuss about the disaster risk reduction and paradigm shift from post disaster relief and pre disaster preparedness for relief to long term risk reduction which is more developmental in nature: how attaching risk reduction lens to development can ensure sustainable development and livelihoods; disaster management structure beginning from National Disaster Management Council down to Union Disaster Management Committee might be given in a box in the form of bullet points. It can discuss the need to form village based volunteer groups citing the example of CPP for village based early warning and other related activities. This might encourage the students and imbue them with the spirit of volunteerism.

vi.a.1. Class Eleven: third chapter of the second part of Geography book for class 11 and 12 is on "Rivers". It describes the topography of the major rivers in Bangladesh, their influence on the environment and on the economy. Then it goes into flood: causes, beneficial and harmful impacts of flood, measures of flood control (hardware or structural), and the problems faced in flood control.

vi.a.2. Class Eleven, Commercial Geography: chapter 9 of second part of Commercial Geography is on "Flood Control and Drainage". It starts with the definition of flood and then discusses some of the major flood in Bangladesh and their impact on our economy. Next it discusses in detail the causes of flood; effects of flood in brief; and finally the flood control measures, their problems, and government's role in flood control. The next section in the chapter is on drainage: it tries to define what drainage is and mentions some of the drainage programmes. Finally the chapter touches on the issue of salinity.

vi.a.3. Class Eleven, Bangla: the Bangla text book for class 11 and 12 has an essay on "Disaster Prone Earth: Bangladesh and the World". The essay gives an overview of different types of disaster occurring in the world, and the devastating ones in Bangladesh; manmade disasters; United Nation's initiatives, and some other common issues of disaster management.

vi.b. Geography—the chapter should give all the detail geographical analysis of the causes of various disasters in Bangladesh, which should include the major ones like flood and cyclone. It should then discuss the measures already taken and the ones needed to be taken to mitigate impact of these disasters at pre, during and post disaster phases and at all levels: GOB, NGOs, and Community.

Commercial Geography: The chapter might be in line with or almost exact copy of the general Geography book with additional paragraphs on the impact of these disasters on our economy in detail.

Bangla: Since not all groups will study geography, it's good that the Bangla book has a chapter on it. It should have an overview of different types of disasters occurring in the world, and the devastating ones in Bangladesh: 1988, 1998, 2004, 2007 floods, 1970, 1991, 2007 cyclones in brief; United Nation's initiatives in DRR; some common issues of disaster management. There should be some insertions with regards to HFA, the SAARC disaster management strategy (2006–2015), draft national policy on disaster reduction, SOD, and roles and responsibilities of DMB, DRR, and CDMP in boxes.

Madrasa Education Board is about to review the text books for the "ebtedayee" or primary level from May 2008. For Dakhil, the existing chapters of class 6 and 9/10 would presumably be revised as and when the NCTB books for general education are revised, since the Madrasa Board reproduces the chapters from NCTB books. Thus DMB should immediately approach Madrasa Education Board to integrate disaster risk reduction into Ebtedayee, Dakhil and Alim levels.

Technical Education Board follows the NCTB books. Therefore the revisions of NCTB books would contribute to the Board's curriculum revision. However, for Diploma education, DMB might approach a few Technical Colleges to assist them to integrate DRR and/or introduce new course on DRR.

With contrast of the amendment by NCTB, it has been seen that it only gives emphasis on disaster risk reduction education at primary to higher secondary level, but it does not consider non-formal as well as in-formal education. But to achieve the best results, formal, non-formal and in-formal education should be partnered together, if possible. Because, Hofstein and Rosenfeld (1996) argued that integration of in-formal learning experiences within the formal school curriculum enriches the learning process. On the other hand, La Belle (1981) suggested that the contribution of non-formal education as a complement and supplement to formal education. Furthermore, Lidstone and Nielsen (1999) suggested that while formal disaster education would remain a systematic and structured learning that progresses over time, the real life context, reflective learning, and situated-learning characteristics of non-formal/informal education are equally important. On the other hand, the Hyogo Framework of Action (HFA) (2005–2015) stresses "use of knowledge, innovation and education to build a culture of safety and resilience at all levels" as one of the five priorities of action with a focus on including disaster education in formal and non-formal education at all levels. Likewise, according to Shaw et al. (2011a, b), "Both formal and non-formal education is indispensible to change people's attitudes so that they have the capacity to assess and address their sustainable development concerns" that suggesting the synergetic effect resulting from the combination of the various types of education. One example from Iran shows that Earthquake Safety Education program is successful there, because this program integrates disaster education into the formal education and performs non-formal education activities at the same time, involving students in the local context as well as engaging participation from the community. Another example from Myanmar highlights that the need of better awareness on DRR at the community level, give emphasis to take action properly to prepare to mitigate disaster risks. Therefore, a unique DRR education project "Mobile Knowledge Resource Centre (MKRC) and Water Knowledge Resource Centre (WKRC)" is getting importance at highest level because it increases the awareness of communities on disaster risks and hazards, and to enhance individual, family and neighborhood level actions. On the contrary, KIDA model developed by Shaw et al. (2009) that also gives emphasize on knowledge, interest, desire to promote action and actual actions are significant outputs of disaster education. Therefore, it can be said that non-formal as well as informal education is imperative to promote disaster education. But it is difficult for the school to have special knowledge on disaster risk reduction. In this regard, NGOs and research institutes can play a role to enhance the awareness of the school students and teachers that further disseminates to the parents and family members.

14.8 Challenges and Gaps to Implement Disaster Education

The country has been experiencing major disasters in increasingly closer time gap: floods of 2000, 2004, 2007, SIDR 2007. This has increased the level of awareness among not only the relevant key stakeholders about hazards but also among the key people in other sectors at the policy level. Since natural disasters create a humanitarian situation, people tend to go beyond political divide and consider it precisely as a humanitarian issue and want to deal with it with utmost sincerity. However, this good will persists during the emergency period only. Soon after the shock is over, people tend to slide back to the usual state of apathy and bureaucracy.

Due to many a seminars and workshops at the national level, organized by various agencies on disasters, the policy makers are more or less aware about the hazards: the types of hazards; why they occur; their impact; who are the responsible agencies: Ministry of Food and Disaster Management (MoFDM), Disaster Management Bureau (DMB), Directorate of Relief and Rehabilitation (DRR); and the existence of disaster management committees. However, there are gaps in the understanding of the issues related to risk reduction among the policy makers of all sectors. The risk reduction program of the MoFDM, has designed primarily to utilize the Ministry's own funding innovatively, to reduce vulnerability of the poorest and the most vulnerable population through long term program, turned into a post disaster rehabilitation program. The concept of risk reduction is not understood and therefore not given any prominence or necessary weight age by the key people in the government. Thus people from within the line Ministry: Ministry of Food and Disaster Management (MoFDM) lack the clarity about the concept; that means the level of understanding of other Ministries in this regard. Over the years, people have come to recognize the issues of disaster management with regards to preparedness for better management of post disaster response. However, with regards to risk reduction which in effect is development effort, keeping disaster scenario in mind, is still being seen as a difficult proposition to all, including the decision makers at the government level.

There is another big gap in the understanding of the people in the helm with regards to people's capacity, indigenous coping capability, and people's empowerment for sustainable risk reduction. The government officials think of the vulnerable people as "victims", "passive recipients" of emergency support, "dependent" on relief' and incapable of taking the right decisions, and "ignorant" bunch. Therefore, they feel obliged to serve them, the poor, weak, and incapable vulnerable community, and thereby patronize them. The inherent flaw of such assumptions is that it is "half truth", which is worse than "full lies". People are vulnerable and poor, but in no way, they are ever completely incapacitated. They rather have the capacity, and resilience to not only survive but also start afresh in the face of all odds. This inherent power, this indomitable fighting spirit of the people of Bangladesh are often undermined by the government people and the policy makers, by considering them as passive victims. As a result all sort of post disaster response and risk reduction programs are designed and undertaken without any consultation with the targeted population, and therefore are often far short than being effective. For example, in SIDR response, donors spent millions of dollar for drinking water while it was proven to be an unnecessary support, because people had repaired the existing tubewells, right after SIDR. In the name of risk reduction, programs are undertaken as per the advice and design of the "experts" "academicians", who are often detached from the vulnerable community and therefore do not understand the needs or the pulse of the people. Many a times there are research or action research programs, studies etc. which seldom reach the people, whom these are meant for.

One of the problems of teaching disaster management let alone disaster risk reduction at the higher level is lack of books and materials in Bengali (bangla), and more importantly, country specific materials. Students mostly have to rely on internet to access international publications and proceedings of international conferences on disasters and environment. It is easier to amend curriculum of university courses than it is for school level education. Here the course teachers can propose the amendments to the syllabus, and unlike school curriculum it does not require any revision of any books. However, as mentioned above, reading material remains an endemic problem in higher education level, which needs to be addressed.

14.9 Recommendations for Strengthening Disaster Education at Different Level

14.9.1 Primary, Secondary and Higher Secondary Levels

- Given the fact that Bangladesh is a high risk country to all sorts of recurrent natural disasters, there should be a clear policy statement of the government for integration of disaster risk reduction into education curriculum at various levels. This should be spelt out in the National Disaster Management Policy. It might also be incorporated in the PRSP.
- Revision of curriculum is a long-winding participatory process oriented activity. Considering the experience of DoE on environmental curriculum revision, it is extremely important that the proposed activity should be undertaken in an appropriate institutional framework involving the relevant institutions and experts, under the leadership of the Ministry of Education. MoFDM should organize an inter-ministerial meeting with the representatives of the relevant Ministries and Institutions to determine the appropriate institutional framework within which the proposed activity can be initiated. With agreement of all, DMB might provide the secretarial support to the group/committee/or the task force.
- Bangladesh, being one of the most vulnerable countries to climate change, the curriculum of disaster management should include climate change, it's likely impacts in Bangladesh, and the linkage between climate change and disaster risk reduction.
- The revision process should leverage international best practices and lessons learnt. Disaster reduction curriculum of other countries could also be consulted: e.g. of Australia, India, Bangkok, UK.
- Risk reduction should be integrated in the non formal education too. A large number of adults are part of adult education program, and they should be included in the knowledge network.
- The issue of differently able people, who constitute 10 % of the population, should be kept in mind. They are the most vulnerable people in any disaster. Therefore the content should have information which would help the physically disable students to help themselves, or help the students to take care of disable family members. In this respect, incorporating this topic in the Braille system might also be considered.
- The cost factor has to be kept in mind while recommending for curriculum revision in terms of colour, paper quality or even significantly increasing the number of pages. The books for the primary schools are all printed with UNICEF money, and distributed to all children in the country free of cost. Ministry of Primary and Mass Education entirely deals with primary education in the country. Only the books of Secondary and Higher Secondary levels are being paid for by the students. Currently NCTB print eight crore books every year on top of the old books which are also put on circulation. This costs on average Tk125 crore per year. One of the reasons why 45 % of the children enroll in primary education level while only 12–16 % enroll in higher secondary level is that the students do not have the financial ability to buy the text books. Therefore, increasing the cost

of the books would mean more drop outs. So, amendments have to be done with careful consideration about the limits as to how much cost hike the students would be able to cope with. However, since the primary level books are provided free of cost and are entirely financed by UNICEF, some improvements like colorful illustrations might be planned for these books for the children.

14.9.2 Higher Education

- DMB might encourage and assist universities, both private and public, to develop post graduation programs on disaster risk reduction. Jahangir Nagar University, IUB, North South Universities might be approached.
- Since disaster reduction is a cross cutting issue, it is important that the inclusion of disaster risk reduction is encouraged in all the relevant subjects/programs/departments beyond science and geography, such as history, sociology, engineering, architecture, environmental management, hydrology, soil science, planning and public health, economics, public administration, business administration, etc.
- National University, that governs the degree and bachelors level education of all colleges and universities in the country should be encouraged to include risk reduction into their curriculum. That way, it would be possible to reach a vast number of students. National University might be encouraged and assisted to incorporate Disaster Risk Reduction as a full-fledged optional course; might also form part of a core course of B.A. as well as B.S.S.
- For Universities appropriate books/reading materials, mostly from international institutions can be adopted/translated, and new materials should be produced/published. DMB might facilitate this.
- Develop linkages with international institutions and universities to harness their knowledge and expertise in teaching disaster reduction. Many organizations like ISDR, Prevention, and others have produced many materials. Internationally a huge amount of work has been done in this area which could be adopted and utilized in teaching disaster risk reduction in Bangladesh at different levels.
- Electronic media, having fast increasing viewer-ship, and becoming more and more accessible in rural areas, could be utilized to provide distant learning on disaster risk reduction.

14.9.3 In Terms of ESD and CCE

In Bangladesh, ESD (Education for Sustainable Development) and CCE (Climate Change Education) is a new concept. ESD has been practiced to some extent in this country, mainly in the form of environmental education. On the other hand, CCE is taught in a form of disaster management. Contrasting with disaster education, ESD and CCE should be reorientation and restructured, so that it should be part of an education for sustainable development that helps to develop attitudes, knowledge to make informed decisions for the benefit of students, communities, and others, now and in the future—and to act upon these decisions (Habiba et al. 2013).

14.10 Conclusion

From this study, it has been concluded that GoB has carried out disaster education at different levels of educational system in Bangladesh. In the recent years, they also made amendments for all system of education up to higher secondary through NCTB in co-operation with CDMP. But, for better awareness on disaster education needs make people more aware at individual, family and community or local level. Because awareness facilitates preparedness and preparedness leads to disaster risk reduction. However, it needs the participation of different groups of stakeholders from students and teachers, to government, non-government organizations, the media and private sectors. Since, disaster education is a process based approach that links school to communities and families, links humans to nature, links different disciplines, links theory to practice, and links indigenous knowledge and its transferability over time (Shaw et al. 2011a, b).

In the recent years, disaster education is getting importance, because, past experiences show that disaster education has a positive effects on disaster management and risk reduction. Therefore, not only giving focus on school educational system, it is noteworthy to link school education with family and community. According to UNISDR (2006), it not only highlights the importance of integrating disaster risk reduction into formal education, but also emphasized the importance of community participation in order to achieve sustainability within the community. The home and community are considered as an important unit of disaster management. Because, children who knows how to react during disaster like cyclone, earthquake, floods; conscious and educated family members who know how to manage family in response to disaster; community leaders who have learned how to warn their neighbors in a timely manner and communities familiar with preparing themselves for natural hazards-all will demonstrate how disaster education can make an important difference in protecting people at the time of calamity. Therefore, it is paramount to consider communities awareness at the local level.

In order to prepare the nation of about 160 million people, Bangladesh requires professionals with background and education in dealing with natural and manmade hazards. Although, professional training course is offered by various training institutes in Bangladesh such as Bangladesh Public Administration Training Centre (BAPTC), Rural Development Academy (RDA), Bangladesh Academy for Rural Development (BARD), National Academy for Educational Management (NAEM). But, the enhancement of the capacity of partners in terms of infrastructure, training, curriculum, syllabus, resource material development and procurement, education and research support are inevitable. Finally it requires political commitment to ensure that disaster risk reduction is incorporated in the education system of Bangladesh, for which it would require allocating the requisite human and financial resources.

Acknowledgements The first author gratefully acknowledges the support from GCOE-ARS post-doctoral program of Kyoto University. The second author gratefully acknowledges the support from the Japan Society for the Promotion of Science (JSPS) of Japan.

References

Disaster risk in public education system in Bangladesh: emphasis on tornado. Available at http://www.iawe.org/WRDRR_Bangladesh/Preprints/S2CDMP.pdf. Accessed 15 Nov 2012

Dufty N (2009) Natural hazards education in Australian schools: how can we make it more effective? Aust J Emerg Manag 24(2): 13–16. Available at: http://www.em.gov.au/Documents/APEC%20Hanoi%20Outcomes%20Report.PDF. Accessed 15 Jan 2011

Habiba U, Abedin M, Shaw R (2013) Desk research to develop policy guidelines on climate change education: a national policy review in Bangladesh. Kyoto University, Kyoto

Haider D (2008) Curriculum review with regards to hazard awareness and risk reduction. Comprehensive Disaster Management Programme (CDMP), Ministry of Food and Disaster Management (MoFDM), Dhaka

Hofstein A, Rosenfeld S (1996) Bridgoing the gap between formal and informal science learning. Available at: http://www.ec.europa.eu/education/lifelong-learning-policy/doc52_en.htm. Accessed 25 Dec 2012

La Belle TZ (1981) An introduction to the nonformal education of children and youth. Comp Educ Rev 25(3):313–329

Lidstone J, Nielsen S (1999) Public education and disaster management: is there any guiding theory? Aust J Emerg Manag 13(3):14–19

Ministry of Education (MoE) (2012) Government of the People's Republic of Bangladesh. Available at http://moedu.gov.bd/index.php. Accessed 7 Dec 2012

OECD (2008) Stocktaking report on risk awareness and education on natural catastrophes 2008. Available at: http://www.oecd.org/dataoecd/24/50/42221829.pdf. Accessed 15 Nov 2012

Pandey B, Okazaki K (2005) Community based disaster management: empowering communities to cope with disaster risks. Reg Dev Dialogue 26(2):52–57

Paton D (2005) Community resilience: integrating hazard management and community engagement. In: Proceedings of the international conference on engaging communities, Queensland Government/UNESCO, Brisbane. Available at http://www.engageingcom.munities 2005.org/abstracts/Paton-Douglas-final.pdf. Accessed 3 Dec 2013

Petal M (2008) Concept note: formal and informal education for disaster risk reduction. Available at http://www.riskred.org/activities/ddredislamabad.pdf. Accessed 3 Dec 2013

Ronan KR, Johnson DM, Daly M, Fairley R (2001) School children's risk perceptions and preparedness: a hazards education survey. Aust J Disast Trauma Stud 1. Available at http://www.massey.ac.nz/~trauma/issues/2001-1/ronan.htm. Accessed 3 Dec 2013

Shaw R, Takeuchi Y, Shiwaku K, Fernandez G, Gwee QR, Yang B (2009) 1-2-3 of Disaster Education. Graduate School of Global Environmental Studies, Kyoto University, Kyoto

Shaw R, Takeuchi Y, Shiwaku K (2011a) 'TSUNAGARU': the essence of disaster education. In: Shaw R, Shiwaku K, Takeuchi Y (eds) Disaster Education: community, environment and disaster risk management, vol 7. Emerald Group Publishing Limited, Bingley, pp 153–162

Shaw R, Takeuchi Y, Gwee QR, Shiwaku K (2011b) Disaster Education: an introduction. In: Shaw R, Shiwaku K, Takeuchi Y (eds) Disaster Education: community, environment and disaster risk management, vol 7. Emerald Group Publishing Limited, Bingley, pp 1–22

Shiwaku K (2009) Essentials of school disaster education: example from Kobe, Japan. In: Shaw R, Krishnamurthy RR (eds) Disaster Management: global challenges and local solutions. University Press, Hyderabad, pp 321–337

Shiwaku K, Shaw R, Kandel RC, Shrestha S, Dixit A (2007) Future perspective of school disaster education in Nepal. Disaster Prev Manag 16(4):576–587

UN/ISDR (2005) In: World Conference on Disaster Reduction, 18–22 January 2005, Kobe, Hyogo. Thematic Discussion Paper Cluster 3. Knowledge, innovation and education: building culture of safety and resilience. Available at: http://www.UN/ISDR.org/wcdr/thematic-sessions/WCDR-discussion-paper-cluster3.pdf

UNCRD (2008) School earthquake safety initiatives. Reducing vulnerability of school children to earthquakes. Available at: http://www.hyogo.uncrd.or.jp/publication/pdf/Report/2008SESI/UNCRD%20SESI%20publication%202008.pdf. Accessed 15 Nov 2012

UNEP (1992) Agenda 21/Chapter 36: Promoting education, public awareness and training. Available at: http://habitat.igc.org/agenda21/a21-36. htm. Accessed 15 Nov 2012

Unicef (2012) Quality Primary Education in Bangladesh. Available at http://www.unicef.org/bangladesh/Quality_Primary_Education%281%29.pdf. Accessed 12 Dec 2012

UNSIDR (2006) Newsletter ISDR Inform-Latin American and the Caribbean. No.13. Available at: http://www.erid.org/eng/revista/no_13_2006/art7.htm. Accessed 15 Nov 2012

Wisner B (2006) Let our children teach us! A review of the role of education and knowledge in disaster risk reduction. Inter-Agency Task Force Cluster Group on Education and Knowledge. Available at http://scholar.google.co.jp/scholar?q=2006+Review+of+the+Role+ofEducation+and+Knowledge+in+Disaster+Risk+Reduction&hl=ja&as_sdt=1&oi=scholar. Accessed 27 Dec 2012

Chapter 15
Disaster Risk Reduction Investment and Reduction of Response Cost in Bangladesh

Ahmadul Hassan, Aminul Islam, Nityananda Chakravorty, and Bhuiya Md. Tamim Al Hossain

Abstract Disaster Risk Reduction (DRR) is one of the prime concerns of development policies throughout the world. Bangladesh, being a disaster prone country affected by severe natural hazards like flood, cyclone, drought, has been undertaking disaster risk reduction initiatives from past few decades. A large amount of investment was made with assistance from development partners. The general assumption that informs these activities is that investment in DRR will result in less damage and loss when a disaster strikes and thus a less costly and timely response and recovery. This paper focuses on the results of a study conducted in Bangladesh based on the available data on disaster events, damage, DRR and response activities to test the hypothesis that disaster risk reduction reduces response costs. Using simple data analysis, taking absolute values of costs on both risk reduction and response measures of the limited data available, the results of the study provide a very clear indication in favour of the hypothesis. The findings of the study indicate that there has been a dramatic downward shift in the cost of responding to disasters since 1984 due to a steady increase in DRR investment.

Keywords Bangladesh • Cyclone • Disaster risk reduction • Investment • Response cost

A. Hassan • N. Chakravorty • B.Md. Tamim Al Hossain
Center for Environmental and Geographic Information Services, Dhaka, Bangladesh

A. Islam (✉)
Sustainable Development, United Nations Development Programme [UNDP], Dhaka, Bangladesh
e-mail: aminul.islam@undp.org

R. Shaw et al. (eds.), *Disaster Risk Reduction Approaches in Bangladesh*, Disaster Risk Reduction, DOI 10.1007/978-4-431-54252-0_15, © Springer Japan 2013

15.1 Background

Bangladesh frequently suffers from devastating natural hazards such as floods, cyclones with accompanying storm surges, tornadoes, river-bank erosion and drought. The significant loss of life and assets caused by natural disasters classify Bangladesh as one of the most disaster prone countries in the world. Evidence also suggests that natural disasters in Bangladesh are occurring with repeated frequency and in short intervals that leave little time for recovery.

These natural disasters greatly hinder the development of the country because of the loss of human lives, loss of livestock, destruction and damage to assets and infrastructure including access roads, protective embankments and agricultural land. This is particularly relevant in the Bangladesh context, where disasters, although random and unpredictable, are not surprising. Following the devastating floods of 1988 and the cyclone of 1991, the Bangladesh government adopted a holistic approach to natural disaster management embracing the processes of hazard identification, disaster mitigation, community preparedness and integrated response efforts. Relief and recovery activities are now planned within an all-risk management framework seeking enhanced capacities of at-risk communities and thereby lowering their vulnerability to specific hazards. In line with this paradigm shift from relief and response to comprehensive disaster management, a range of disaster risk reduction activities have been initiated.

Although there is conventional wisdom around the idea that disaster preparedness will improve disaster response in terms of fewer losses and more effective response there is limited empirical evidence to demonstrate this assumption. The Stern report (2007) found that 1 US$ of investment in DRR saves US $7 in disaster response. A limited number of studies have demonstrated that disaster prevention can pay high dividends, one study found that for every Euro invested in risk management, broadly 2–4 Euros are returned in terms of avoided or reduced disaster impacts on life, property, the economy and the environment (Mechler et al. 2008). It is also true that despite assumed benefits, disaster risk management (DRM) measures are not widely implemented and there continues to be, for the most part, a reliance on reactive approaches that only address the impact of disasters. Bilateral and multilateral donors allocate 90 % of their disaster management funds for relief and reconstruction and only 10 % for disaster risk management. This low level of investment in prevention can be explained by a lack of understanding and concrete evidence regarding the types and extent of the cost and benefits of preventive disaster risk management measures.

Bangladesh has shown itself to be a leading proponent of DRR, gaining recognition at the global level by pioneering a number of interventions and innovative approaches for reducing the impact of disasters on people, infrastructure and property. DRR activities have become part of the development culture in Bangladesh as a necessity because of the country's frequent exposure to natural disasters. This paper aims at initiated with the objective to interpreting and establishing relation between DRR investment and post-disaster resource requirements for response and recovery. The core concept behind this paper is to examine the effectiveness of DRR investment in terms of reduced response requirements and to identify the gaps that presently prevent achieving the Hyogo Framework for Action (HFA) goals (including inadequate DRR investment, inadequate implementation and processes and weaknesses in planning).

15.2 Vulnerability to Cyclones

Coastal region of Bangladesh is one of the most vulnerable regions to cyclonic storms (Fig. 15.1). The region comprises the 16 coastal districts covering 40,919 km^2 area and hosts about 38.2 million populations (estimated in 2009). About 5.5 % of global total tropical storms (with wind speed 62 km/h) form in the Bay of Bengal. Of these, Bangladesh is hit by about 1 % of them (Ali 1999). Figure 15.2 shows the

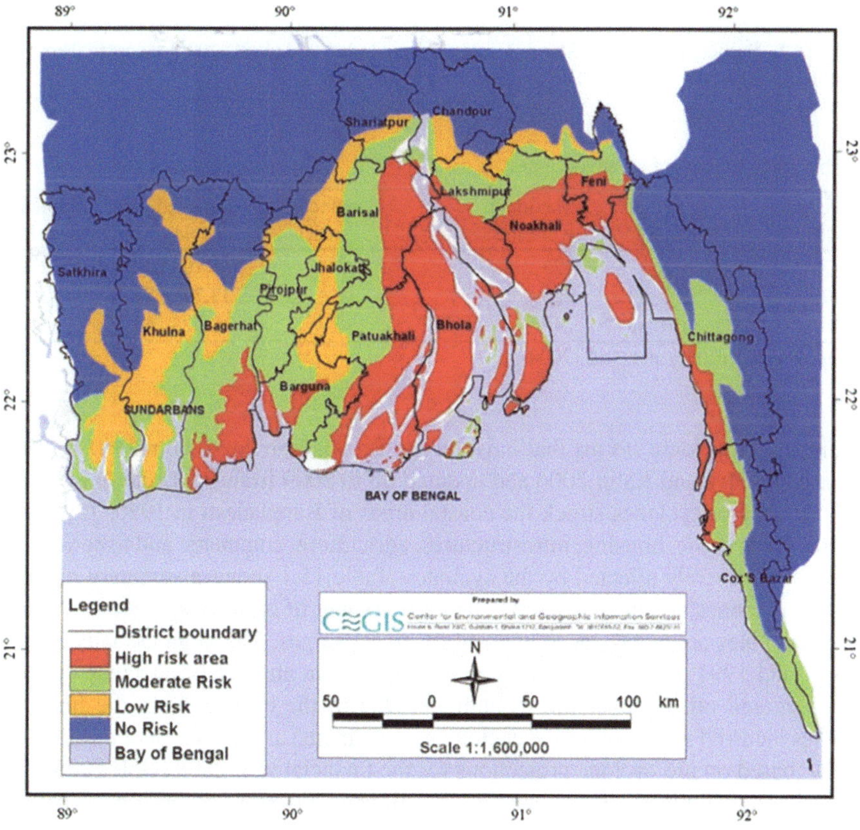

Fig. 15.1 Coastal area of Bangladesh showing cyclone risk zones

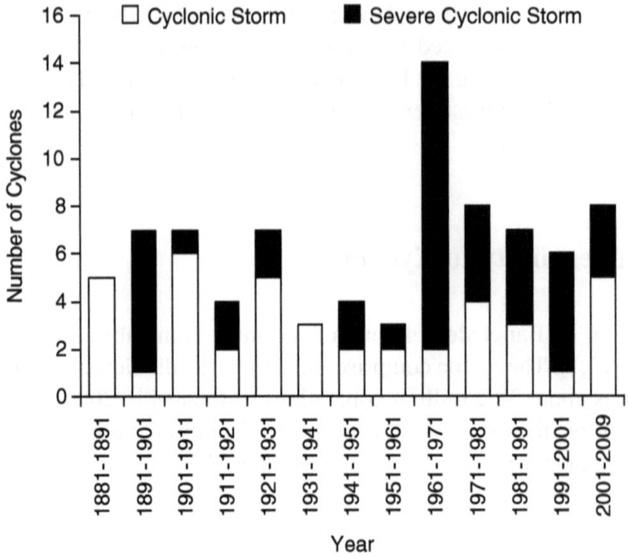

Fig. 15.2 Number of Cyclonic Storms Hitting Bangladesh Coast (1881–2009) (adopted from Hye and Kabir 2006 and BMD data)

Table 15.1 Impacts of severe cyclones

Cyclone events	Storm surge	Maximum wind speed (km/h)	Affected districts	Affected people	Human casualties
1970	6–9 m	223	5	1,100,000	470,000
1991	6–7.5 m	225	19	13,798,275	138,958
2007	Up to 10 m	240	30	8,923,259	3,363

Source: DMB (2008) and GoB (2008a, b)

number of cyclonic storms that have hit the Bangladesh coast from 1881 to 2009 (based on Hye and Kabir 2006 and updated up to 2009 from BMD data). In recent history, major cyclones struck the coastal areas of Bangladesh in 1970, 1991, and 2007. Human life, housing, infrastructures, agriculture, commerce and fisheries sectors were severely affected by the cyclones. Table 15.1 shows a summary of these major cyclones and their impacts. The high number of casualties is due to the fact that cyclones are always accompanied by storm surges. The large casualty rate of 1971 and 1991 cyclones is generally attributed to an absence of adequate cyclone shelters and other preparedness facilities. Due to the cyclone Sidr, the financial losses incurred as a result of the cyclone was estimated around 0.5 % of the national GDP based on pre-cyclone projections for the financial year 2008 (GoB 2008a, b).

Although impacts of cyclones were estimated in terms of human casualties, house and infrastructure destroyed or agriculture crop damaged for cyclonic events, a comprehensive vulnerability and exposure of human life, houses, infrastructures and economic activities to cyclone hazards is not well known in Bangladesh. In this study, we have estimated the vulnerability and exposure of coastal areas of Bangladesh to cyclonic storms considering four elements or sectors—housing, agriculture, livestock and infrastructure. Historical records of each of the four elements of 19 coastal districts were collected from Statistical Yearbooks published by the Bangladesh Bureau of Statistics (BBS 1992–2007), which were used to assess the potential exposure to cyclones and storm surges. The vulnerability and exposure of an element to cyclone is calculated based on the specific location within the "risk zones" of coastal region defined by IWM (2008) and distribution of the element within the risk zones. IWM (2008) defined four risk zones in the coastal region of Bangladesh: wind/no risk (no inundation), low risk (inundation 0–1 m), Moderate risk (inundation 1–3 m) and high risk (inundation >3 m) (Fig. 15.1). The historical data of each element was available for each coastal district. The distribution of the four elements within the risk zones was estimated as proportional area of a district falling in risk zones multiplied by the density of each element within the district. Then the potential damage of each element in each risk zone was calculated by the amount of the element present in the risk zone multiplied by the damage factor of that element. Finally, total potential damage of four elements due to cyclone was computed by combining damage values of four individual elements. The study assumes various damage factors for each element based on risk zones and infrastructure type as follows.

Housing: Household damage is normally doubled when surge height increases approximately 2 m. Based on this observation, damage percentage area have been assumed as 80 % for High risk, 40 % for Moderate risk, 5 % for Low risk and 0 % for Wind/no risk area. Besides, house types are also important in its vulnerability to damage, so a damage percentage of 10 % and 80 % for pakka and kacha house respectively have been considered. Finally, the number of potentially exposed households for different years has been calculated multiplying the number of houses by all the factors.

Agriculture: The potential exposure of agricultural production has been calculated using an estimated "percentage of crops damaged" for the different risk zones (25 % for wind/no risk, 50 % for low risk and 100 % for moderate and high risk areas). The height of crops in field is normally below 1 m, whereas the expected inundation depth in high risk area is more than 3 m and 1–3 m in moderate risk area. As a result, damage factor for crop is assumed as 100 % for high and moderate risk, 50 % for low risk and 25 % for wind risk area. After assigning these zones, values of agricultural products had been estimated using yearly crop procurement prices and inflation rates.

Livestock: As there are not enough safe haven facilities for livestock in the coastal areas, they normally stay either at homestead or fields during a cyclone. The potential damage to livestock has been assumed to be 80 % for high risk, 40 % for moderate risk, 5 % for low risk and 0 % for wind/no risk areas. Based on these values the damage and exposure of livestock assets are calculated using livestock prices and inflation rates.

Table 15.2 Potentially exposed asset worth and actual damage occurred by severe cyclones

	Potentially exposed asset worth (in billion Taka)			Actual damage (in billion Taka)		
Year	1991	1997	2007	1991	1997	2007
Housing	83	94	113	59	23	47
Agriculture	61	45	41	48	22	28
Livestock	90,779	131,434	26,435	40	0.44	18
Infrastructure	30,200	301,14	41,959	6	2	25

Infrastructure: The potential exposure for roads has been calculated incorporating damage percent for different risk zone (0 % for wind/no risk, 5 % for low risk and 40 % for moderate risk and 80 % for high risk areas). The potential exposure for embankments have been calculated incorporating damage percent similar to roads along with a factor considering the exposure of different types of embankments (0.4 for interior dyke, 0.7 for marginal dyke and 1 for Sea dyke). Based on the calculations, values of these assets are calculated.

The worth of potential exposure of these sectors is presented in Table 15.2. This table also presents the actual damage worth for different cyclones which was calculated using the damage figures available from the disaster management bureau (DMB 2008).

15.3 Investment in Disaster Risk Reduction

Disaster risk reduction is a complex process. Effective DRR needs to address social, institutional, technical, environmental, financial and economic issues in a way which is cohesive and interconnecting. Calculating total investment in DRR is challenging because the scope of DRR is so broad. Not only does DRR cut across a range of different issues, it also takes place at different levels and in conjunction with other processes, at household level, community level, local government, district and national level. The disaster risk reduction measures to reduce impacts of cyclones in Bangladesh includes cyclone early warning system, Cyclone Protection Program (CPP volunteers trained to assist those affected by cyclones), construction of cyclone shelters and embankment for surge protection, afforestation and green belt development. Recent data show that CPP program holds 42,000 trained volunteers in 2008 (about 14 volunteers per 1,000 vulnerable households); about 2,500 cyclone shelters were constructed by the year 2007 (served about 2.7 million people, 17 % of total vulnerable population); about 4,600 km embankment were constructed in the coastal areas by 2007 (approximately 11,000 km^2 area is benefited); coastal afforestation program benefited 88 km^2 area of coastal region in 2007; and the cyclone early warning system was improved significantly with high accuracy of prediction and dissemination throughout the coastal region. Government of Bangladesh (GoB) as well as international NGOs (INGO) has been implementing these types of DRR measures

Table 15.3 Investment in disaster risk reduction projects

Year	DRR investment (in million Taka)		
	GoB	INGO	Total
1984–1991	53	13	66
1992–1997	829	481	1,310
1998–2007	11,068	321	11,389

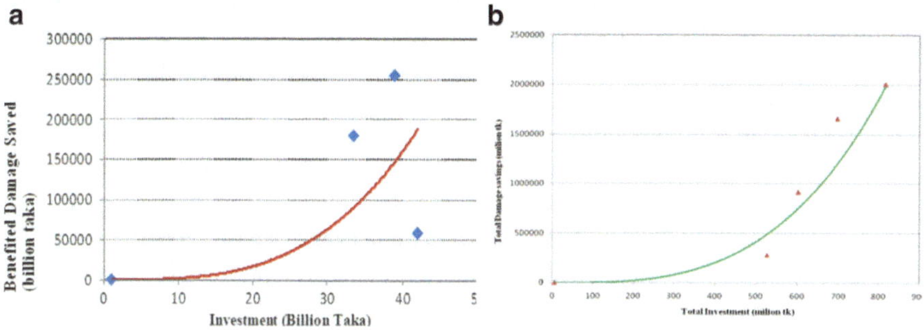

Fig. 15.3 (**a**) Relationship of benefited damage saved and investment in coastal embankments. (**b**) DRR cost in afforestation and corresponding savings of damage

for past few decades. Both GoB and INGOs invested a huge amount of money in those projects. A summary of investments in DRR projects in different periods is presented in Table 15.3, which show that total investment in DRR projects was increased about ten times after 1997 cyclone till cyclone Sidr in 2007.

The benefits of cyclone shelter, together with CPP pool and other supports are reflected in Fig. 15.3a, b for a gross understanding. The figures show that, the investment in cyclone shelter construction is increasing and this is contributing to serve more people at the risk due to cyclone. The percentage of served population at risk has grown from merely 2 % to around 17 %.

The purpose of embankments is to protect agricultural land, household assets, infrastructure and livestock from all water related hazards including cyclone. Table 15.4 presents the potential cost of damage and the benefit incurred in terms of savings due to embankments. The percentage of damage savings to potential cost of damage estimated to be was 15 % in 1997 and has increased to 85 % in 2007 as more embankments have been constructed.

The value of investment in embankments and estimated savings against damage that this investment has resulted in when cyclones occurred are presented in Fig. 15.3a, b. This figures show that the ratio of potential savings to investment is increasing with time. In 1985, investing one Taka resulted in a saving of 326 Taka while in 2007 this saving is 1386 Taka.

Afforestation can reduce damage to houses, crops, livestock and infrastructures because of the protection a green belt can provide in reducing the severity of tidal surges during disasters. The degree to which damage is reduced will depend on

Table 15.4 Benefited damage savings and potential damage worth

Year	Potential damage worth (Taka)	Benefited damage savings (Taka)	Percentage (%)
1997	161,687	24,134	15
2007	68,549	58,086	85

when the cyclone makes landfall (at high tide or low tide), the specific nature of the cyclone (in terms of wind-speed, proximity to origin) and the surge height that accompanies the storm. Historical events from 1984 to 2007 (cyclone Sidr) are recorded in the following table with an estimation of the savings incurred as a result of afforestation.

Afforestation is a cumulative process which multiplies the benefits gradually to sectors. The information above demonstrate that cumulative investment in afforestation at three stages (1991, 1997 and 2007) has increased at an average rate of 1.6 % per annum and damage has reduced at an average rate of 1.9 %.

The relation between investment and damage savings is represented in Fig. 15.3a, b.

15.4 Relief and Response

Relief activities take place in immediate aftermath of a disaster and typically include distributions of cash and of commodities essential for survival in the post disaster context including food, water, clothes, blankets as well as provision of medical support and temporary water supply and sanitation.

After the immediate relief phase has prevented further loss of life in the aftermath of a disaster, reconstruction/rehabilitation activities begin. The focus of rehabilitation activities are the longer term needs for affected communities to recover. This includes the repair and reconstruction of houses, roads, bridges, embankments and water supply system as well as technical support for the rehabilitation of agriculture, livestock and fisheries and other measures to ensure food security which might include creation of alternative income sources for the affected community.

Relief and rehabilitation (R&R) are the two major components of disaster response costs in Bangladesh. Table 15.5 illustrates the spending on these activities over the period.

15.5 Assessment: Ratio of Preparedness and Response

Establishing a direct relationship between DRR investments and Response costs is a complex exercise. Because of an absence of systematic data it is very difficult to arrive at a conclusive statement that reflects the relationship between DRR investment and response costs. The emergence of DRR as a sector globally is relatively new. The limited data available to disasters is sufficient to give some indicative notions of the effect of investment in DRR has on the cost of responding on the

Table 15.5 Response costs for disasters in Bangladesh (Taka in millions)

	GoB response cost		INGOs response cost		
Year	Relief	Rehabilitation	Relief	Rehabilitation	Total response cost
1984–1991	6.67	64	2.26	0.24	9.57
1992–1997	2.98	63.25	5.24	2	71.49
1998–2007	105.52	38.43	2.75	3.99	150.68

Source: GoB (1990–2007), Ministry of Food and Disaster Management, INGOs

Table 15.6 Relation between DRR and Response costs (Taka in millions)

Year	DRR investment	Increase in cost	Response cost	Increase in cost	Percentage (response/DRR)
1984–1991	66	1	957	1	1,446
1992–1997	1,310	20	7,149	8	545
1998–2007	11,389	172	15,068	16	132

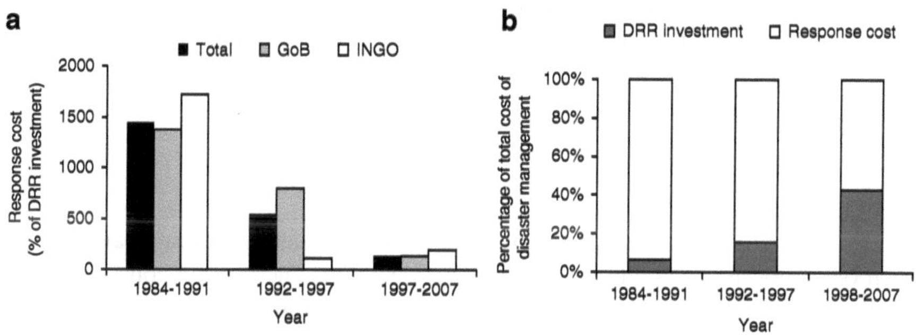

Fig. 15.4 (**a**) Relation between DRR and response total costs, government investment and INGO investment. (**b**) Relation between percentage of DRR investment and response cost

major cyclones in Bangladesh. As more information becomes available over time and is more systematically managed there is scope for reviewing and testing this tentative conclusion.

Table 15.6 presents the relationship among DRR and response costs. Table 15.6 indicates that, both the investment in DRR and the cost of responding to cyclones has increased over the period reviewed for this study. Investment in DRR has increased by 172 times from 1991 to 2007, at the same time; response cost has increased by 16 times. In Table 15.6, "increase in cost" indicates the ratio of DRR or Response cost of a specific time period to that of 1984–1991 period. The "Percentage (Response/DRR)" indicates the percent of response cost to DRR investment for the time period. Over the years, the percentage of response cost to DRR investment has been decreased rapidly. In 1991 this percent was 1,446 % and in 2007 for the cyclone Sidr it had reduced to 132 %.

Table 15.6 gives a summary relationship which is quantified in Fig. 15.4a, b.

A general inference from the information available is that as investment in DRR increases, the relative cost of response decreases. In order to establish a more

Table 15.7 Benefits in different sectors due to DRR investments

DRR investment	Household	Agriculture	Livestock	Infrastructure	Life saving
Early warning	+++	0	++++	0	++++
CPP	+++	0	++++	0	++++
Cyclone shelter	+	0	++++	0	+++
Embankment	++++	++++	+++	+++	++
Afforestation	++	++	+	++	0

Note: "+" indicate positive impact, "0" indicate less/no relation

specific relationship between DRR investment and resultant response cost a more detailed analysis would ideally involve a greater breakdown of the types of DRR investment over more disasters. Table 15.7 shows that investment in early warning and CPP activities has a positive impact on preserving household assets, protecting livestock and in reducing casualties as a result of cyclones. Investment in cyclone shelter is found to result in preserving the lives of people and livestock in the affected areas. Investing in coastal embankments can benefit in all sectors and the benefit of coastal afforestation is mainly seen in household, agriculture and livestock sectors.

15.6 Conclusion

Some aspects of DRR and preparedness are easier to measure and calculate the cost than of others; for example infrastructure investments such as the construction of cyclone shelters, building embankments and raising homesteads are straight forward indicators to account for, non-structural measures including increasing community knowledge about early warning and the use of cyclone shelters are more difficult to measure but may also play an important part in saving lives and avoiding injuries during cyclones and should be factored in to analysis of investments and savings. One important lesson that can be learned from this study is that detailed information on disasters should be collected and maintained in a comprehensive information management system.

Another way to understand the benefits of DRR investments is to include a cost–benefit analysis (CBA). The interest in economic aspects of DRM has been increasing with high profile disaster events due to climate change impacts already being observed and projected. However, the short timeframe in which the study was carried out and the limited information in Bangladesh on DRR spending and disaster response spending have limited the scope of this study only on cyclonic storm surges.

Acknowledgements The authors would like to express their utter gratitude to CEGIS authority and all the people involved in making this study successful. Special thanks go to the UNDP for the financial support and continuous encouragement during the study. Acknowledgement is also due to the relevant GoB agencies (particularly, BMD, DMB and BWDB) for their cooperation and support. The International NGOs working in DRR activities in Bangladesh also deserve appreciation for their data support.

References

Ali A (1999) Climate change impacts and adaptation assessment in Bangladesh. Clim Res 12(2/3): 109–116

BBS (1992–2007) Statistical Pocket Book Bangladesh 1992–2007. Bangladesh Bureau of Statistics, Government of Bangladesh, Dhaka

DMB (2008) List of past disasters and corresponding damages in Bangladesh. Retrieved from http://www.dmb.gov.bd/pastdisaster

GoB (1990–2007) Annual development Program, 1990–1991 to 2006–2007. Planning Commission, Government of Bangladesh, Dhaka

GoB (2008a) Cyclone Sidr in Bangladesh: damage, loss and needs assessment for disaster recovery and reconstruction. Government of Bangladesh/the World Bank/the European Commission, Dhaka

GoB (2008b) Super cyclone SIDR: impacts and strategies for interventions. Ministry of Food and Disaster Management, Government of Bangladesh, Dhaka

Hye JMA, Kabir MM (2006) Cyclonic storm surge modelling in designing coastal embankment. In: Workshop on climate change impact modeling, climate change cell, Department of Environment, Componenet 4b, Comprehensive Disaster Management Programme (CDMP), Government of Bangladesh, Dhaka

IWM (2008) Preparation of Inundation Map Considering the Decay Factors for Landuse, Geomorphology and Slope, Comprehensive Disaster Management Programme (CDMP), Ministry of Food and Disaster Management. Government of Bangladesh, Dhaka

Mechler R, Hochrainer S, Kull D, Chopde S, Singh P, Wajih S, the Risk to Resilience Study Team (2008) Uttar Pradesh drought cost-benefit analysis, from risk to resilience working paper no. 5. Moench M, Caspari E, Pokhrel A (eds) ISET, ISET-Nepal and ProVention, Kathmandu

Stern N (2007) The economics of climate change – the Stern Review. Cambridge University Press, London

Chapter 16
Gender and Social Exclusion Analysis in Disaster Risk Management

A.K.M. Mamunur Rashid and Hasan Shafie

Abstract Gender is one of the key perspectives in disaster risk management in Bangladesh. The vulnerabilities to disasters are different for male and female; the scale of the vulnerabilities also differs by sex and the sexual construction of the culture. While the whole of the disaster management discourse is emphasizing the shift of relief dominated disaster management towards a more comprehensive disaster risk reduction culture, the gender remained a very important aspect of risk reduction because of the cultural process of causation of vulnerability to disasters. Discrimination, exclusion and marginalization based on the sexual identity is a pre-disaster characteristic of a society, and such characteristics is a root cause of vulnerability, which made a particular sex more exposed to disasters than other. In different agro-ecological zone based on diversified livelihood pattern and social division of labour, the vulnerability to hazards varies by sex and experiences different degree of disasters. This is also observed that due to relative high vulnerability of women than men in disasters also made other cycles of disasters as the response to disasters also fail to address the gender needs in emergencies. One of the key roles of disaster managers should be to identify the gender needs and address the needs in a sensitive way. There is ample scope to address the gender perspectives in defining and redefining the risk environment, managing risk environment and responding to emergencies. Gender and social exclusion analysis could be a key start in the process of planning for disaster risk management and relevant policy, programmes and institutions could be responsive to address the key analysis imperatives into a comprehensive manner. Addressing gender and social inclusion framework will therefore make the disaster risk management more comprehensive in Bangladesh.

A.K.M.M. Rashid (✉)
Poverty Environment and Climate Mainstreaming Project, UNDP, Dhaka, Bangladesh
e-mail: nemoibsru@gmail.com

H. Shafie
Department of Anthropology, University of Dhaka, Dhaka, Bangladesh

R. Shaw et al. (eds.), *Disaster Risk Reduction Approaches in Bangladesh*, Disaster Risk Reduction, DOI 10.1007/978-4-431-54252-0_16,

Keywords Gender • Inclusion of excluded • Resilience • Risk and vulnerability • Social exclusion

16.1 Issues and Concern

The present paper aims at exploring the problem of a more equal integration of women and different excluded people in the disaster management in Bangladesh. The processes of discrimination, exclusion and marginalization produce asymmetric experience of disaster impacts for women and certain groups of people. Evidences from Bangladesh suggest that individuals and communities are differentially exposed and vulnerable to disasters depending on factors such as wealth, education, race, ethnicity, religion, gender, age, class, caste, disability, and health status. This is because society transforms natural hazards of different intensities or magnitudes through social processes that increase the exposure and vulnerability of population groups, their livelihoods, production, support infrastructure, and services (Cannon 1994; Wisner et al. 2004; Wisner and Cluster 2006; Carreno et al. 2007; Chambers 2009; ICSU-LAC 2010). The social distribution of vulnerabilities tends to be concentrated on socially excluded people including women, children, elderly, indigenous people and religious minorities in the context of Bangladesh. Therefore, the present paper makes the point that the analysis of socio-cultural factors can be made relevant at policy level to devise necessary and appropriate strategies to counteract exclusion processes in disaster management for inclusive outcomes.

The promotion of resilient and adaptive societies requires a paradigm shift away from the primary focus on natural hazards and extreme climatic events toward the understanding of vulnerability. The reduction of vulnerability has become the core element of both climate change adaptation and disaster risk management and, therefore, constitutes an important common ground between these two areas of policy and practice. The severity of the impacts of disasters and extreme climate events depends strongly on the level of vulnerability and exposure to these events. Therefore, understanding the multi-faceted nature of vulnerability and exposure is a prerequisite for determining how disasters and climate events contribute to the occurrence of disasters, and for designing and implementing effective adaptation and disaster risk management strategies. But, owing to the diversities in conceptual frameworks, definitions and disciplinary views, approaches to address the causes of vulnerability also differ (Blaikie 1994; Twigg 2001; Adger and Brooks 2003; Cardona 2003; Turner et al. 2003a, b; Schroter et al. 2005; Adger 2006; Fussel and Klein 2006; Cutter et al. 2008; Cutter and Finch 2008). Moreover, there is high agreement in disaster literatures that the women, children, elderly, or indigenous people as to be the "most vulnerable", but there are hardly any indications as to what these groups are vulnerable to and why (Wisner and Luce 1993; Enarson and Morrow 1998; Morrow 1999; Cardona 2003, 2011; Bankoff 2004).

The exclusion and enduring undervaluing of women, persons with disabilities and socially excluded groups perpetuate a cycle of poverty and isolation. Moreover, their vulnerabilities are accelerated by the extensive presence of inequality characterized by poor health, disease, hunger, malnourishment, illiteracy, restricted access to resources, poor housing, unemployment, and the lack of access to basic services such as water, sewage, electricity and so forth (Psacharopoulos and Patrinos 1994; Partridge et al. 1996). At the same time, they often remain "invisible" in disaster reduction or emergency response programmes, even though their number statically makes up a significant portion of any population. These most vulnerable categories are, most often, not included in the various stages of disaster response, neither as recipients of aid to meet their basic as well as specific needs, nor as active stakeholders and designers or planners of aid measures, voicing their own needs and opinions. The realities on the ground exhibit that these groups of people suffer the most from the disaster damages and have least capacities to resist and recover from the losses sustained from a hazard or other threats, without any external assistance. The situation turns even worse under policy and performance breakdown in the planning for and response to the needs of the most vulnerable categories. Therefore, understanding vulnerability of the excluded groups is a prerequisite for understanding disaster risk and the development of risk reduction and adaptation strategies to extreme natural events.

The subsequent discussion proceeds onward with clarification of the problem-field from the analytical perspectives of exclusion, deprivation and social construction of vulnerabilities in the context of disaster management in Bangladesh. Pursuant to such aspiration, the paper would try to shed light on the causal and constitutive relations between social exclusion and vulnerability from the analytic perspective of freedom and capability deprivation. This would introduce the issues and concerns of Rights Based Approach in disaster management. Thereafter, the specific ways of addressing the needs of gender and excluded people at successive phases of disaster risk management would be discussed towards the end of this chapter.

16.2 Setting the Context: Place and People

Ecological Profile of Bangladesh: Agriculture is the mainstay of Bangladesh economy and of the people at large. This sector also accounts for 48.1 % of labor force (BBS 2007). The contribution of agriculture to Gross Domestic Product (GDP) averages 22 % of GDP (BBS 2007). Bangladesh is endowed with potential agricultural resources and characterized by diverse physical features that make up 30 major agro-ecological zones (AEZ) with high diversity indices of genetic resources. These ecological zones resemble homogeneity in agricultural and ecological features, while differences are accounted on fourfold characteristic classifications in relation to physiography, soil types, land types by inundation levels and agro-climatology. Physiographic forms the primary element in delineating the agroecological regions, which comprises broadly of three regions including floodplain area, Pleistocene terrace, and the hilly regions. The floodplain covers 80 % (NAPA 2005) which are located in the north-western, central, south-central and north-eastern regions. Pleistocene terrace covers 8 % (ibid) which consists of Madhupur and Barind Tract.

The hills cover 12 % of the total land area which include the low hills and hillocks of Sylhet and the Chittagong Hill Tracts.

Bangladesh, known as the land of rivers, is geographically positioned at the receiving end of perhaps the largest river system, and formed at confluence of the Ganges (Padma), Brahmaputra (Jamuna), and Meghna rivers and their tributaries-the number would exceed 310 (BBS 2007). While the Tropic of Cancer divides Bangladesh and has endowed the country with a tropical monsoon climate characterized by heavy seasonal rainfall, moderately warm temperatures and high humidity. Three distinct seasons can be recognized in Bangladesh from climatic point of view: the dry winter season from November through February; the pre-monsoon hot summer season from March through May, and the rainy monsoon season which lasts till October.

Hazard Portfolio: Bangladesh is said to be the most disaster prone region on earth owing to the factors like geographical positioning, deltaic formation history and low-lying coastal morphology. The country is exposed to natural hazards of all possible sorts, such as, floods, river bank erosion, cyclones, droughts, water logging, arsenic contamination, salinity intrusion, tornadoes, heat waves, cold waves, earthquakes etc. The (co)occurrence of these natural events are often coupled with and multiplied by the high base vulnerabilities of the individuals, households and communities resulting in disasters that further drive the country towards greater environmental degradation, hunger, poverty, social deprivation and political conflicts, and thereby impeding the socio-economic development of the country. The country is highly susceptible to natural hazards and climate change impacts. The livelihood base of the people significantly suffers from erosion resulting from recurrent and over exposure to diverse natural hazards. Underlying the impact of a disaster is the issue of vulnerability. Vulnerability, at the national level, to natural hazards is determined by a complex and dynamic set of influences, such as the economic structure of a country, its stage of development and prevailing economic conditions and policy.

Vulnerability Scenarios: The high population density, widespread poverty, lack of awareness and education, enormous pressure on rural land, and agriculture based economic system cause and accelerate the vulnerabilities of the general population of Bangladesh. The impacts of any disasters are not equally distributed over the population, since some groups of people are more vulnerable than others. Hence the most vulnerable people suffer the most from the disaster damages and have least capacities to resist and recover from the losses sustained from a hazard or other threat. Women and children, Persons with disabilities (PWDs) as well as the socially excluded people are most vulnerable people in these hazards due to their high degrees of base vulnerabilities. When determining vulnerability by establishing the capability of these groups and their encompassing environment to anticipate, cope with, and recover from emergencies, it is important to consider the level of volatility and the potential rates of change that may exist. A variety of models exist to identify or describe the cumulative impact of risks and the causal relationship between them.

Vulnerability scenarios in Bangladesh, like elsewhere (Lavell et al. 2003; Cannon 2006; Cutter et al. 2008), are situation and hazard specific. Yet there are

Table 16.1 Comparative profiles of groups by vulnerability indicators

Indicators	Less vulnerable	More vulnerable
Ethnicity	Groups with sufficient knowledge of Bangla language, socially cohesive members of supporting groups	Groups with no or insufficient knowledge of Bangla language (in the context of CHTs), socially not cohesive, nonmembers of supporting groups
Government policies and planning	In place and effective	Not in place or not effective
Local economic production and employment opportunities	Robust and protected, little unemployment	Frail and exposed, substantial unemployment
Medical and emergency services	Robust and resilient	Frail and not resilient
Response and recovery capability	Tested and adequate	Untested or inadequate
Community based DRR planning process and implementation	Stakeholders and communities participate in planning process, effective mitigation strategies	Stakeholders and communities not involved in planning process, no or ineffective mitigation strategies
Critical infrastructure	Robust and protected	Frail and exposed

generic underlying factors of vulnerabilities, such as poverty, and the lack of social networks and social support mechanisms aggravating vulnerability levels irrespective of hazard types. These types of generic factors are different from the hazard specific factors and assume a different position in the intervention actions and the nature of risk reduction and adaptation processes (ICSU-LAC 2010). Vulnerability of human settlements and ecosystems is intrinsically tied to different socio-cultural and environmental processes (Kasperson et al. 1988; Cutter 1994; Adger 2006; Cutter et al. 2008; Cutter and Finch 2008; Williams et al. 2008; Dawson et al. 2011). Again, from a climate change perspective, basic environmental conditions are changing progressively and thereby posing new risk conditions for local societies. In fact, future vulnerability is embedded in the present conditions of the communities that may be exposed in the future (Patt et al. 2005, 2010; Klein and Patt 2012); that is, new hazards in areas not previously subject to them will reveal, not necessarily create, underlying vulnerability factors (Alwang et al. 2001; Cardona et al. 2003; International Strategy for Disaster 2009; Lopez-Calva and Ortiz-Juarez 2009; UNISDR 2009a, b). Table 16.1 represents a comparative scenario of people in Bangladesh by selected vulnerability indicators:

16.3 Social Exclusion and Vulnerability

Concepts like social exclusion, deprivation, multiple marginalization, social capital, social capability, and civic engagement have gained wide acceptance in much of the recent development literature. The concept of social exclusion has been defined as "the process through which individuals or groups are wholly or partially excluded from full participation in the society within which they live" (Haan and Maxwell 1998; Haan 2009). There has been extensive empirical work done focusing on the range of social, political, institutional, and cultural processes that lead to the exclusion of marginal groups in society. Contrarily, such research enterprises on social exclusion and the concept as such have been challenged for being limited in theoretical underpinning (Oyen 1997; Sen 2000). A conceptual framework to understand the macro processes of social exclusion is yet to be developed and, it has still no precise definition and means different things to different people (Atkinson and Hills 1998). If the advocates have been vocal, so have been the critics. I shall, however, try to grasp the idea and relevance of "social exclusion" and "capability deprivation" in the context of disaster risk management in Bangladesh.

The central focus, here, is on the phenomenon of exclusive groups membership, which entitles the members for certain benefits and excludes non-members from such benefits. This membership may be based on interest group, political party, ethnicity, class etc. (Jordan 1996; Evans 1998; Haan and Maxwell 1998). The idea of social exclusion has drawn significant attention to the role of social and institutional processes that lead to exclusion, deprivation and, marginalization of individuals or social groups, certain pre-existing (age, gender, strata, class etc.) categories, from employment, markets, training, power etc. Atkinson and Hills stated of three aspects of social exclusion (Atkinson and Hills 1998). Firstly, social exclusion is relative, i.e., excluded from a particular society, at a particular place and time. Second, it implies an act of exclusion and hence an agent or agency. Finally, exclusion has a dynamic aspect that entails people are excluded not just because they are without a job or income, at present, but because they have little prospect for the future even though the excluded people have the same productive endowment in relation to those who exclude them. The study of social encounter in Bangladesh exhibits that the salient features of social exclusion have dynamic outcome for individual choices, opportunities and strategies. It affects both the quality of life of individuals and the equity and cohesion of society at large.

Social exclusion has sanctioned restriction on their freedom, in Aristotelian sense of the term, to undertake necessary activities. Social exclusion is a complex and multi-dimensional process of restricting accesses to resources, rights, goods and services, while it is exhibited in the inability to participate in the normal relationships and activities, available to the majority of people in a society, whether in economic, social, cultural or political arenas. Some of these dimensions of social exclusion are cited in Table 16.2:

Social exclusion occurs where different factors combine to trap individuals and areas in a spiral of disadvantage (Department of Social Security 1999). An individual is socially excluded if (a) he or she is geographically resident in a society but (b) for reasons beyond his or her control, he or she cannot participate in the normal activities of citizens in that society, and (c) he or she would like to so participate (Burchardt et al. 2002). A lack or denial of access to the kinds of social relations, social customs and activities in which the great majority of people are engaged (Gordon et al. 2000). In current usage, social exclusion is often regarded as a

Table 16.2 Different dimensions of social exclusion

Dimensions of social exclusion	Characteristics	Sources
Individual	Physical health, education, and events in the individual's life, disability	Burchardt (2000) and Room (2000)
Social	Inequality and injustice, gender, age, ethnicity, cultural practices	Phipps and Curtis (2001) and Curran et al. (2007)
Economic	Financial problems, unemployment, scarcity of resources	Hammer (2003) and Pleace (2003)
Socioeconomic	Poverty	Dowling (1999)
Political	Political marginalization, lack of political participation	Nabatchi et al. (2012) and Beland (2007)
Institutional and organizational	Rules, policies, processes	Beall and Piron (2005)

Source: Tambulasi (2009)

Table 16.3 Different processes whereby social exclusion operates

Interactional/associative exclusion	Institutional/organizational exclusion	Market exclusion
• Corporate or group interests • Stigmatization • Weak leadership among them • Restrained Interactions • Mistrust by others • Identity exclusion • Lack of mutually exchangeable resources or skills	• Discriminatory institutional mission or objectives • Low status entitlements • Lack of competitive skills of the excluded people	• Few resources/assts to exchange • Little mobility • Distorted markets arrangements • High transaction costs of capital and others • Labor and financial market segmentation and discrimination

"process" rather than a "state" and this helps in being constructively precise in deciding its relationship to poverty (Howarth et al. 1998). Social exclusion is a broader concept than poverty, encompassing not only low material means but the inability to participate effectively in economic, social, political and cultural life and in some characterizations alienation and distance from mainstream society (Duffy 1995). In other words, social exclusion is an accumulation of confluent processes with successive ruptures arising from the heart of the economy, politics and society, which gradually distances and places persons, groups, communities and territories in a position of inferiority in relation to centres of power, resources and prevailing values (Estivill 2003). There are different processes whereby social exclusion operates on the ground and accelerate the vulnerabilities of the excluded groups (Table 16.3).

The structural character of social exclusion, relating it to wider economic processes and inequalities, has been a feature of much academic writing (Brown 1994; Byrne 2005; Levitas 2005). While some others' have suggested that the focus should exclusively be on social exclusion rather than poverty (Oppenheim 1998).

Because social exclusion is multi-causal, relational, and it includes less tangible aspects than poverty such as the loss of status, power, self-esteem and expectations. Again some work has focused on the persistence over time is an integral aspect of social exclusion (Room 2000; Barnes 2005) on the ground that the continuity of poverty, deprivation and multiple disadvantage exacerbates their vulnerabilities, especially on the quality of life as well as future life chances. Vulnerability refers to the propensity of exposed elements such as human beings, their livelihoods, and assets to suffer adverse effects when impacted by hazard events (UNDRO 1980; Liverman 1990; Maskrey 1993; Blaikie 1994; Cannon 1994; Weichselgartner 2001; Bogardi and Birkmann 2004; UNISDR 2004; Birkmann 2006; Cannon 2006; Janssen et al. 2006; Thywissen 2006; UNISDR 2009a, b).

Vulnerability in the context of disaster risk management is the most palpable manifestation of the social construction of risk (Aysan 1993; Blaikie 1994; Wisner et al. 2004; ICSU-LAC 2010). This notion emphasizes that society, in its interaction with the changing physical world, constructs disaster risk by transforming physical events into hazards of different intensities or magnitudes through social processes that increase the exposure and vulnerability of population groups, their livelihoods, production, support infrastructure, and services (Cannon 1994; Wisner et al. 2004; Wisner and Cluster 2006; Carreno et al. 2007; Chambers 2009; ICSU-LAC 2010). It is evident and largely agreed upon that high vulnerability and exposure are mainly an outcome of skewed development processes, including those associated with environmental mismanagement, demographic changes, rapid and unplanned urbanization, and the scarcity of livelihood options (Maskrey 1993, 1994; Lavell et al. 2003; Cannon 2006; Cardona 2011; ICSU-LAC 2010).

Effective management of risk requires understanding of how vulnerability is generated, how it increases, and how it builds up (Maskrey 1989; Fernandez 1999; Cardona et al. 2003; O'Brien et al. 2004; Cardona 2011). While vulnerability is a key concept for both disaster risk and climate change adaptation, the term is employed in numerous other contexts, for instance to refer to epidemiological and psychological fragilities, ecosystem sensitivity, or the conditions, circumstances, and drivers that make people vulnerable to natural and economic stressors (Kasperson et al. 1988; Cutter 1994; Wisner et al. 2004; Brklacich and Bohle 2006; Haines et al. 2006; Villagrán de León 2006). Despite various frameworks developed for defining and assessing vulnerability, there are at least some common causal factors of vulnerability across societies (Cardona and Hurtado 2000; McCarthy et al. 2001; Cardona 2003, 2011; Cardona et al. 2003; Gallopin 2006; Carreno et al. 2007, 2009; Parry et al. 2007; ICSU-LAC 2010):

- **Susceptibility/Fragility (in DRM) or Sensitivity (in CCA)**: physical attributions of human beings, infrastructure, and environment to be affected by a hazard events due to lack of resistance and predisposition of society and ecosystems.
- **Lack of Resilience (in DRM) or Lack of Adaptive Capacities (in CCA)**: restricted accessibility to and limited capacity to mobilize the resources, and

incapacity to anticipate, adapt, and respond in absorbing the socio-ecological and economic impact.

16.4 Socially Excluded Groups in Bangladesh

The enduring experience of living with disasters in Bangladesh suggests that the women, children, PWDs and socially excluded groups constitute the most vulnerable segments of the society and are disproportionately affected by the negative impacts of any disaster. This paper defines the situation of these groups as a form of social exclusion on the ground that: (1) these groups are, in various ways, kept away from full participation in the wider economic, political, cultural, and social life; (2) the enduring discrimination and historical social features entrapped these groups in a situation below the minimum threshold of well-being while hindering their full participating in the society; and (3) these groups are lacking in power and access to decision-making that could influence policies or create opportunities for improving their standard of living. These groups have some additional vulnerability against different hazards and often remain "invisible" in disaster reduction or emergency response programmes. Some of these most vulnerable groups in Bangladesh are:

• Women (Including Female Headed Households & Victims of Domestic Violence)	• Religious Minorities Groups • Indigenous and Ethnic Minorities • Occupational Minority Groups
• Children	• Very Poor/Homeless
• Adolescents	• Victims of Trafficking
• People with Disabilities (PWDs)	• Refugees (Rohingga, Bihari etc.)
• Older Persons/Elderly Citizens	• HIV Positive Individuals

This paper recognizes that all these aforesaid groups entail certain degrees of alienation in the disaster reduction or emergency response programmes. This distancing is also an outcome of the wider social processes within which these groups usually receive inadequate attention and are systematically disadvantaged owing to different reasons. The socially excluded groups are systematically disadvantaged because they are discriminated against on the basis of their ethnicity, race, religion, sexual orientation, caste, descent, gender, age, disability, HIV status, migrant status or where they live. Such discrimination occurs in public institutions, such as the legal system or education and health services, as well as social institutions like the household, and in the community (DFID 2005; Gaynor et al. 2007).

The degree of discrimination, however, varies from one society to another, as do the forms that social exclusion takes. Understanding and measuring the disaster risks accelerated by social exclusion is quite difficult because of the limitations of any basic indicator sets on social exclusion. However, the development of deprivation index as well as quality of life and well-being indicators could be operationalized at an individual rather than an aggregate level. Based on these sets of indicators, the Bristol Social Exclusion Matrix (B-SEM) has identified three domains of

Table 16.4 Domain and dimension of gender and exclusion in Bangladesh

Domain	Dimension	Vulnerability components
Resources	Material/economic resources	1. Lack of earning opportunities and low income
		2. Low possession of necessities
		3. No home ownership
		4. Restricted access to other assets
		5. No savings and high debt
		6. Higher incidence of subjective poverty
		7. Limited access to common property resources
	Access to public and private services	1. Limited access to public services
		2. Restricted access to utility services
		3. Limited access to public transportation
		4. Discriminated access to private services
		5. Limited or no access to financial services
	Social resources	1. Institutionalization/separation from family
		2. Limited or no access to social support system
		3. Limited access to safety net programmes
		4. Low frequency and quality of social contact
Participation	Economic participation	1. Differential access to labor market
		2. Gender differentials in labor market outcomes
		3. Prevalence of bonded labor
		4. Undertaking unpaid work
		5. Delayed and discounted payment for wage labor
		6. Low quality of working life
	Social participation	1. Limited participation in common social activities
		2. Limited or no participation in decision making
	Cultural participation	1. Limited or discriminated access to educational attainments
		2. Limited or discriminated access to basic skills
	Political and civic participation	1. Restricted and conditional citizenship status
		2. Lack of political participation
		3. Limited participation in civic and voluntary activities
		4. Limited membership in civil society organizations
Quality of life	Health and well-being	1. Weak physical health and exercise
		2. Poor mental health situation
		3. Low life expectancies
		4. Limited opportunities for personal development
		5. Low self-esteem
		6. Higher susceptibility to stigma and prejudice
	Living environment	1. Low housing quality
		2. Higher incidence of homelessness
		3. Lower neighborhood safety
		4. Low neighborhood satisfaction
		5. Limited access to open space
	Crime, harm and criminalization	1. Low social safety
		2. Victim of crime
		3. Low subjective safety, i.e. under threat and fear of crime
		4. Exposed to eve-teasing, bullying and harassment
		5. Victim of discrimination
		6. Unaware of legal rights & victim of imprisonment

resources, participation and quality of life; and a total of ten dimensions within these domains (Levitas and Britain 2007). The domains and dimensions, and their corresponding vulnerability components for the excluded people in Bangladesh are set out in Table 16.4.

The interactive nature of social exclusion means that these domains and dimensions are often both outcomes and risk factors. The relationships between the dimensions and domains of social exclusion are quite complex. The very interactional processes recognized in this paper of multivariate analysis imply that many, if not all, these dimensions are simultaneously exclusionary outcomes and causal factors for other dimensions of exclusion in the context of Bangladesh.

16.4.1 Exclusion of Women and Children

Gender is an informal yet important institution based on socially approved role differentiations, behaviors and activities of men and women. Women's position is society often makes them more vulnerable than men. They tend to have less access to livelihood resources and income earning opportunities. Women tend to be underrepresented in decision-making processes due to various factors including active exclusion by men, lack of time to participate due to domestic responsibilities, and lack of confidence to express their views. At the same time, the women in Bangladesh are engaged in bearing social and economic burden: taking responsibility for domestic tasks such as producing and cooking food, fetching water, looking after children and caring for the sick and elderly. Despite these challenges, experience has shown that women often play a proactive role in the restoration of household and community functions when hazards and stresses occur. During hazards' exposure, women play important roles in ensuring the safety of younger or older members of the family.

Disasters in Bangladesh, according to recent histories, have always impacted the women and children more adversely because of built-in societal norms. However, owing to different social norms in Bangladesh, the women are made reluctant to move from their home to a safer area, or they may be less aware than men of how to respond. Women and children are rather confined to homes and are less mobile than the males, which increase their vulnerabilities and sufferings during disasters. Different studies on 1988 and 1998 floods in Bangladesh reveals that the women became more severely affected than men because of their responsibilities and everyday activities including food processing and cooking, cleaning, collecting water and fuel, looking after the livestock and belongings (Kafi 1992; Lovekamp 2003). In a similar study on 1991 cyclone, carried out in Chakaria and Kutubdia, exhibits that the children under 10 years and women over 40 years sustained the highest mortality of 26 % and 31 % respectively (Kabir 1995). Besides, the cultural practice of wearing *sharee* and long hair of the women have turned out to increase vulnerability for them during storm surge and wind blow of cyclone SIDR 2007 (Shafie 2009). Women are often vulnerable to sexual harassment in pre and post disaster situation.

In some areas, it was reported that the hooligans and looters took the opportunity of the people's distressed situation for robbery and sexual violence after the 1991 cyclone (Kafi 1992; Kabir 1995; Lovekamp 2003). As a result, in anticipation of potential sexual offence against them, young women and adolescent girls are not preferred to go to the cyclone shelters in order to avoid staying with males (strangers). Moreover, as cultural sanctions, they have restricted access to resources, services and decision-making. Therefore, the vulnerability of women and children is extremely high and thus they are most at risk among the community.

16.4.2 People with Disabilities in Disaster

Persons with disabilities (PWDs), as being one of the most vulnerable groups, suffer most during the natural hazards' exposure. Similarly, cultural beliefs or community attitudes toward the persons with disabilities contribute to either the vulnerability or the capacity of this category depending on the extent to which they reduce or increase the risk of disaster. Almost 6 % of Bangladesh's 147 million inhabitants have some form of disabilities and are largely absent from national development initiatives (Foley and Chowdhury 2007; Hosain et al. 2011) and disaster reduction programmes. Persons with disabilities face many obstacles in their normal life which increases during disaster episodes.

The persons with disabilities are highly vulnerable to natural disasters because there are: (1) deprived of equitable access to resources and services; (2) lacking in the ownership of productive resources, money, education, jobs, confidence, skills, and ability to access service systems; (3) limited with relatively fewer coping mechanisms with which to recover from disaster; (4) excluded from normal social interactions and placed in low social rank; and (5) not having adequate access to early warning dissemination mechanisms. Following from these reasoning, the persons with disabilities are perpetually dependent on support services which remain to be inadequate during different phases of disasters.

16.4.3 Exclusion of Indigenous, Religious and Occupational Minorities

Indigenous religious and occupational minorities, particularly those living in geographically remote areas, face injury, loss and damage from the effects of natural disasters each year. The existing policy and planning framework of the Government in disaster management are not readily inclusive of these communities, as this guideline does not adequately cater for the special needs of such excluded groups. Consequently, the capacity of these communities to prepare for, deal with and mitigate the effects of any future disasters is severely limited. Most of these excluded groups have little or no financial capacity for mitigation and encounter considerable disadvantages in the key areas of health, education, employment and housing. These special need people have been accorded much lower priorities in disaster risk management activities of the Government and Non-Government organizations in Bangladesh. Owing to the presence of more pressing concerns of the wider communities, the additional needs of these excluded communities are always compromised at both policies and programmes levels.

The indigenous, religious and occupational communities, particularly those in Chittagong Hill Tracts, are located in remote and risk prone areas and are highly susceptible to disaster impacts. There are higher cost structures and additional demands associated with the delivery of relief and recovery assistance to remote communities. The entrenched level of socioeconomic disadvantage experienced by these communities also means they require more assistance than other less remote communities in achieving similar disaster recovery outcomes. Three are some additional challenges for an equitable achievement of disaster risk management objectives by addressing the needs of these excluded communities. Some of these challenges in the context of Bangladesh include: (1) geographic isolation in hilly areas and limited accessibility during different seasons of the year; (2) community infrastructure and services that are well below national standards, and high repair and maintenance cost for infrastructure due to remoteness; (3) low economic base and absence of capacity to raise funds; (4) unique cultural, linguistic and communication issues; and (5) deprivation of equitable access to education, health and other facilities.

Improved risk reduction for these indigenous, religious and occupational minority communities cannot be attained without the active participation and involvement of the communities and strategies that take into account the needs and diversities of these communities. This requires long-term commitment to the inclusion of these excluded communities in mainstream planning and service delivery from all levels of Government and other agencies.

16.5 Including the Excluded in Disaster Management

The Government of Bangladesh is highly committed to develop institutional premises and mechanisms to ensure that appropriate policies are taken to foster a culture of prevention at all levels of our societies. All the Government efforts are being objectified in the line of establishing hazard-resilient communities and the protection of people from the threat of disasters. The Government, therefore, has adopt and implement policy measures at the regional, sub-regional, national and local levels aimed at reducing the vulnerability of the poor people to both natural and technological hazards through proactive rather than reactive approaches. The disaster management model in Bangladesh is based on three major elements: (1) defining and redefining risk environment; (2) managing risk environment; and (3) responding to emergencies. However, at policy level, this model acknowledges the women and other excluded groups as the most vulnerable groups in disasters. Yet, at the

Table 16.5 Defining and redefining risk environment

Elements of DM	Prospects for including gender and exclusion
Developing criteria	The specific needs of women and socially excluded groups are to be considered while developing the risk evaluation criterion. Women, girls, person with disability as well as indigenous, religious and occupational minorities are endowed with different vulnerability features to hazards. Recognizing these differences, the risk environment should be defined by multiple criteria instead of any single criterion
Identifying hazards	The perception of hazards (ranking, severity) largely varies according to the viewpoints of male and female, able and differently able persons, rich and landless people, ethnic minority and majority people, girls and boys and so on. Therefore, the identification of hazards should consider all these viewpoints of different social segments especially the excluded people
Assessing vulnerabilities	Vulnerability and exposure are dynamic, varying across temporal and spatial scales, and depend on economic, social, geographic, demographic, cultural, institutional, governance, and environmental factors such as wealth, education, race, ethnicity, religion, gender, age, class, caste, disability, and health status Vulnerabilities are dependent on the nature of gender relationships, entitlements, hazard types, social identities as majority or minority, attitudes towards disable people, social stigmatizations, and so on. Therefore these are important factors for assessing vulnerabilities
Analyzing risks	The analysis of risks needs to be done based on segregation by gender and socially excluded people
Evaluating risks	Conventionally, risk evaluation criteria are developed with gender and social biasness, while priorities are ordered according the majoritarian, male and risk peoples' perspectives. Therefore, risk evaluation criteria are to be developed in light of the practical and strategic needs of women and socially excluded people
Risk prioritization	During risk prioritization, the high risk portfolios of the women, person with disabilities and excluded people are often ignored and receive less priority. Usually the needs of the powerful people get priorities in decision-making. Therefore, the voices of the women and socially excluded peoples are to be amplified and accounted during risk prioritization

Table 16.6 Managing risk environment

Elements of DM	Prospects for including gender and exclusion
Participation	The effectiveness of managing risk environment depends on the spontaneous participation of all cross-cutting social categories including women and men, landlord and landless, ethnic minority and majority, person with disabilities, higher caste/class and lower caste/class. Hence, the participation of female, poor, person with disability, landless, lower caste people needs to be encouraged and ensured
Risk reduction options	The risk reduction options are usually generated to fulfill the needs of the socially dominant groups e.g. the male, rich, ethnic or religious majority and higher caste people. Besides, the generated risk reduction options may have detrimental effects on the existing social inequalities by gender, class and exclusion. These problems can be addressed by cultivating the voice and quality participation of the women and excluded people
Prioritization of options	Prioritizing the risk reduction measures often reflects the asymmetric gender relations and existing social power structure. Like aforesaid steps, prioritizing risk reduction options under managing risk environment should also be more sensitive towards the needs of the women, poor, person with disability and socially excluded people
Implementation and monitoring	The women and socially excluded people should have their respective stakes in the implementation process. The risk reduction programmes need to be inclusive of gender and social equity based monitoring and evaluation. The monitoring framework and tools are to be sensitive towards gender and social exclusion issues

Table 16.7 Responding to emergencies

Elements of DM	Prospects for including gender and exclusion
Warning dissemination	Most of the women, person with disability, lower caste and poor people have less education and mostly warning system are not friendly to women and socially excluded people. The messages of warning signals are not always comprehendible to the women, person with disabilities and socially excluded people. Therefore, the community based warning dissemination system should effectively and sensitively warn the women and socially excluded people
Evacuation and shelter	The evacuation route planning should include the special needs of the women, person with disabilities and socially excluded people. The shelter management and planning need adequate consideration of the requirements of these excluded groups. The safety and privacy of the women and girls are also of important concerns to be addressed
Search and rescue	The volunteers and rescuers teams should adequately include women members. They are to be trained to deal with the specific needs of women during search and rescue sensitively
Need assessment	During humanitarian emergency situations, different aid agencies come up with very generic relief goods. Usually the relief items or goods are not selected to address the need of lactating mother, breast-feeding children, reproductive and hygienic needs of women. While conducting need assessment, often these specific needs are ignored
Post hazard shelter	According to local cultural sanctions, women are not quite willing to share same shelter centre along with male. They need privacy in shelter centre. Hence, the special needs of women and excluded people are to be considered during post hazard shelter management

programme and implementation level, these excluded groups remain to be invisible in terms of their respective needs and priorities. The below discussion aims at identifying some possible ways of including these excluded people in disaster management in Bangladesh (Tables 16.5–16.7).

16.6 Gender and Social Exclusion Analysis

Gender and social exclusion analysis is the study of the different roles of women, men and excluded groups to understand what they do, what resources they have and what their needs and priorities are. The analysis of gender and social exclusion can be explored in the context of the linkages of different development and disaster paradigms. Currently, the approaches to gender issues are addressed as cross-cutting issues in the economic growth, poverty alleviation and disaster reduction paradigms. In general, enabling environment programmes pay little direct attention to the empowerment of women, while it is expected to occur naturally as an offshoot of economic growth. To be effective, however, mainstreaming of gender policies in the context of economic reform has to draw on all these paradigms, with gender analysis taking place at all stages of the programme cycle, from appraisal and design to monitoring and long-term assessment of impact. It is also necessary to accommodate social, economic and cultural diversity within that analysis. Besides, the women are not homogenous category because in some situations poor women may have more in common with their male compatriots than they do with rich women in their particular sector of economic activity, whilst in other contexts, women's might have similar special needs. Detailed gender and social exclusion analysis should be undertaken at all stages of appraisal and design to avoid the type of unforeseen and unintended impacts.

Addressing gender issues must also be a multi-level process, so that gender perspectives are incorporated into formal policies and institutions relating to disasters and livelihoods from local to national levels. Actions very often need to be tailored to ensure that women are able to participate equally, for example training courses should be held at times of day when women are able to attend. Providing child-care can improve attendance. Training may need to be tailored to the educational level of women. It is important to address the deep-rooted social inequalities between women, men and groups as a long-term process, rather than a one-off project activity. However, the analysis gender and social exclusion requires describing the key issues to consider different areas of enquiry e.g. roles and responsibilities, assets and livelihoods, power and decision-making and needs analysis.

16.7 Conclusion and Policy Implications

The analysis in this paper sheds light on the prevalence and persistence of social exclusion experienced in Bangladesh with regard to disaster risk management. A significant portion of the population, including the women, in Bangladesh experience social exclusion at some level. However, individuals and households experience multiple barriers to social and economic participation. This paper makes the point that there is a growing need for policies to address the issues of social exclusion in disaster risk management in Bangladesh. With equally importance, it supports the argument for a more inclusive model of support aimed at women, excluded people and communities.

The multidimensional nature of gender discrimination and social exclusion requires policy reform taking a more holistic perspective in devising appropriate effective strategies to counteract such processes. The design and calculation of multidimensional measures of social exclusion have become quite inevitable to transcend the inherent weaknesses of our conventional over reliance on the measures of income-based poverty. Disaster risk management programmes in Bangladesh, at present, address gender needs based on conventional approaches by prioritizing the selection of women among the programme beneficiaries. Needless to state that, this approach is not only inadequate but also creating new forms of vulnerabilities for them in many respects. The practical and strategic needs of the gender cannot be addressed without proper understanding of gender differentials of vulnerabilities.

The gender and social exclusion analysis would provide a more complete understanding of the social barriers faced by individuals, households and communities to full participation. In addition to the individual behavior explanations, this analysis also accounts the external or systemic barriers of equitable access to both resources and participation. Conceptual understandings of the pathways to social and economic participation explain structural or systemic barriers beyond the control of individuals and households. While substantial advances have been made in the fields of disaster risk management in Bangladesh, yet we need to have better integrated and inclusive approaches based on social exclusion understandings. Gender and social exclusion analysis would address barriers to participation concurrently with building self esteem, foundational capabilities, skills and training for improved quality of life, wellbeing, safety and life satisfaction. However adopting gender and social exclusion analysis would drive public investments anchored in social inclusion principles and thereby the integrated approaches will place the individual or household at the centre of assistance.

References

Adger WN (2006) Vulnerability. Glob Environ Chang 16(3):268–281

Adger WN, Brooks N (2003) Does global environmental change cause vulnerability to disaster. In: Pelling M (ed) Natural disaster and development in a globalizing world. Routledge, London, pp 19–43

Alwang J, Siegel PB et al (2001) Vulnerability: a view from different disciplines. In: Social Protection Discussion Paper Series

Atkinson AB, Hills J (1998) Exclusion, employment and opportunity. LSE STICERD Research Paper No. CASE004

Aysan Y (1993) Vulnerability assessment. In: Merriman PA, Browitt CWA (eds) Natural disasters: protecting vulnerable communities. Thomas Telford, Londres

Bankoff G (2004) The historical geography of disaster: 'Vulnerability' and 'Local Knowledge' in Western Discourse. In: Bankoff G, Frerks G, Hilhorst D (eds) Mapping vulnerability: disasters, development and people. Earthscan, London, pp 25–37

Barnes M (2005) Social exclusion in Great Britain: an empirical investigation and comparison with the EU. Ashgate Publishing Company, Aldershot

BBS (2007) Statistical Year Book. Bangladesh Bureau of Statistics, Bangladesh Government, Dhaka

Beall J, Piron LH (2005) Social exclusion review. Department for International Development, London

Beland D (2007) The social exclusion discourse: ideas and policy change. Policy Politics 35(1):123–139

Birkmann J (2006) Measuring vulnerability to promote disaster-resilient societies: conceptual frameworks and definitions. In: Birkmann J (ed) Measuring vulnerability to natural hazards. Towards disaster resilient societies. United Nations University Press, Tokyo, pp 9–54

Blaikie PM (1994) At risk: natural hazards, people's vulnerability, and disasters. Taylor & Francis Group, London

Bogardi J, Birkmann J (2004) Vulnerability assessment: the first step towards sustainable risk reduction. In: Malzahn D, Plapp T (eds) Disaster and society: from hazard assessment to risk reduction. Logos Verlag Berlin, Berlin, pp 75–82

Brklacich M, Bohle HG (2006) Assessing human vulnerability to global climatic change. In: Krafft T, Ehliers E (eds) Earth system science in the Anthropocene: emerging issues and problems. Springer, New York, pp 51–61

Brown P (1994) Economic restructuring and social exclusion: a new Europe? Routledge, London
Burchardt T (2000) Social exclusion: concepts and evidence. In: Gordon D, Townsend P (eds) Breadline Europe: the measurement of poverty. The Policy Press, Bristol, pp 385–406
Burchardt T, Le Grand J et al (2002) Degrees of exclusion: developing a dynamic, multidimensional measure. In: Hills J, Le Grand J, Piachaud D (eds) Understanding social Exclusion. Oxford University Press, Oxford
Byrne D (2005) Social exclusion. Open University Press, Maidenhead
Cannon T (1994) Vulnerability analysis and the explanation of natural disasters. In: Varey A (ed) Disasters, development and the environment. Wiley, Chichester, pp 13–30
Cannon T (2006) Vulnerability analysis, livelihoods and disasters. In: Dannemann S (ed) Coping with Risks due to natural hazards in the 21st century. Taylor & Francis, London, pp 41–49
Cardona OD (2003) The need for rethinking the concepts of vulnerability and risk from a holistic perspective: a necessary review and criticism for effective risk management. In: Bankoff G, Frerks G, Hilhorst D (eds) Mapping vulnerability: disasters, development and people. Earthscan, London, p 17
Cardona OD (2011) Disaster risk and vulnerability: notions and measurement of human and environmental insecurity. In: H.G. Brauch, U. Oswald Spring, C. Mesjasz, J. Grin, P. Kameri-Mbote, B. Chourou, P. Dunay, and J. Birkmann, (eds.) Coping with Global Environmental Change, Disasters and Security: Threats, Challenges, Vulnerabilities and Risks. Vol. 5, Hexagon Series on Human and Environmental Security and Peace. Springer, Berlin, pp. 107–122
Cardona OD, Hurtado JE (2000) Holistic seismic risk estimation of a metropolitan center. In: Proceedings of the 12th World Conference of Earthquake Engineering, January 30 – February 4, 2000. New Zealand Society for Earthquake Engineering, Auckland, New Zealand
Cardona OD, Hurtado JE et al (2003) The notion of disaster risk: conceptual framework for integrated management. IDB/IDEA Program of Indicators for Disaster Risk Management, National University of Colombia, Manizales
Carreno ML, Cardona OD et al (2007) Urban seismic risk evaluation: a holistic approach. Nat Hazards 40(1):137–172
Carreno, Martha Liliana et al (2009) Holistic urban seismic risk evaluation of megacities: application and robustness. In: L.A. Mendes-Victor, C.S. Oliveira, J. Azevedo, and A. Ribeiro, (eds.) The 1755 Lisbon Earthquake: Revisited. Geotechnical, Geological, and Earthquake Engineering, Vol. 7. Springer, Netherlands, pp. 167–183
Chambers R (2009) Vulnerability, coping and policy (editorial introduction). IDS Bull 37(4):33–40
Curran C, Burchardt T et al (2007) Challenges in multidisciplinary systematic reviewing: a study on social exclusion and mental health policy. Soc Pol Admin 41(3):289–312
Cutter SL (1994) Environmental risks and hazards. Prentice Hall, Englewood Cliffs
Cutter SL, Finch C (2008) Temporal and spatial changes in social vulnerability to natural hazards. Proc Natl Acad Sci USA 105(7):2301–2306
Cutter SL, Barnes L et al (2008) A place-based model for understanding community resilience to natural disasters. Glob Environ Chang 18(4):598–606
Dawson TP, Jackson ST et al (2011) Beyond predictions: biodiversity conservation in a changing climate. Science 332(6025):53–58
DFID (2005) Reducing poverty by tackling social exclusion. GB, Department for International Development, London
Dowling M (1999) Social exclusion, inequality and social work. Soc Pol Admin 33(3):245–261
Department of Social Security (1999) Opportunity for all: Tackling poverty and social exclusion. Stationery Office Limited, London
Duffy K (1995) Social exclusion and human dignity in Europe. Council of Europe, Strasbourg
Enarson EP, Morrow BH (1998) The gendered terrain of disaster: through women's eyes. Praeger, Westport
Estivill J (2003) Concepts and strategies for combating social exclusion: an overview. ILO, Geneva
Evans M (1998) Behind the rhetoric: the institutional basis of social exclusion and poverty. IDS Bull 29(1):42–49

Fernandez MA (1999) Cities at Risk: Environmental Degradation, Urban Risks and Disasters in Latin America. A/H Editorial, La RED, USAID, Quito, Ecuador
Foley D, Chowdhury J (2007) Poverty, social exclusion and the politics of disability: care as a social good and the expenditure of social capital in Chuadanga, Bangladesh. Soc Pol Admin 41(4):372–385
Fussel HM, Klein RJT (2006) Climate change vulnerability assessments: an evolution of conceptual thinking. Clim Change 75(3):301–329
Gallopin GC (2006) Linkages between vulnerability, resilience, and adaptive capacity. Glob Environ Chang 16(3):293–303
Gaynor C, Watson S et al (2007) Evaluating DFID's policy on tackling social exclusion: baseline, framework and indicators. Department for International Development, London
Gordon D, Adelman L et al (2000) Poverty and social exclusion in Britain. Joseph Rowntree Foundation, York
Haan A (2009) Social exclusion: an alternative concept for the study of deprivation? IDS Bull 29(1):10–19
Haan A, Maxwell S (1998) Editorial: poverty and social exclusion in north and south. IDS Bull 29(1):1–9
Haines A, Kovats RS et al (2006) Climate change and human health: impacts, vulnerability, and mitigation. Lancet 367(9528):2101–2109
Hammer T (2003) Youth unemployment and social exclusion in Europe: a comparative study. Policy Press, Bristol
Hosain GM, Atkinson D et al (2011) Impact of disability on quality of life of rural disabled people in Bangladesh. J Health Popul Nutr 20(4):297–305
Howarth C, Kenway P et al (1998) Monitoring poverty and social exclusion: labour's inheritance. Joseph Rowntree Foundation, York
ICSU-LAC (2010) Science for a better life: developing regional scientific programs in priority areas for Latin America and the Caribbean. ICSU-LAC/CONACYT, Rio de Janeiro
International Strategy for Disaster Reduction (2009) Global Assessment Report on disaster risk reduction: risk and poverty in a changing climate invest today for a safer tomorrow. United Nations, Geneva
Janssen MA, Schoon ML et al (2006) Scholarly networks on resilience, vulnerability and adaptation within the human dimensions of global environmental change. Glob Environ Chang 16(3):240–252
Jordan B (1996) A theory of poverty and social exclusion. Polity Press, Cambridge
Kabir R (1995) Bangladesh: surviving the cyclone is not enough. Stop Disasters 24:5–6
Kafi S (1992) Disasters and destitute women: twelve case studies. Bangladesh Development Partnership Centre, Dhaka
Kasperson RE, Renn O et al (1988) The social amplification of risk: a conceptual framework. Risk Anal 8(2):177–187
Klein RJTJT, Patt AGG (2012) Assessing vulnerability to global environmental change: making research useful for adaptation decision making and policy. Taylor & Francis, London
Lavell A, Mansilla E et al (2003) Local level risk management: concept and practices. CEPREDENAC-UNDP, Quito
Levitas R (2005) The inclusive society?: social exclusion and new labour. Palgrave Macmillan, Basingstoke
Levitas R, Britain G (2007) The multi-dimensional analysis of social exclusion. Cabinet Office, London
Liverman DM (1990) Vulnerability to global environmental change. In R. Kasperson et al (eds) Understanding Global Environmental Change: The Contributions of Risk Analysis and Management. Clark University, Worcester, MA, pp 27–44
López-Calva LF, Ortiz-Juárez E (2009) Evidence and policy lessons on the links between disaster risk and poverty in Latin America: methodology and summary of country studies. Research for Public Policy, MDGs and Poverty, MDG-01-2009, RBLAC-UNDP, New York

Lovekamp W (2003) Gender and disaster: a synthesis of flood research in Bangladesh. The International Sociological Association, Research Committee on Disasters. Exlibris publication, USA
Maskrey A (1989) Disaster mitigation: a community based approach. Oxfam, Oxford, United Kingdom
Maskrey A (1993) Vulnerability accumulation in peripheral regions in Latin America: the challenge for disaster prevention and management. In: Merriman MA, Browitt CWA (eds) Natural disasters: protecting vulnerable communities. Thomas Telford, London, pp 461–472
Maskrey A (1994) Disaster mitigation as a crisis of paradigms: reconstruction after the Alto Mayo Earthquake, Peru. In: Varley A (ed) Disasters, development and environment. Wiley, New York, pp 109–123
McCarthy JJ, Canziani OF et al (2001) Climate change 2001: impacts, adaptation, and vulnerability: contribution of Working Group II to the third assessment report of the Intergovernmental Panel on Climate Change. Cambridge University Press, Cambridge
Morrow BH (1999) Identifying and mapping community vulnerability. Disasters 23(1):1–18
Nabatchi T, Gastil J et al (2012) Democracy in motion: evaluating the practice and impact of deliberative civic engagement. OUP, New York
NAPA (2005) National Adaptation Programme of Action. Ministry of Environment and Forest, Government of the People's Republic of Bangladesh, Dhaka
O'Brien K, Leichenko R et al (2004) Mapping vulnerability to multiple stressors: climate change and globalization in India. Glob Environ Chang 14(4):303–313
Oppenheim C (1998) An inclusive society: strategies for tackling poverty. Institute for Public Policy Research, London
Oyen E (1997) The contradictory concepts of social exclusion and social inclusion, International Institute for Labour Studies, United Nations Development Programme, Research Series 110:63–66
Parry ML, Canziani OF et al (2007) IPCC, 2007: climate change 2007: impacts, adaptation and vulnerability. Contribution of working group II to the fourth assessment report of the intergovernmental panel on climate change. Cambridge University Press, Cambridge
Partridge, William L, and Jorge E. Uquillas (1996) Including the Excluded: Ethnodevelopment in Latin America. Annual World Bank Conference on Development in Latin America and the Caribbean, 1996: Poverty and Inequality, Proceedings of a Conference Held in Bogota, Colombia. World Bank Publications, New York
Patt A, Klein RJT et al (2005) Taking the uncertainty in climate-change vulnerability assessment seriously. CR Geosci 337:411–424
Patt A, Schroter D et al (2010) Assessing vulnerability to global environmental change: Making Research Useful for Adaptation Decision Making and Policy. Earthscan, London
Phipps S, Curtis L (2001) The social exclusion of children in North America. Dalhousie University, Halifax
Pleace N (2003) Single homelessness as social exclusion: the unique and the extreme. Soc Pol Admin 32(1):46–59
Psacharopoulos G, Patrinos HA (1994) Indigenous people and poverty in Latin America: an empirical analysis. World Bank, Washington
Room G (2000) Trajectories of social exclusion: the wider context for the third and first worlds. In: Gordon D, Townsend P (eds) Breadline Europe: the measurement of poverty. The Policy Press, Bristol, pp 407–440
Schroter D, Polsky C et al (2005) Assessing vulnerabilities to the effects of global change: an eight step approach. Mitig Adapt Strat Glob Chang 10(4):573–595
Sen AK (2000) Social exclusion: concept, application, and scrutiny. Office of Environment and Social Development, Asian Development Bank, Manila
Shafie H (2009) Voices from afar: assessing the inclusion and integration of disable people in emergency response to SIDR. ADD, Dhaka
Tambulasi RIC (2009) Pushed to the abyss of exclusion: ICT and social exclusion in developing countries. J Inform Commun Ethics Soc 7(2/3):119–127

Thywissen K (2006) Core terminology of disaster reduction. Measuring vulnerability to natural hazards: towards disaster resilient societies. United Nations University Press, Hong Kong
Turner BL, Kasperson RE et al (2003a) A framework for vulnerability analysis in sustainability science. Proc Natl Acad Sci USA 100(14):8074–8079
Turner BL, Matson PA et al (2003b) Illustrating the coupled human-environment system for vulnerability analysis: three case studies. Proc Natl Acad Sci USA 100(14):8080–8085
Twigg J (2001) Sustainable livelihoods and vulnerability to disasters. Benfield Greig Hazard Research Center, London
UNDRO, O. o. t. U. N. D. R., Co-ordinator (1980) Natural Disasters and Vulnerability Analysis: Report of Expert Group Meeting (9–12 July 1979), Office of the United Nations Disaster Relief Co-ordinator
UNISDR (2009a) Risk and Poverty in a Changing Climate: Invest Today for a Safer Tomorrow - 2009 Global Assessment Report on Disaster Risk Reduction. United Nations International Strategy for Disaster Reduction (UN/ISDR), Geneva
UNISDR (2009b) Terminology on disaster risk reduction. United Nations International Strategy for Disaster Reduction (UN/ISDR), Geneva
UNISDR (2004) Living with Risk: A Global Review of Disaster Reduction Initiatives. United Nations International Strategy for Disaster Reduction (UN/ISDR), Geneva
Villagrán de León JC (2006) Vulnerability: a conceptual and methodological review. United Nations University, Institute for Environment and Human Security, Bonn
Weichselgartner J (2001) Disaster mitigation: the concept of vulnerability revisited. Disaster Prev Manag 10(2):85–95
Williams SE, Shoo LP et al (2008) Towards an integrated framework for assessing the vulnerability of species to climate change. PLoS Biol 6(12):e325
Wisner, B (2006) Let our children teach us: A Review of the Role of Education and Knowledge in Disaster Risk Reduction. UNISDR System Thematic Cluster/Platform on Knowledge and Education, Geneva, Switzerland
Wisner B, Luce HR (1993) Disaster vulnerability: scale, power and daily life. GeoJournal 30(2):127–140
Wisner B, Blaikie P et al (2004) At risk: natural hazards, people's vulnerability and disasters, 2004. Routledge, London

About the Editors

Rajib Shaw is an Associate Professor in the Graduate School of Global Environmental Studies of Kyoto University, Japan. He worked closely with the local communities, NGOs, governments and international organization, including United Nations, especially in the Asian countries. He is currently the Chair of the United Nations Asia Regional Task Force for Urban Risk Reduction, and the President of Asian University Network of Environment and Disaster Management (AUEDM). His research interests are: community based disaster risk management, climate change adaptation, urban risk management, and disaster and environmental education. He has published several books in the field of disaster and environmental management. He is also the Chief Editor of Asian Journal of Environment and Disaster Management.

Fuad Mallick is the Professor and Chairperson of Department of Architecture and Director of the Postgraduate Programs in Disaster Management (PPDM) at BRAC University. He did pioneering research on thermal comfort for Bangladesh and published extensively on it. He is the founder of the department of architecture of BRAC University and the author of its curriculum. He is also the founder director of the university's Post Graduate Programs in Disaster Management, the first of its kind in the region. Fuad's research background is on environmental aspects of architectural design, low income and rural housing, architectural education and in the last few years disaster management and climate change. He has written book chapters and published journal articles on disaster management and climate change issues. He has been a reviewer for the Aga Khan Award for Architecture and a member of Aga Khan Education Program's thinking group for conceptualizing a School of Architecture and Human Settlements for the Aga Khan University. He is the Vice President of the Asian Universities Network for Environment and Disaster Management (AUEDM).

Aminul Islam is specialized in climate change, environment and disaster management and contributed as an academic, researcher, practitioner, manager and policy adviser. He is associated with United Nations Development Programme since 1996

Disaster Risk Reduction, DOI 10.1007/978-4-431-54252-0, © Springer Japan 2013 and served different positions in Bangladesh Country Office started as the Sustainable Development Adviser and served as the Assistant Country Director and lead the Climate change, Environment and Disaster Management Cluster and currently Senior Adviser. Dr. Islam also served as the Senior Technical Adviser and Team Leader of the Water Disaster Risk Reduction Project in Central Vietnam supported by UNDP Vietnam. As an academic, he taught research methodology for the post graduate programme in the Institute of Bangladesh Studies in Rajshahi University, founder Director and Associate Professor of the School of Environment Science and Management of Independent University, Bangladesh and also contributed as an Adjunct Professor of the Population Science Department in Dhaka University. He was Environmental Scientist of the Flood Action Plan's Component 16 on Environment implemented by the ISPAN (Virginia, US based) supported by USAID.

MIX
Papier aus verantwortungsvollen Quellen
Paper from responsible sources
FSC® C105338

If you have any concerns about our products,
you can contact us on
ProductSafety@springernature.com

In case Publisher is established outside the EU,
the EU authorized representative is:
Springer Nature Customer Service Center GmbH
Europaplatz 3, 69115 Heidelberg, Germany

Printed by Libri Plureos GmbH
in Hamburg, Germany